Symmetrical Components for Power Systems Engineering

ELECTRICAL ENGINEERING AND ELECTRONICS

A Series of Reference Books and Textbooks

1. Rational Fault Analysis, *edited by Richard Saeks and S. R. Liberty*
2. Nonparametric Methods in Communications, *edited by P. Papantoni-Kazakos and Dimitri Kazakos*
3. Interactive Pattern Recognition, *Yi-tzuu Chien*
4. Solid-State Electronics, *Lawrence E. Murr*
5. Electronic, Magnetic, and Thermal Properties of Solid Materials, *Klaus Schröder*
6. Magnetic-Bubble Memory Technology, *Hsu Chang*
7. Transformer and Inductor Design Handbook, *Colonel Wm. T. McLyman*
8. Electromagnetics: Classical and Modern Theory and Applications, *Samuel Seely and Alexander D. Poularikas*
9. One-Dimensional Digital Signal Processing, *Chi-Tsong Chen*
10. Interconnected Dynamical Systems, *Raymond A. DeCarlo and Richard Saeks*
11. Modern Digital Control Systems, *Raymond G. Jacquot*
12. Hybrid Circuit Design and Manufacture, *Roydn D. Jones*
13. Magnetic Core Selection for Transformers and Inductors: A User's Guide to Practice and Specification, *Colonel Wm. T. McLyman*
14. Static and Rotating Electromagnetic Devices, *Richard H. Engelmann*
15. Energy-Efficient Electric Motors: Selection and Application, *John C. Andreas*
16. Electromagnetic Compossibility, *Heinz M. Schlicke*
17. Electronics: Models, Analysis, and Systems, *James G. Gottling*
18. Digital Filter Design Handbook, *Fred J. Taylor*
19. Multivariable Control: An Introduction, *P. K. Sinha*
20. Flexible Circuits: Design and Applications, *Steve Gurley, with contributions by Carl A. Edstrom, Jr., Ray D. Greenway, and William P. Kelly*
21. Circuit Interruption: Theory and Techniques, *Thomas E. Browne, Jr.*
22. Switch Mode Power Conversion: Basic Theory and Design, *K. Kit Sum*
23. Pattern Recognition: Applications to Large Data-Set Problems, *Sing-Tze Bow*
24. Custom-Specific Integrated Circuits: Design and Fabrication, *Stanley L. Hurst*
25. Digital Circuits: Logic and Design, *Ronald C. Emery*
26. Large-Scale Control Systems: Theories and Techniques, *Magdi S. Mahmoud, Mohamed F. Hassan, and Mohamed G. Darwish*
27. Microprocessor Software Project Management, *Eli T. Fathi and Cedric V. W. Armstrong (Sponsored by Ontario Centre for Microelectronics)*
28. Low Frequency Electromagnetic Design, *Michael P. Perry*
29. Multidimensional Systems: Techniques and Applications, *edited by Spyros G. Tzafestas*
30. AC Motors for High-Performance Applications: Analysis and Control, *Sakae Yamamura*
31. Ceramic Motors for Electronics: Processing, Properties, and Applications, *edited by Relva C. Buchanan*

32. Microcomputer Bus Structures and Bus Interface Design, *Arthur L. Dexter*
33. End User's Guide to Innovative Flexible Circuit Packaging, *Jay J. Miniet*
34. Reliability Engineering for Electronic Design, *Norman B. Fuqua*
35. Design Fundamentals for Low-Voltage Distribution and Control, *Frank W. Kussy and Jack L. Warren*
36. Encapsulation of Electronic Devices and Components, *Edward R. Salmon*
37. Protective Relaying: Principles and Applications, *J. Lewis Blackburn*
38. Testing Active and Passive Electronic Components, *Richard F. Powell*
39. Adaptive Control Systems: Techniques and Applications, *V. V. Chalam*
40. Computer-Aided Analysis of Power Electronic Systems, *Venkatachari Rajagopalan*
41. Integrated Circuit Quality and Reliability, *Eugene R. Hnatek*
42. Systolic Signal Processing Systems, *edited by Earl E. Swartzlander, Jr.*
43. Adaptive Digital Filters and Signal Analysis, *Maurice G. Bellanger*
44. Electronic Ceramics: Properties, Configuration, and Applications, *edited by Lionel M. Levinson*
45. Computer Systems Engineering Management, *Robert S. Alford*
46. Systems Modeling and Computer Simulation, *edited by Naim A. Kheir*
47. Rigid-Flex Printed Wiring Design for Production Readiness, *Walter S. Rigling*
48. Analog Methods for Computer-Aided Circuit Analysis and Diagnosis, *edited by Takao Ozawa*
49. Transformer and Inductor Design Handbook: Second Edition, Revised and Expanded, *Colonel Wm. T. McLyman*
50. Power System Grounding and Transients: An Introduction, *A. P. Sakis Meliopoulos*
51. Signal Processing Handbook, *edited by C. H. Chen*
52. Electronic Product Design for Automated Manufacturing, *H. Richard Stillwell*
53. Dynamic Models and Discrete Event Simulation, *William Delaney and Erminia Vaccari*
54. FET Technology and Application: An Introduction, *Edwin S. Oxner*
55. Digital Speech Processing, Synthesis, and Recognition, *Sadaoki Furui*
56. VLSI RISC Architecture and Organization, *Stephen B. Furber*
57. Surface Mount and Related Technologies, *Gerald Ginsberg*
58. Uninterruptible Power Supplies: Power Conditioners for Critical Equipment, *David C. Griffith*
59. Polyphase Induction Motors: Analysis, Design, and Application, *Paul L. Cochran*
60. Battery Technology Handbook, *edited by H. A. Kiehne*
61. Network Modeling, Simulation, and Analysis, *edited by Ricardo F. Garzia and Mario R. Garzia*
62. Linear Circuits, Systems, and Signal Processing: Advanced Theory and Applications, *edited by Nobuo Nagai*

63. High-Voltage Engineering: Theory and Practice, *edited by M. Khalifa*
64. Large-Scale Systems Control and Decision Making, *edited by Hiroyuki Tamura and Tsuneo Yoshikawa*
65. Industrial Power Distribution and Illuminating Systems, *Kao Chen*
66. Distributed Computer Control for Industrial Automation, *Dobrivoje Popovic and Vijay P. Bhatkar*
67. Computer-Aided Analysis of Active Circuits, *Adrian Ioinovici*
68. Designing with Analog Switches, *Steve Moore*
69. Contamination Effects on Electronic Products, *Carl J. Tautscher*
70. Computer-Operated Systems Control, *Magdi S. Mahmoud*
71. Integrated Microwave Circuits, *edited by Yoshihiro Konishi*
72. Ceramic Materials for Electronics: Processing, Properties, and Applications, Second Edition, Revised and Expanded, *edited by Relva C. Buchanan*
73. Electromagnetic Compatibility: Principles and Applications, *David A. Weston*
74. Intelligent Robotic Systems, *edited by Spyros G. Tzafestas*
75. Switching Phenomena in High-Voltage Circuit Breakers, *edited by Kunio Nakanishi*
76. Advances in Speech Signal Processing, *edited by Sadaoki Furui and M. Mohan Sondhi*
77. Pattern Recognition and Image Preprocessing, *Sing-Tze Bow*
78. Energy-Efficient Electric Motors: Selection and Application, Second Edition, *John C. Andreas*
79. Stochastic Large-Scale Engineering Systems, *edited by Spyros G. Tzafestas and Keigo Watanabe*
80. Two-Dimensional Digital Filters, *Wu-Sheng Lu and Andreas Antoniou*
81. Computer-Aided Analysis and Design of Switch-Mode Power Supplies, *Yim-Shu Lee*
82. Placement and Routing of Electronic Modules, *edited by Michael Pecht*
83. Applied Control: Current Trends and Modern Methodologies, *edited by Spyros G. Tzafestas*
84. Algorithms for Computer-Aided Design of Multivariable Control Systems, *Stanoje Bingulac and Hugh F. VanLandingham*
85. Symmetrical Components for Power Systems Engineering, *J. Lewis Blackburn*

Additional Volumes in Preparation

Digital Filter Design and Signal Processing, *Glen Zelniker and Fred Taylor*

ELECTRICAL ENGINEERING-ELECTRONICS SOFTWARE

1. Transformer and Inductor Design Software for the IBM PC, *Colonel Wm. T. McLyman*
2. Transformer and Inductor Design Software for the Macintosh, *Colonel Wm. T. McLyman*
3. Digital Filter Design Software for the IBM PC, *Fred J. Taylor and Thanos Stouraitis*

Symmetrical Components for Power Systems Engineering

J. Lewis Blackburn
Consultant
Bothell, Washington

CRC Press is an imprint of the
Taylor & Francis Group, an **informa** business

Published in 1993 by
CRC Press
Taylor & Francis Group
6000 Broken Sound Parkway NW, Suite 300
Boca Raton, FL 33487-2742

Printed in the United States of America on acid-free paper
20 19 18 17

International Standard Book Number-10: 0-8247-8767-6 (Hardcover)
International Standard Book Number-13: 978-0-8247-8767-7 (Hardcover)

Library of Congress Cataloging-in-Publication Data

Catalog record is available from the Library of Congress

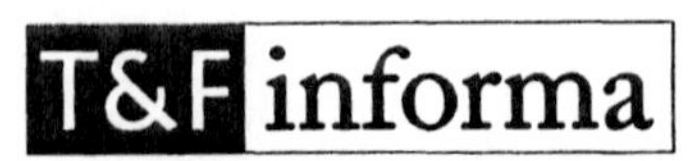

Taylor & Francis Group
is the Academic Division of T&F Informa plc.

Visit the Taylor & Francis Web site at
http://www.taylorandfrancis.com

and the CRC Press Web site at
http://www.crcpress.com

To my wife PEGGY
for more than fifty years
of patience, loving care, and understanding

Preface

The method of symmetrical components is a powerful tool for understanding and determining unbalanced currents and voltages in three-phase electrical power systems. In a sense it is the "language" of those associated with relay protection. For almost all faults, the intolerable conditions that require isolation of the problem area involve unbalances. Thus the quantities that operate the protection are directly or indirectly related to symmetrical components.

This book presents the fundamental concepts of symmetrical components along with a review of per unit (percent), phasors, and polarity. Typical examples are solved throughout the text, and an additional problem section is included for further studies. The book is intended as a text for students and as a reference manual for practicing engineers and technicians—all who are involved or associated with relaying and power system analysis.

The modern computer provides large volumes of fault and related data but without any understanding or appreciation. Thus the aim of this book is to provide (1) a practical understanding of system unbalances, the basic circuits, and calculations, (2) the techniques of making calculations when a computer or program is unavailable, (3) an overview for visualization of faults, unbalances, and the sequence quantities, (4) the determination of system parameters for manual calculations or computer programs.

This text has been developed from notes used over many years of presenting symmetrical components. Originally they were based on the classic book by Wagner and Evans, *Symmetrical Components* (1933). Associates, students and friends within Westinghouse, the IEEE, CIGRE, many utilities, and industrial and consulting companies around the world have directly or indirectly added contributions over the last fifty-five years.

Special thanks are extended to William M. Strang for his great work on the figures and to Ruth A. Dawe and Lila Harris, Marcel Dekker, Inc., editors, for their wonderful assistance and encouragement.

J. Lewis Blackburn

Contents

1

Introduction and Historical Background

1.1 INTRODUCTION AND GENERAL AIMS

The *method of symmetrical components* provides a practical technology for understanding and analyzing electric power system operation during unbalanced conditions. Typical unbalances are those caused by faults between the phases and/or ground (phase to phase, double phase to ground, phase to ground), open phases, unbalanced impedances, and combinations of these. Balanced three-phase faults are included. Also, many protective relays operate from symmetrical component quantities. For example, all ground relays operate from zero-sequence quantities, which are normally not present in the power system. Therefore, a good understanding of this subject is of great importance and is a very important "tool" in system protection.

This discussion is for three-phase electric systems, which are assumed from a practical standpoint to be balanced or symmetrical up to a point or area of unbalance. A normal area or point of unbalance in a power system will usually be down in the low-voltage or distribution area, where single-phase loads are connected or where nonsymmetrical equipment is used.

In a symmetrical or balanced system the source voltages (generators) are equal in magnitude and are in phase, with their three phases displaced 120° in phase relations. Also, the impedances of the three-phase circuits and equipment are of equal magnitude and phase angle.

In a sense, *symmetrical components* can be called the "language" of the protection engineer or technician. Its great value is both in thinking or visualizing system unbalances, and as a means of detailed analysis of them from the system parameters. In this simile it is like a language, in that it requires experience and practice for easy access and application. Fortunately, faults and unbalances occur infrequently, and many do not require detailed analysis. Thus it becomes difficult to "practice the language." This difficulty has increased significantly with the ready availability of fault studies via computers. These provide rapid access to voluminous data, frequently with little user understanding of the information, the background, or the method that provides the data.

The goal of this book is to provide (1) a practical understanding and appreciation of the fundamentals, basic circuits, and calculations; (2) an overview directed toward a clear visualization of faults and system unbalances; (3) where access to a computer or a proper fault program may not be available, the means of making fault and unbalanced calculations "by hand"; and (4) determination of the necessary system parameters for calculations or computer programs. Throughout, the math is kept as simple as possible.

Although computers and hand calculators provide great accuracy, fault and unbalance calculations will not be the same as will be experienced for real-life occurrences. The principal reasons are (1) approximations involved in the determination of many of the system parameters, especially for lines; (2) parameter changes resulting from temperature variations; (3) highly variable and relatively unknown fault resistance; and (4) variable generator impedances with time. Thus high accuracy for fault and unbalance calculations is not possible and should not be expected, and from a practical standpoint is not really needed.

Surprisingly, the calculated values often are quite close to actual values, and in general the calculations are very practical for system design and equipment selection and for the application and setting of protective relays. The general practice is to calculate only solid faults. This provides maximum values important for system design. Fault resistance will reduce the current values. However, as indicated above, it is highly variable and relatively unknown. Thus it becomes impossible practically to select the "correct" value for fault and unbalance studies.

In protective relaying, using solid fault data, overcurrent relays should be set so that their minimum pickup is at least one-half of the minimum fault currents and, hopefully, more sensitive as long as the phase relays do not operate on the maximum short-time load current and ground relays do not operate on the maximum tolerable zero-sequence unbalance. This generally provides good protection for system arcing faults, except for possible very high resistance faults in low-voltage distribution.

The method of symmetrical components is applicable to multiphase systems, but only symmetrical components for three-phase systems are covered in this book.

1.2 HISTORICAL BACKGROUND

The method of symmetrical components was developed late in 1913 by Charles L. Fortescue of Westinghouse when investigating mathematically the operation of induction motors under unbalanced conditions. At the 34th Annual Convention of the AIEE on June 28, 1918, in Atlantic City, he presented a paper entitled "Method of Symmetrical Co-ordinates Applied to the Solution of Polyphase Networks." This was published in the *AIEE Transactions*, Volume 37, Part II, pages 1027–1140. This published paper was 89 pages (5 by 8 in.) in length, with 25 pages of discussion by six well-known "giants" of electric power engineering: J. Slepian, C. P. Steinmetz, V. Karapetoff, A. M. Dudley, Charles F. Scott, and C. O. Mailloux.

Application of the method "to the study of short circuits and system disturbances," and the method as we know it today, was made by C. F. Wagner and R. D. Evans. They began a series of articles in the Westinghouse magazine *The Electric Journal* that ran for 10 issues, from March 1928 through November 1931. This series was enlarged by the two authors and published in the classic and still very useful book *Symmetrical Components*, published by McGraw-Hill Book Company, New York, 1933.

Just as the Wagner–Evans book was about to be printed, another Westinghouse engineer, W. A. Lewis, developed the concept of splitting the line reactance into components: one associated with the conductor reactance (X_a), one associated with the spacing to the return conductor(s) (X_d), and one associated with the depth of the earth return (X_e). This was added to the book as an appendix (VII) and is covered in this book in Chapter 11.

Another Westinghouse engineer, E. L. Harder, provided very useful tables of fault and unbalance connections that were presented in his paper "Sequence Network Connections" published in the December 1937 issue of *The Electric Journal*. This is covered in Chapter 5.

During this time, Edith Clarke of General Electric was developing notes and lecturing in this area. However, formal publication of her work did not occur until 1943.

On a personal note it was my privilege to meet Dr. Fortescue and Mr. Evans several times during my student assignment on the AC Network Analyzer in 1937. I knew C. F. Wagner through his son Chuck (C. L. Wagner). Chuck, past President of the IEEE Power Engineering Society, and I have had a long association together in protective relaying.

I took my first course in symmetrical components under Dr. Lewis and then applied it while working for Dr. Harder. In fact, Ed Harder was instrumental in my obtaining a permanent job in relaying. At that point I was totally unfamiliar with relays, but it was a job and in the New York area where I wanted to be. And 55 years have quickly slipped by!

2

Per Unit and Percent Values

2.1 INTRODUCTION

Power systems operate at voltages where kilovolt (kV) is the most convenient unit for expressing voltage. Also, these systems transmit large amounts of power, so that kilovolt-ampere (kVA) and megavolt-ampere (MVA) are used to express the total (general or apparent) three-phase power. These quantities, together with kilowatts, kilovars, amperes, ohms, flux, and so on, are usually expressed as a per unit or percent of a reference or base value. The per unit and percent nomenclatures are widely used because they simplify specification and computations, especially where different voltage levels and equipment sizes are involved.

2.2 PER UNIT AND PERCENT DEFINITIONS

Percent is 100 times per unit. Both are used as a matter of convenience or of personal choice and it is important to designate either percent (%) or per unit (pu).

The per unit value of any quantity is the ratio of that quantity to its base value, the ratio expressed as a nondimensional decimal number. Thus actual quantities, such as voltage (V), current (I), power (P), reactive power (Q), volt-amperes (VA), resistance (R), reactance (X), and impedance (Z), can be expressed in per unit or percent as follows:

$$\text{quantity in per unit} = \frac{\text{actual quantity}}{\text{base value of quantity}} \tag{2.1}$$

$$\text{quantity in percent} = (\text{quantity in per unit}) \times 100 \tag{2.2}$$

where "actual quantity" is the scalar or complex value of a quantity expressed in its proper units, such as volts, amperes, ohms, or watts. "Base value of quantity" refers to an arbitrary or convenient reference of the same quantity chosen and designated as the base. Thus per unit and percent are dimensionless ratios that may be either scalar or complex numbers.

As an example for a chosen base of 115 kV, voltages of 92, 115, and 161 kV become 0.80, 1.00, and 1.40 pu or 80%, 100%, and 140%, respectively.

2.3 ADVANTAGES OF PER UNIT AND PERCENT

Some of the advantages of using per unit (or percent) are:

1. Its representation results in more meaningful data where the relative magnitudes of all similar circuit quantities can be compared directly.
2. The per unit equivalent impedance of any transformer is the same when referred to either the primary or the secondary side.
3. The per unit impedance of a transformer in a three-phase system is the same regardless of the type of winding connections (wye–delta, delta–wye, wye–wye, or delta–delta).
4. The per unit method is independent of voltage changes and phase shifts through transformers, where the base

voltages in the windings are proportional to the number of turns in the windings.

5. Manufacturers usually specify the impedance of equipment in per unit or percent on the base of its nameplate rating of power (usually kVA) and voltage (V or kV). Thus the rated impedance can be used directly if the bases chosen are the same as the nameplate ratings.
6. The per unit impedance values of various ratings of equipment lie in a narrow range, while the actual ohmic values may vary widely. Therefore, where actual values are not known, a good approximate value can be used. Typical values for various types of equipment are available from many sources and reference books. Also, the correctness of a specified unit can be checked knowing the typical values.
7. There is less chance of confusion between single-phase power and three-phase power, or between line-to-line voltage and line-to neutral voltage.
8. The per unit method is very useful for simulating the steady-state and transient behavior of power systems on computers.
9. The driving or source voltage usually can be assumed to be 1.0 pu for fault and voltage calculations.
10. With per unit the product of two quantities expressed in per unit is expressed in per unit itself. However, the product of two quantities expressed in percent must be divided by 100 to obtain the result in percent. For this reason it is desirable to use per unit rather than percent in computations.

2.4 GENERAL RELATIONSHIPS BETWEEN CIRCUIT QUANTITIES

Before continuing the discussion of the per unit method, a review of some general relationships between circuit quantities applicable to all three-phase power systems is in order. This will focus on the two basic types of connections, wye and delta, as shown in Fig. 2.1. For either of these the following basic equations apply*:

$$S_{3\phi} = \sqrt{3}\, V_{LL} I_L \tag{2.3}$$

$$V_{LL} = \sqrt{3}\, V_{LN}\angle{+30^\circ} \tag{2.4}$$

$$I_L = \frac{S_{3\phi}}{\sqrt{3}\, V_{LL}} \tag{2.5}$$

From these three equations the value of the impedances and the delta current can be determined.

Wye-connected impedances (Fig. 2.1a)

$$\begin{aligned} Z_Y &= \frac{V_{LN}}{I_L} = \frac{V_{LL}\angle{-30^\circ}}{\sqrt{3}} \times \frac{\sqrt{3}\, V_{LL}}{S_{3\phi}} \\ &= \frac{V_{LL}^2\angle{-30^\circ}}{S_{3\phi}} \end{aligned} \tag{2.6}$$

Delta-connected impedances (Fig. 2.1b)

$$I_D = \frac{I_L\angle{+30^\circ}}{\sqrt{3}} \tag{2.7}$$

$$\begin{aligned} Z_D &= \frac{V_{LL}}{I_D} = \frac{\sqrt{3}\, V_{LL}\angle{-30^\circ}}{I_L} \\ &= \frac{\sqrt{3}\, V_{LL}\angle{-30^\circ}}{} \times \frac{\sqrt{3}\, V_{LL}}{S_{3\phi}} \\ &= \frac{3V_{LL}^2\angle{-30^\circ}}{S_{3\phi}} \end{aligned} \tag{2.8}$$

$$I_D = \frac{V_{LL}}{Z_D} = \frac{S_{3\phi}\angle{+30^\circ}}{3V_{LL}} \tag{2.9}$$

* S is the apparent or complex power in volt-amperes (VA, kVA, MVA), P is the active power in watts (W, kW, MW), and Q is the reactive power in vars (var, kvar, Mvar). Thus $S = P + jQ$.

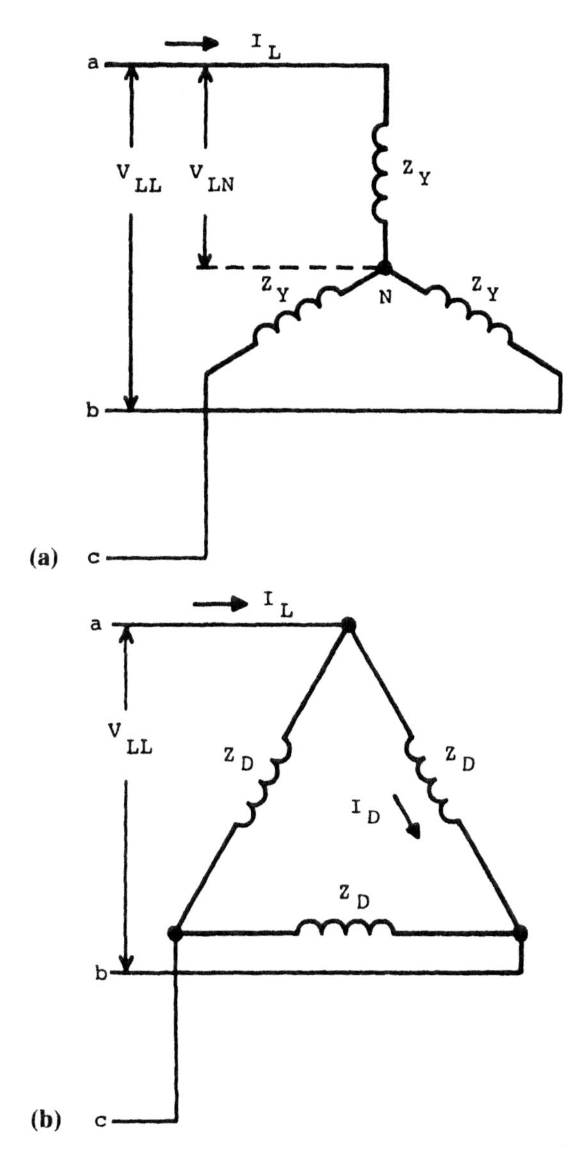

Figure 2.1 Impedances in three-phase wye and delta circuits. (a) Wye-connected impedances. (b) Delta-connected impedances.

These equations show that the circuit quantities S, V, I, and Z are so related that the selection of any two of them determines the values of the remaining two quantities. Usually, the wye connection is assumed, so Eqs. (2.3) through (2.6) are most commonly used for power system calculations. A great deal of confusion can be avoided by clearly remembering that wye connections are assumed and not delta connections, or vice versa. If a delta connection is given, it can be converted into an equivalent wye connection for calculation purposes. Alternatively, Eqs. (2.8) and (2.9) can be used directly, if the need arises, to express the impedance and current in terms of delta circuit quantities.

2.5 BASE QUANTITIES

In the following chapters it is more convenient to use the notation kVa or MVA instead of S, and kV instead of V. The base quantities are scalar quantities, so that phasor notation is not required for the base equations. Thus equations for the base values can be expressed from Eqs. (2.3), (2.5), and (2.6), with the subscript B to indicate a base quantity as follows:

$$\text{For base power: } \text{kVA}_B = \sqrt{3}\,\text{kV}_B I_B \tag{2.10}$$

$$\text{For base current: } I_B = \frac{\text{kVA}_B}{\sqrt{3}\,\text{kV}_B} \quad \text{amperes} \tag{2.11}$$

$$\text{For base impedance: } Z_B = \frac{\text{kV}_B^2 \times 1000}{\text{kVA}_B} \quad \text{ohms} \tag{2.12}$$

and since

$$1000 \times \text{MVA} = \text{kVA} \tag{2.13}$$

the base impedance can also be expressed as

$$Z_B = \frac{\text{kV}_B^2}{\text{MVA}_B} \quad \text{ohms} \tag{2.14}$$

In three-phase electric power systems the common practice is to use the standard or nominal system voltage as the voltage base, and a convenient MVA or kVA quantity as the power base. One widely used power base is 100 MVA. The system voltage commonly specified is the voltage between the three phases (i.e., the line-to-line voltage). This is the voltage used as a base in Eqs. (2.10) through (2.14). As a shortcut and for convenience, the line-to-line subscript designation (LL) is omitted. With this practice it is always understood that the voltage is the line-to-line value unless indicated otherwise.

The *major exception* is in the method of symmetrical components, where line-to-neutral phase voltage is used. This should always be specified carefully, but there is sometimes a tendency to overlook this step. Similarly, current is always the phase or line-to-neutral current unless otherwise specified.

Power is always understood to be three-phase power unless otherwise indicated. General power, also known as complex or apparent power, is designated by MVA or kVA, as indicated above. Three-phase power is designated by MW or kW. Three-phase reactive power is designated by RMVA or RkVA.

2.6 PER UNIT AND PERCENT IMPEDANCE RELATIONSHIPS

Per unit impedance is the ratio of Eq. (2.1), using actual ohms (Z_{Ω}) and the base ohms of Eq. (2.14):

$$Z_{\text{pu}} = \frac{Z_{\Omega}}{Z_B} = \frac{\text{MVA}_B Z_{\Omega}}{\text{kV}_B^2} \quad \text{or} \quad \frac{\text{kVA}_B Z_{\Omega}}{1000\ \text{kV}_B^2} \tag{2.15}$$

or in percent notation,

$$\%Z = \frac{100\ \text{MVA}_B Z_{\Omega}}{\text{kV}_B^2} \quad \text{or} \quad \frac{\text{kVA}_B Z_{\Omega}}{10\ \text{kV}_B^2} \tag{2.16}$$

Where the ohm values are desired from per unit, percent values, the equations are

$$Z_{\Omega} = \frac{\text{kV}_B{}^2 Z_{\text{pu}}}{\text{MVA}_B} \quad \text{or} \quad \frac{1000\ \text{kV}_B{}^2 Z_{\text{pu}}}{\text{kVA}_B} \tag{2.17}$$

$$Z_{\Omega} = \frac{\text{kV}_B{}^2 (\%Z)}{100\ \text{MVA}_B} \quad \text{or} \quad \frac{10\ \text{kV}_B{}^2 (\%Z)}{\text{kVA}_B} \tag{2.18}$$

The impedance values may be either scalars or phasors. The equations are also applicable for resistance and for reactance calculations.

Per unit is recommended for calculations involving division, as it is less likely to result in a decimal-point error. However, the choice of per unit or percent is personal. It is often convenient to use both, but care should be used.

Careful and overredundant labeling of all answers is *recommended very strongly*. This is valuable in identifying a value or answer, particularly later, when you or others refer to the work. Too often, answers such as 106.8, for example, are indicated without any label. To others, or later when memory is not fresh, questions can arise, such as: "What is this? amperes? volts? per unit? what?" Initially, the proper units were obvious, but to others, or later, they may not be. A little extra effort and the development of the good habit of labeling leaves no frustrating questions, doubts, or tedious rediscovery later.

Currents in amperes and impedances in ohms should be referred to a specific voltage base or to primary or secondary windings of transformers. Voltages in volts should be clear as to whether they are primary, secondary, high, low, and so on, quantities.

When per unit or percent values are specified for impedances, resistance, or reactance, *two bases* must be indicated. These are the MVA (or kVA) and the kV bases using Eqs. (2.15) through (2.18). Without the two bases the per unit or percent values are meaningless. For electrical equipment these two bases are the rated values cited on the equipment nameplate or on the manufacturer's drawings or other data supplied. Where several ratings are specified, generally it is correct to assume that the normal rated values were used to determine the per unit or percent values specified. Fundamentally, the manufacturer should specifically indicate the bases where several ratings exist.

System drawings should clearly indicate the MVA (or kVA) base with the base voltages indicated for the various voltage levels shown, where all the impedance components have been reduced to one common base value. Otherwise, the per unit or percent impedances with their two bases must be indicated for every piece of equipment or circuit on the drawing.

For per unit or percent voltages, only the voltage base is required. Thus a 90% voltage on a 138-kV system would be 124.2 kV. For per unit or percent currents, one or two bases are required. If the base current is specified, that is sufficient. A 0.90-pu current with a 1000-A base specifies that the current is 900 A. Where the more common MVA (or kVA) and kV bases are given, Eq. (2.1i), with Eq. (2.13), provides the base current required. Thus with 100-MVA 138-kV bases, the base current is

$$I_B = \frac{1000 \times 100}{\sqrt{3} \times 138} = 418.37 \text{ A at } 138 \text{ kV} \tag{2.19}$$

Thus 418.37 A is 1 pu or 100% current in the 138-kV system.

2.7 PER UNIT AND PERCENT IMPEDANCES OF TRANSFORMER UNITS

As indicated in Section 2.3, a major advantage of the per unit (percent) system is its independence of voltage and phase shifts through transformer banks, *where the base voltages on the different terminals of the transformer are proportional to the turns in the corresponding windings.*

This can be demonstrated by the following analysis. From basic fundamentals, the impedance on one side of a transformer is reflected through the transformer by the square of the turns ratio, or where the voltages are proportional to the turns by the square of the voltage ratio. Thus for one phase of a transformer

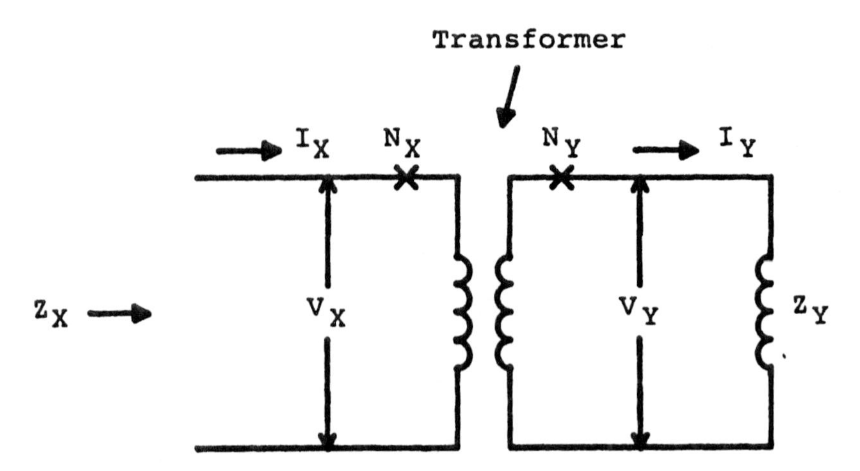

Figure 2.2 Impedances through one phase of a three-phase transformer.

as shown in Fig. 2.2, the impedance Z_y on the N_y turns winding appears as Z_x on the N_x turns-winding side, as

$$Z_x = \left(\frac{N_x}{N_y}\right)^2 Z_y = \left(\frac{V_x}{V_y}\right)^2 Z_y \tag{2.20}$$

The impedance bases on the two sides of the transformer are, from Eq. (2.14),

$$Z_{xB} = \frac{\mathrm{kV}_x^{\,2}}{\mathrm{MVA}_B} \quad \text{where kV}_x \text{ is the } x\text{-side base} \tag{2.21}$$

$$Z_{yB} = \frac{\mathrm{kV}_y^{\,2}}{\mathrm{MVA}_B} \quad \text{where kV}_y \text{ is the } y\text{-side base} \tag{2.22}$$

Taking the ratio of Z_{xB} and Z_{yB} yields

$$\frac{Z_{xB}}{Z_{yB}} = \frac{\mathrm{kV}_x^{\,2}}{\mathrm{kV}_y^{\,2}} = \left(\frac{N_x}{N_y}\right)^2 \tag{2.23}$$

where the turns are proportional to the voltages.

The per unit impedances are, from Eqs. (2.1), (2.20), and (2.24),

$$
\begin{aligned}
Z_x(\text{pu}) &= \frac{Z_x\ (\text{ohms})}{Z_{xB}} = \left(\frac{N_x}{N_y}\right)^2 \left(\frac{N_y}{N_x}\right)^2 \frac{Z_y\ (\text{ohms})}{Z_{yB}} \\
&= \frac{Z_y\ (\text{ohms})}{Z_{yB}} = Z_y(\text{pu})
\end{aligned}
\tag{2.24}
$$

Thus the per unit impedance is the same on either side of the bank.

2.7.1 Example: Transformer Bank

Consider a transformer bank rated 50 MVA with 34.5- and 161-kV windings connected to a 34.5- and a 161-kV power system. The bank reactance is 10%. Now when looking at the bank from the 34.5-kV system, its reactance is

10% on a 50-MVA 34.5-kV base (2.25)

and when looking at the bank from the 161-kV system, its reactance is

10% on a 50-MVA 161-kV base (2.26)

This equal impedance in percent or per unit on either side of the bank is independent of the bank connections: wye–delta, delta–wye, wye–wye, or delta–delta.

This means that the per unit (percent) impedance values throughout a network can be combined independently of the voltage levels as long as all the impedances are on a common MVA (kVA) base and the transformer windings ratings are compatible with the system voltages. This is a great convenience.

The actual transformer impedances in ohms are quite different on the two sides of a transformer with different voltage levels.

This can be illustrated for the example. Applying Eq. (2.18), we have

$$jX = \frac{34.5^2 \times 10}{100 \times 50} = 2.38\ \Omega \text{ at } 34.5\text{ kV} \tag{2.27}$$

$$= \frac{161^2 \times 10}{100 \times 50} = 51.84\ \Omega \text{ at } 161\text{ kV} \tag{2.28}$$

This can be checked by Eq. (2.20), where for the example x is the 34.5-kV winding side and y is the 161-kV winding side. Then

$$2.38 = \frac{34.5^2}{161^2} \times 51.84 = 2.38 \tag{2.29}$$

2.8 CHANGING PER UNIT (PERCENT) QUANTITIES TO DIFFERENT BASES

Normally, the per unit or percent impedances of equipment are specified on the equipment base, which generally will be different from the power system base. Since all impedances in the system must be expressed on the same base for per unit or percent calculations, it is necessary to convert all values to the common base selected. This conversion can be derived by expressing the same impedance in ohms on two different per unit bases. From Eq. (2.15) for a MVA_1, kV_1 base and a MVA_2, kV_2 base,

$$Z_{1\text{pu}} = \frac{\text{MVA}_1 Z_\Omega}{\text{kV}_1^2} \tag{2.30}$$

$$Z_{2\text{pu}} = \frac{\text{MVA}_2 Z_\Omega}{\text{kV}_2^2} \tag{2.31}$$

By ratioing these two equations and solving for one per unit value, the general equation for changing bases is

$$\frac{Z_{2\text{pu}}}{Z_{1\text{pu}}} = \frac{\text{MVA}_2}{\text{kV}_2^2} \times \frac{\text{kV}_1^2}{\text{MVA}_1} \tag{2.32}$$

$$Z_{2pu} = Z_{1pu} \frac{MVA_2}{MVA_1} \times \frac{kV_1^2}{kV_2^2} \tag{2.33}$$

Equation (2.33) is the general equation for changing from one base to another base. In most cases the turns ratio of the transformer is equivalent to the different system voltages, and the equipment-rated voltages are the same as the system voltages, so that the voltage-squared ratio is unity. Then Eq. (2.33) reduces to

$$Z_{2pu} = Z_{1pu} \frac{MVA_2}{MVA_1} \tag{2.34}$$

It is very important to emphasize that the voltage-square factor of Eq. (2.33) is used *only* in the same voltage level and where slightly different voltage bases exist. It is *never* used where the base voltages are proportional to the transformer bank turns, such as going from the high to the low side across a bank. In other words, Eq. (2.33) has nothing to do with transferring the ohmic impedance value from one side of a transformer to the other side.

Several examples will illustrate the applications of Eqs. (2.33) and (2.34) in changing per unit and percent impedances from one base to another.

2.8.1 Example: Base Conversion with Eq. (2.34)

The 50-MVA 34.5:161-kV transformer with 10% reactance is connected to a power system where all the other impedance values are on a 100-MVA 34.5-kV or 161-kV base. To change the base of the transformer, Eq. (2.34) is used since the transformer and system base voltages are the same. This is because if the fundamental equation (2.33) were used,

$$\frac{kV_1^2}{kV_2^2} = \left(\frac{34.5}{34.5}\right)^2 \quad \text{or} \quad \left(\frac{161}{161}\right)^2 = 1.0 \tag{2.35}$$

so Eq. (2.34) results, and the transformer reactance becomes

$$jX = 10\% \times \frac{100}{50} = 20\% \quad \text{or} \quad 0.20 \text{ pu} \tag{2.36}$$

on a 100-MVA 34.5-kV base from the 34.5-kV side or on a 100-MVA 161-kV base from the 161-kV side.

2.8.2 Example: Base Conversion Requiring Eq. (2.33)

A generator and transformer, shown in Fig. 2.3, are to be combined into a single equivalent reactance on a 100-MVA 110-kV base. With the transformer bank operating on its 3.9-kV tap, the low-side base voltage corresponding to the 110-kV high-side base is

$$\frac{\text{kV}_{LV}}{110} = \frac{3.9}{115} \quad \text{or} \quad \text{kV}_{LV} = 3.73 \text{ kV} \tag{2.37}$$

Since this 3.73-kV base is different from the specified base of the generator subtransient reactance, Eq. (2.33) must be used:

$$jX''_d = 25\% \times \frac{100 \times 4^2}{25 \times 3.73^2} = 115\% \text{ or } 1.15 \text{ pu} \quad \begin{array}{r}\text{on 100-MVA 3.73-kV base}\\ \text{or on 100-MVA 110-kV base}\end{array} \tag{2.38}$$

Similarly, the transformer reactance on the new base is

$$\begin{aligned} jX_T &= 10\% \times \frac{100 \times 3.9^2}{30 \times 3.73^2} = 10\% \times \frac{100 \times 115^2}{30 \times 110^2} \\ &= 36.43\% \text{ or } 0.364 \text{ pu} \quad \begin{array}{r}\text{on 100-MVA 3.73-kV base}\\ \text{or on 100-MVA 110-kV base}\end{array} \end{aligned} \tag{2.39}$$

Now the generator and transformer reactances can be combined into one equivalent source value by adding:

$$115\% + 36.43\% = 151.43\%$$

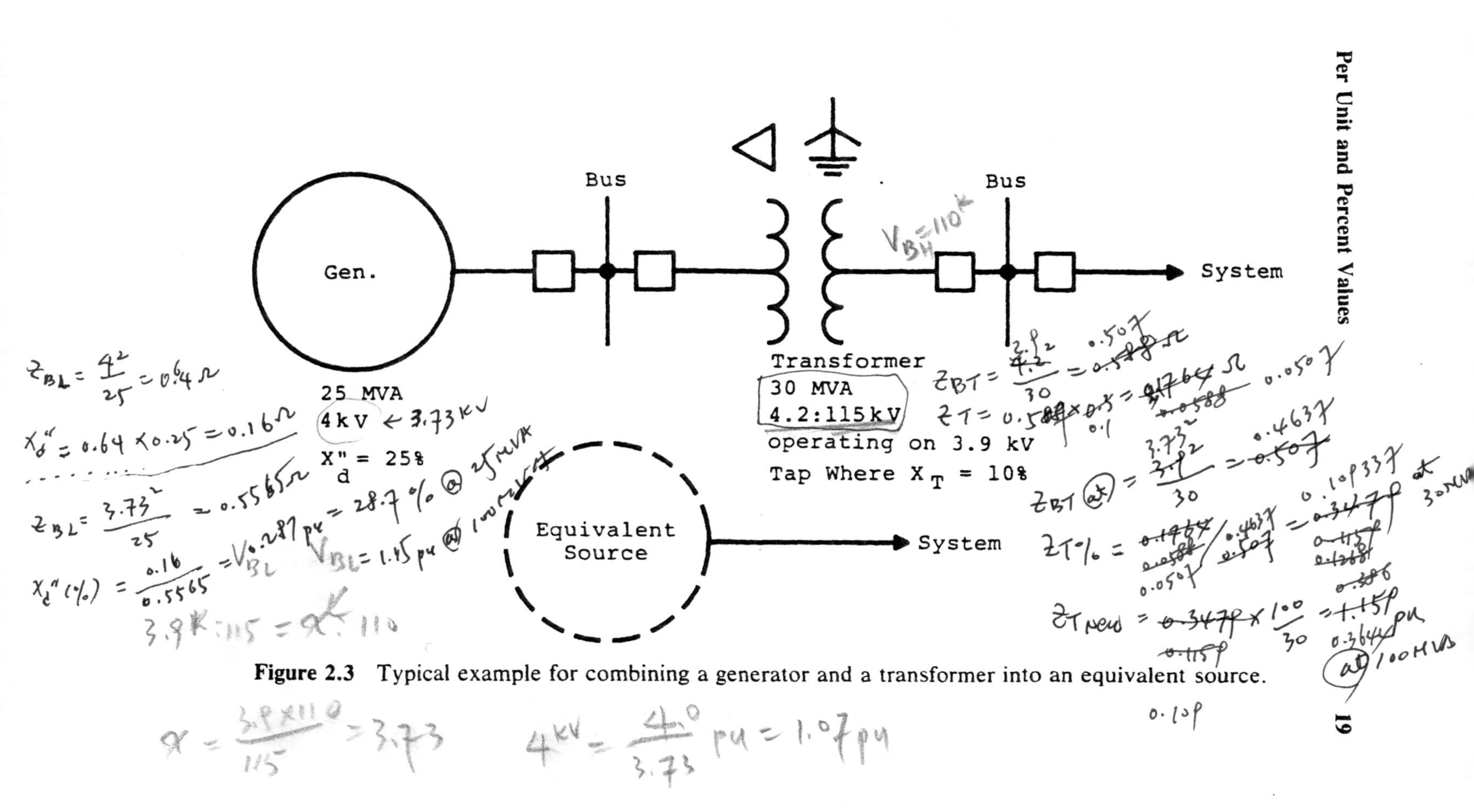

Figure 2.3 Typical example for combining a generator and a transformer into an equivalent source.

or

$$1.15 \text{ pu} + 0.3643 \text{ pu} = 1.514 \text{ pu, both on a 100-MVA 110-kV base} \tag{2.40}$$

The previous warning bears repeating and emphasizing. Never, *never*, NEVER use Eq. (2.33) with voltages on the opposite sides of transformers. Thus the factors $(115/3.9)^2$ and $(110/3.73)^2$ in Eq. (2.33) are *incorrect*.

3
Phasors, Polarity, and System Harmonics

3.1 INTRODUCTION

Phasors and transformer polarity are important in analyzing, documenting, and understanding power system operation and the currents and voltages associated with faults and unbalances. Therefore, a sound theoretical and practical knowledge of these is a fundamental and valuable resource.

3.2 PHASORS

The *IEEE Dictionary* (IEEE 100-1984) defines a phasor as "a complex number. Unless otherwise specified, it is used only within the context of steady state alternating linear systems." It continues: "the absolute value (modulus) of the complex number corresponds to either the peak amplitude or root-mean-square (rms) value of the quantity, and the phase (argument) to the phase angle at zero time. By extension, the term 'phasor' can also be applied to impedance, and related complex quantities that are not time dependent."

In this book, phasors will be used to document various ac voltages, currents, fluxes, impedances, and power. For many years phasors were referred to as "vectors," but this use is deprecated to avoid confusion with space vectors. However, the former use lingers on, so occasionally a lapse to "vectors" may occur.

3.2.1 Phasor Representation

The common pictorial form for representing electrical and magnetic phasor quantities uses the cartesian coordinates with x (the abscissa) as the axis of real quantities and y (the ordinate) as the axis of imaginary quantities. This is illustrated in Fig. 3.1. Thus a point c on the complex plane x-y can be represented as shown in this figure, and documented mathematically by the several alternative forms given in Eq. (3.1).

Phasor form		Rectangular form		Complex form		Exponential form		Polar form
c	$=$	$x + jy$	$=$	$\lvert c \rvert(\cos\phi + j\sin\phi)$	$=$	$\lvert c \rvert e^{j\phi}$	$=$	$\lvert c \rvert \angle{+\phi}$
								(3.1)

Sometimes useful is the conjugate form:

$$c^* = = x - jy = \lvert c \rvert (\cos\phi - j\sin\phi)$$
$$= \lvert c \rvert e^{-j\phi} = \lvert c \rvert \angle{-\phi} \quad (3.2)$$

where

c = the phase
c^* = its conjugate
x = the real value (alternate: Re c or c')
y = the imaginary value (alternate: Im c or c'')
$\lvert c \rvert$ = the modulus (magnitude or absolute value)
ϕ = the phase angle (argument or amplitude) (alternate: arg c)

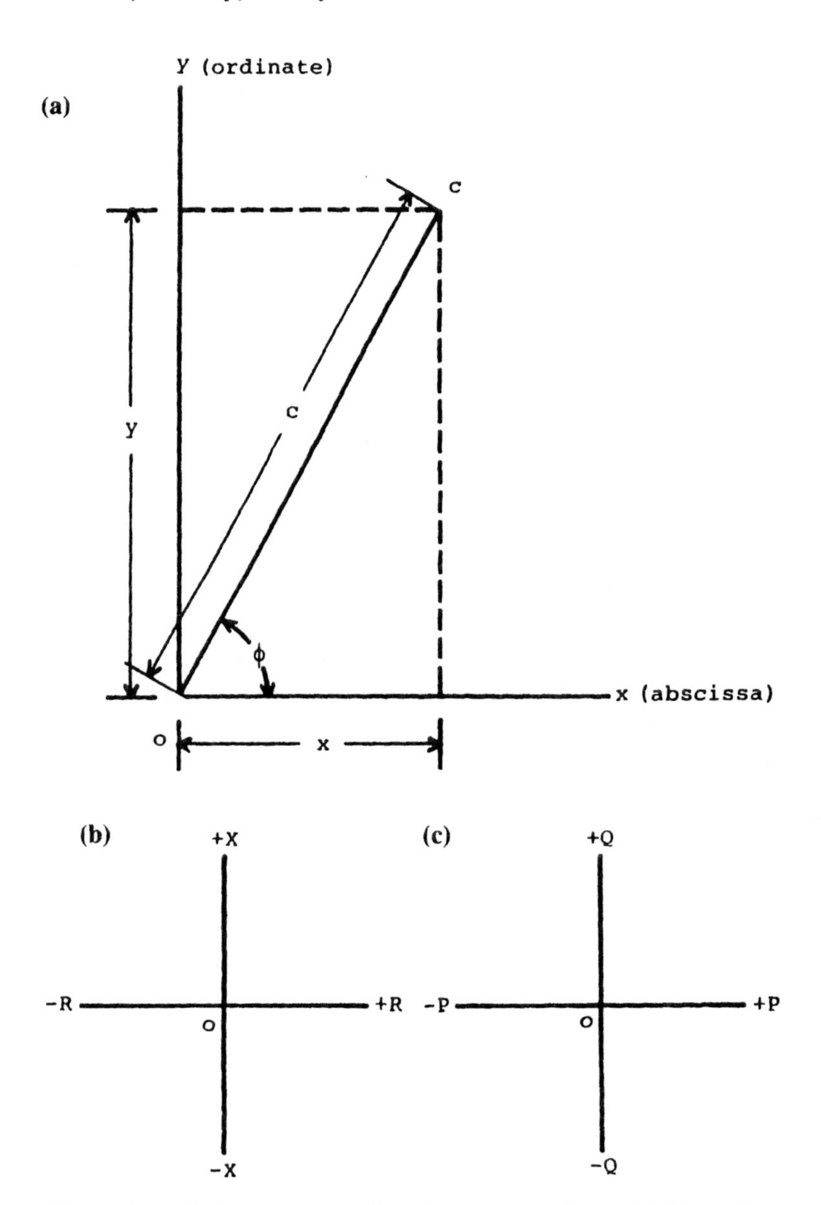

Figure 3.1 Reference axes for phasor quantities: (a) Cartesian x − y coordinates; (b) impedance phasor axes; (c) power phasor axes.

The modulus (magnitude or absolute value) of the phasor is

$$|c| = \sqrt{x^2 + y^2} \tag{3.3}$$

From Eqs. (3.1) and (3.3),

$$x = \tfrac{1}{2}(c + c^*) \tag{3.4}$$

$$y = \frac{1}{2j}(c - c^*) \tag{3.5}$$

3.2.2 Phasor Diagrams for Sinusoidal Quantities

In applying the notation above to sinusoidal (ac) voltages, currents, and fluxes, the axes are assumed fixed, with the phasor quantities rotating at constant angular velocity. The international standard is that *phasors always rotate in the counterclockwise direction.* However, as a convenience, on the diagrams the phasor is always shown "fixed" for the given condition. The magnitude of the phasor (c) can be either the maximum peak value or the rms value of the corresponding sinusoidal quantity. In normal practice it represents the rms maximum value of the positive half-cycle of the sinusoid unless otherwise specifically stated.

Thus a phasor diagram shows the respective voltages, currents, fluxes, and so on, that exist in the electrical circuit. It should document *only* the magnitude and relative phase-angle relations between these various quantities. Thus all phasor diagrams require a scale or complete indications of the physical magnitudes of the quantities shown. The phase-angle reference usually is between the quantities shown, so that the zero (or reference angle) may vary with convenience. As an example, in *fault calculations* using reactance (X) only, it is convenient to use the voltage (V) reference at $+90°$. Then $I = jV/jX$ and the j value cancels, so the fault current does not involve the j factor. On the other hand, in *load calculations* it is preferable to use the voltage (V) at 0° or along the x axis so that the angle of the current (I) represents its actual lag or lead value.

Other reference axes that are in common use are shown in Fig. 3.1b and c. For plotting impedance, resistance, and reactance, the R–X axis of Fig. 3.1b is used. Inductive reactance is $+X$ and capacitive reactance is $-X$.

For plotting power phasors, Fig. 3.1c is used. P is the real power (W, kW, MW) and Q is the reactive power (var, kvar, Mvar). These impedance and power diagrams are discussed in later chapters. While represented as phasors, the impedance and power "phasors" do not rotate at system frequency.

3.2.3 Combining Phasors

The various laws for combining phasors are present for general reference.

Multiplication

The magnitudes are multiplied and the angles added:

$$VI = |V||I| \angle \phi V + \phi I \tag{3.6}$$

$$VI^* = |V||I| \angle \phi V - \phi I \tag{3.7}$$

$$II^* = |I|^2 \tag{3.8}$$

Division

The magnitudes are divided and the angles subtracted:

$$\frac{V}{I} = \frac{|V|}{|I|} \angle \phi V - \phi I \tag{3.9}$$

Powers

$$(I)^n = (|I| e^{j\phi})^n = |I|^n e^{j\phi n} \tag{3.10}$$

$$\sqrt[n]{I} = \sqrt[n]{|I|} \, e^{j(\phi/n)} \tag{3.11}$$

3.2.4 Phasor Diagrams Require a Circuit Diagram

The phasor diagram defined above has a indeterminate or vague meaning unless it is accompanied by a circuit diagram. The circuit diagram identifies the specific circuit involved with the location and assumed direction for the currents and the location and assumed polarity for the voltages to be documented in the phasor diagram. The assumed directions and polarities are not critical, as the phasor diagram will confirm if the assumptions were correct and provide the correct magnitudes and phase relations. These two complementary diagrams (circuit and phasor) are preferably kept separate to avoid confusion and errors in interpretation. This is discussed further in Section 3.3.

3.2.5 Nomenclature for Current and Voltage

Unfortunately, there is no standard nomenclature for current and voltage, so confusion can exist among various authors and publications. The nomenclature used throughout this book has proven to be flexible and practical over many years of use, and it is compatible with power system equipment polarities.

Current and Flux

In the circuit diagrams, current or flux is shown by either (1) a letter designation, such as I or θ, with an arrow indicator for the assumed direction of flow; or (2) a letter designation with double subscripts, the order of the subscripts indicating the assumed direction. The direction is that assumed to be the flow during the positive half-cycle of the sine wave. This convention is illustrated in Fig. 3.2a. Thus in the positive half-cycle the current in the circuit is assumed to be flowing from left to right, as indicated by the direction of the arrow used with I_s, or denoted by subscripts, as with I_{ab}, I_{bc}, and I_{cd}. The single subscript, such as I_s, is a convenience to designate currents in various parts of a circuit and has no directional indication, so an arrow for the direction must be associated with these. Arrows are not required with I_{ab}, I_{bc}, or I_{cd} but are often used for added clarity and convenience.

It is very important to appreciate that in these circuit designations, the arrows do *not* indicate phasors. They are only assumed directional and location indicators.

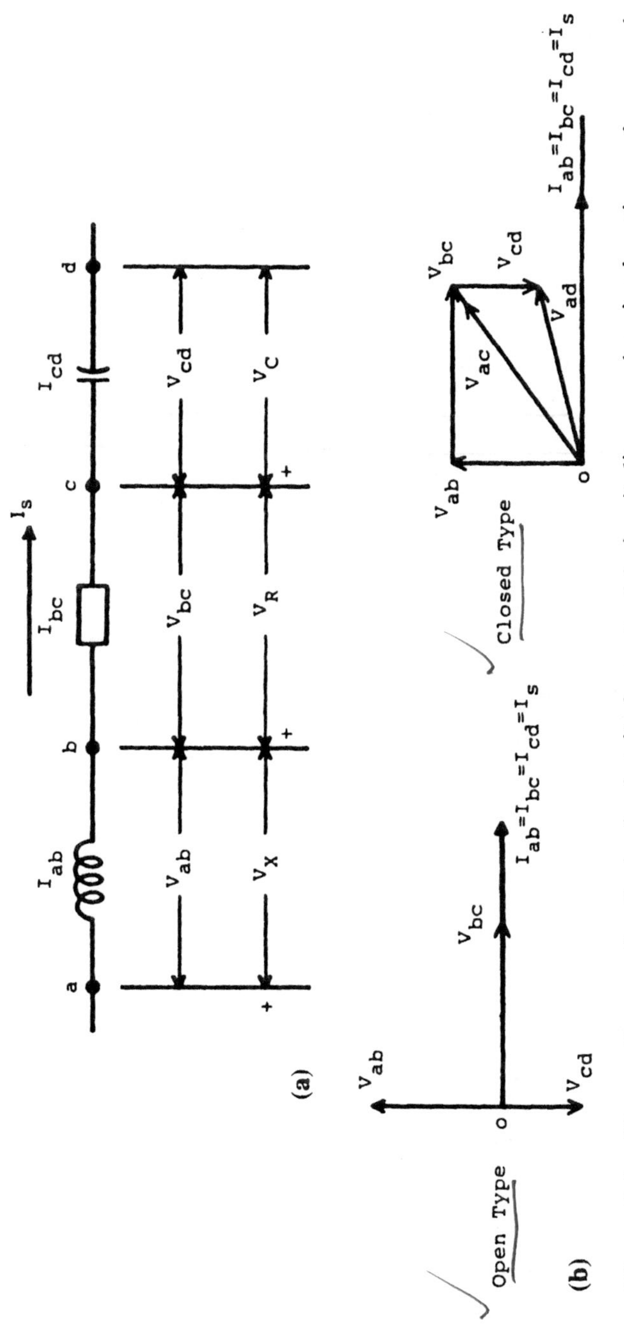

Figure 3.2 Phasor diagram for the basic circuit elements: (a) circuit diagram showing location and assumed directions of current and of voltage drops; (b) phasor diagrams showing current and voltage magnitudes and phase relations.

Voltage

Voltages can be either drops or rises. Much confusion can result by not indicating clearly which is intended or by mixing the two practices in circuit diagrams. This can be avoided by standardizing on one, and *only one*, practice. As voltage drops are far more common throughout the power system, *all* voltages are shown and *always* considered to be *drops* from a higher voltage to a lower voltage. This convention is independent of whether the letter V or E is used for the voltage.

The consistent adoption of *only drops* throughout need cause no difficulties. A generator or source voltage becomes a minus drop since current flows from a lower voltage to a higher voltage. This practice does not conflict with the polarity of equipment, such as transformers, and it is consistent with fault calculations using symmetrical components.

Voltages (always drops) are indicated by either (1) a letter designation with double subscripts, or (2) a small plus (+) indicator shown at the point assumed to be at a relatively high potential. Thus during the positive half-cycle of the sine wave, the voltage drop is indicated by the order of the two subscripts when used, or from the "+" indicator to the opposite end of the potential difference. This is illustrated in Fig. 3.2a, where both methods are shown. It is preferable to show arrows at both ends of the voltage drop designations, to avoid possible confusion. Again, it is most important to recognize that both of these designations, especially where arrows are used, in the circuit diagrams are only location and direction indicators, *not phasors*.

It may be helpful to consider current as a "through" quantity and voltage as an "across" quantity. In this sense, in the representative Fig. 3.2a, the same current flows *through* all the elements in series, so that $I_{ab} = I_{bc} = I_{cd} = I_s$. By contrast, voltage V_{ab} applies only *across* nodes a and b, voltage V_{bc} *across* nodes b and c, and V_{cd} *across* nodes c and d.

3.2.6 The Phasor Diagram

With the proper identification and assumed directions established in the circuit diagram, the corresponding phasor diagram can be drawn from calculated or test data. For the circuit diagram of Fig. 3.2a, two types of phasor diagrams are shown in Fig. 3.2b. The top diagram is referred to as an "open-type" diagram, where all the phasors originate from a common origin. The bottom diagram is referred to as a "closed type," where the voltage phasors are summed together from left to right for the same series circuit. Both types are useful, but the open type is preferred to avoid the confusion that may occur with the closed type. This is amplified in Section 3.3.

3.3 CIRCUIT AND PHASOR DIAGRAMS FOR A BALANCED THREE-PHASE POWER SYSTEM

A typical section of a three-phase power system is shown in Fig. 3.3a. Optional grounding impedances (Z_{Gn}) and (Z_{Hn}) are omitted with solid grounding. (R_{sg}) and (R_{ssg}) represent the ground-mat resistance in the station or substation. Ground g or G represents the potential of the true earth, remote ground plane, and so on. The system neutrals n', n or N, and n'' are not necessarily the same unless a fourth wire is used, as in a four-wire three-phase system. Upper- or lowercase N and n are used interchangeably as convenient for the neutral designation.

The various line currents are assumed to flow through this series section as shown, and the voltages are indicated for a specific point on the line section. These follow the nomenclature discussed previously. To simplify the discussion at this point, symmetrical or balanced operation of the three-phase power system is assumed. Therefore, no current can flow in the neutrals of the two transformer banks, so that with this simplification there is no difference of voltage between n', n or N, n'' and the ground plane g or G. As a result, $V_{an} = V_{ag}$, $V_{bn} = V_{bg}$, and $V_{cn} = V_{cg}$. Again, this is true only for a balanced or symmetrical system. With this the respective currents and voltages are equal in magnitude and 120° apart in phase, as shown in the phasor diagram (Fig. 3.3b), in both the open and closed types. The phasors for various unbalanced and fault conditions are discussed later.

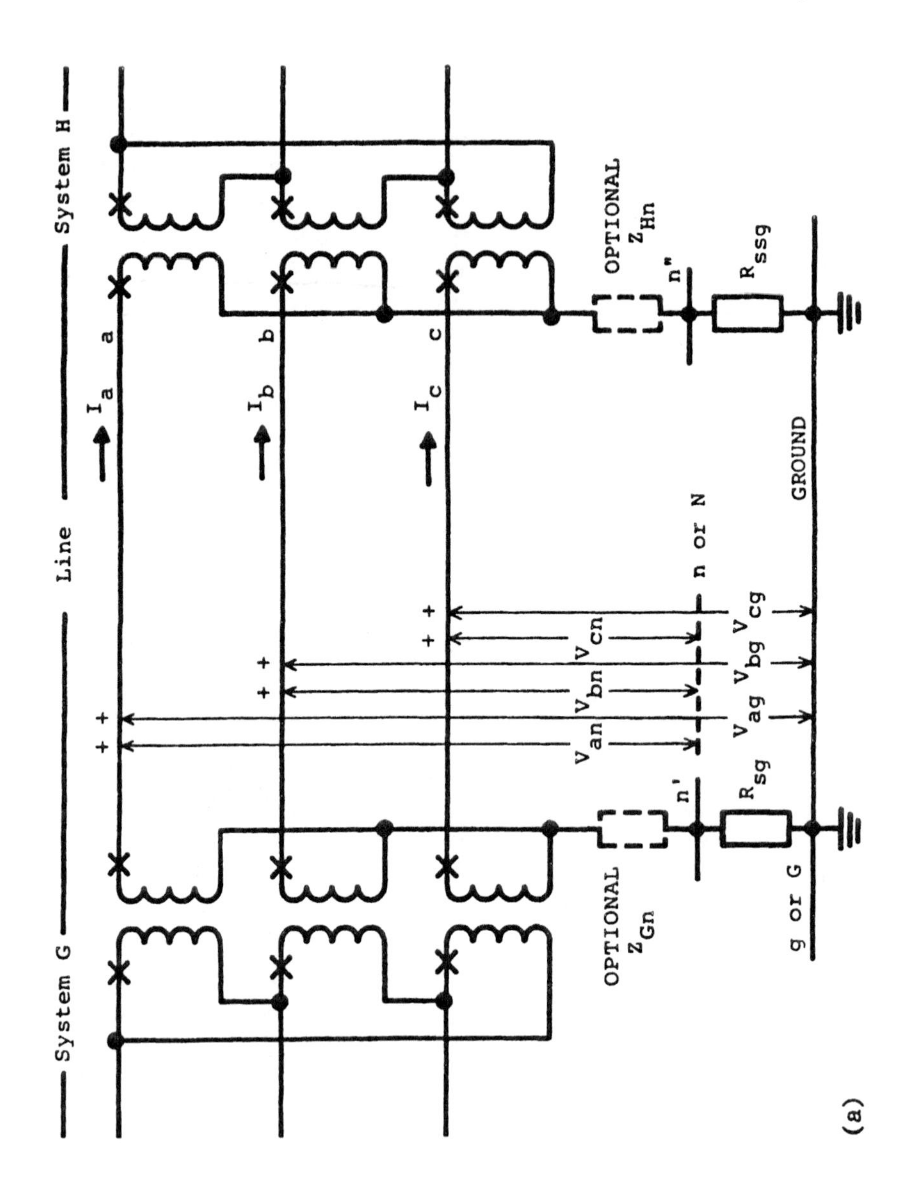
System G
Line
System H
I_a
I_b
I_c
a
b
c
OPTIONAL Z_{Gn}
OPTIONAL Z_{Hn}
n'
n"
n or N
R_{sg}
R_{ssg}
V_{an}
V_{bn}
V_{cn}
V_{ag}
V_{bg}
V_{cg}
GROUND
g or G

(a)

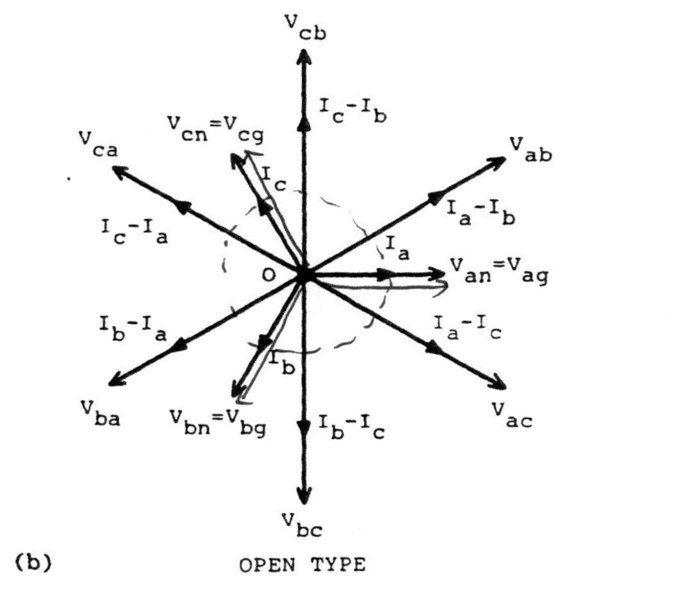

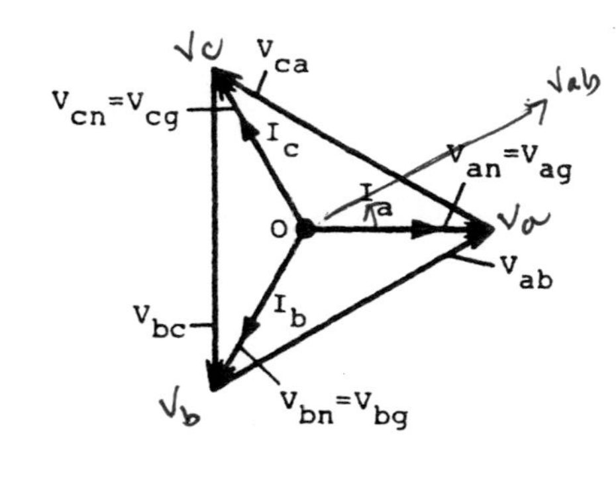

Figure 3.3 Phasor diagram for a typical three-phase circuit operating with balanced or symmetrical quantities: (a) circuit diagram showing location and assumed directions of current and voltage drops; (b) phasor diagrams showing current and voltage magnitudes and phase relations.

The open-type phasor diagram permits easy documentation of all possible currents and voltages, some of which are not convenient in the closed-type phasor diagram. The delta voltage V_{ab}, representing the voltage (drop) from phase *a* to phase *b*, is the same as $V_{an} - V_{bn}$. Similarly, $V_{bc} = V_{bn} - V_{cn}$ and $V_{ca} = V_{cn} - V_{an}$.

As indicated, the closed-type phasor diagram can lead to difficulties. As seen in Fig. 3.3b, its shape lends itself mentally to an assumption that the three vertices of the triangle represent the *a*, *b*, and *c* phases of the power system, and that the origin *O* represents $n = g$. Questions arise with this closed type of phasor diagram as to why $V_{an} = V_{ag}$ has its phasor arrow as shown since the voltage drop is from phase *a* to neutral, and similarly for the other two phases; also why V_{ab}, V_{bc}, and V_{ca} are pointing as shown, as they are drops from phase *a* to phase *b*, phase *b* to phase *c*, and phase *c* to *a*, respectively. It would appear that they should be pointing in the opposite direction.

The phasors shown on this closed phasor diagram (Fig. 3.3b) are absolutely *correct* and must not be changed. The difficulty is in the combining a circuit diagram with the phasor diagram by the mental association of *a*, *b*, and *c* to the closed triangle. The open type avoids this difficulty. This also emphasizes the desirability of having *two separate* diagrams: a circuit diagram and a phasor diagram. Each serves particular but quite different functions.

3.4 PHASOR AND PHASE ROTATION

"Phasor" and "phase rotation" are two entirely different terms, although they look almost alike. *Ac phasors* are always rotating counterclockwise at the system frequency. The fixed diagrams, plotted such as in Fig. 3.3b, represent what would be seen if a stroboscopic light of system frequency were imposed on the system phasors. The phasors would appear fixed in space as plotted.

In contrast, *phase rotation* or *phase sequence* refers to the order in which the phasors occur as they rotate counterclockwise. The standard sequence today is: *a*, *b*, *c*; *A*, *B*, *C*; 1, 2, 3;

or in some areas *r*, *s*, *t*. In Fig. 3.3b the sequence is *a*, *b*, *c*. The *IEEE Dictionary* (IEEE 100-1984) only defines phase sequence, so this is preferred. However, phase rotation has been used over many years and is still used in practice.

Not all power systems operate with phase sequence *a*, *b*, *c* or its equivalent. There are several large electric utilities in the United States that operate with *a*, *c*, *b* phase sequence. In some cases this sequence is used throughout the system; in others, one voltage level may be *a*, *b*, *c* and another voltage level, *a*, *c*, *b*. The specific phase sequence is only a name designation that was established arbitrarily early in the history of a company, and it is difficult to change after many years of operation.

Knowing the existing phase sequence is very important in three-phase connections of relays and other equipment, so should be clearly indicated on the drawings and information documents. This is especially true if it is not *a*, *b*, *c*. The connections from *a*, *b*, *c* to *a*, *c*, *b*, or vice versa, can generally be made by completely interchanging phases *b* and *c* for both the equipment and the connections.

3.5 POLARITY

Polarity is important in all transformers and a discussion of this follows. Polarity is also very important in protection equipment and this is covered in Blackburn's *Protective Relaying: Principles and Applications* (see the Bibliography).

3.5.1 Transformer Polarity

The polarity indications for transformers are well established by standards that apply to all types of transformers. There are two varieties of polarity: substractive and additive. Both follow the same rules. Power and instrument transformers are subtractive, while some distribution transformers are additive. The polarity marking can be a dot, a square, or an X, or it can be indicated by the standardized transformer terminal markings, the practices having varied over the years. It is convenient to designate polarity by an X in this book.

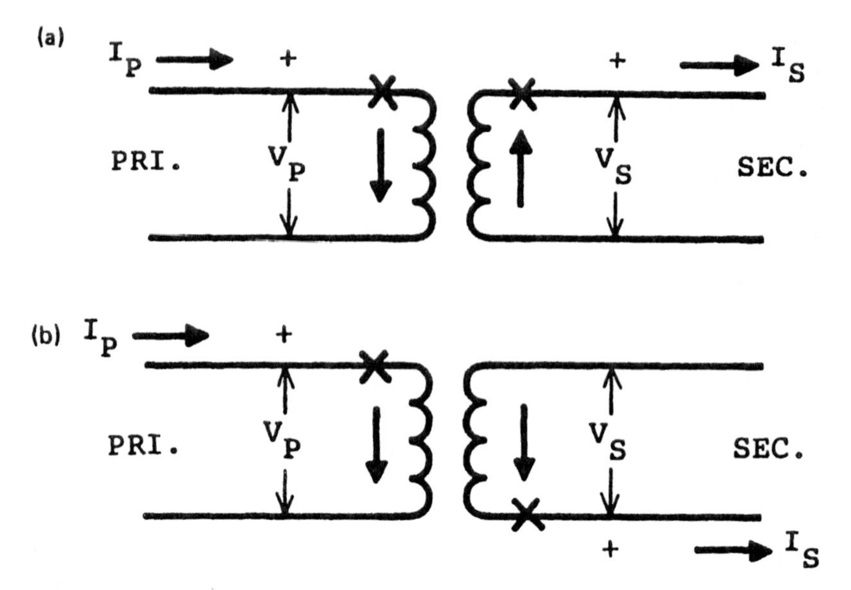

Figure 3.4 Polarity definitions for transformers: (a) subtractive polarity; (b) additive polarity.

The two fundamental rules of transformer polarity are illustrated in Fig. 3.4 and apply to both varieties. These are:

1. Current flowing in at the polarity mark of one winding flows out of the polarity mark of the other winding. Both currents are substantially in phase.
2. The voltage drop from polarity to nonpolarity across one winding is essentially in phase with the voltage drop from polarity to nonpolarity across the other winding(s).

The currents through, and the voltages across, the transformers are substantially in phase since the magnetizing current and the impedance drop through the transformers is very small and can be considered negligible. This is normal and practical for these definitions.

The current transformer polarity markings are shown in Fig. 3.5. It will be noted that the direction of the secondary current is the same independent of whether the polarity marks are together on one side or on the other. For current transformers (CTs) associated with circuit breakers and transformer banks, it is the common practice for the polarity marks to be located on the side away from the associated equipment.

The voltage drop rule is often omitted in the definition of transformer polarity, but it is an extremely useful tool to check the phase relations through wye–delta transformer banks, or in connecting up a transformer bank for a specific phase shift required by the power system. The ANSI/IEEE standard for transformers states that the high voltage should lead the low voltage by 30° with wye–delta or delta–wye banks. Thus different connections are required if the high side is wye than if the high side is delta. The connections for these two cases are shown in Fig. 3.6. The diagrams below the three-phase transformer connection illustrate the use of the voltage drop rule to provide or check the connections. Arrows on these voltage drops have been omitted (preferably not used), as they are not necessary and can cause confusion.

In Fig. 3.6a the check is made by noting that *a* to *n* from polarity to nonpolarity on the left-side winding is in phase with *A* to *B* from polarity to nonpolarity on the right-side winding. Similarly, *b* to *n* (polarity to nonpolarity) is in phase with *B* to

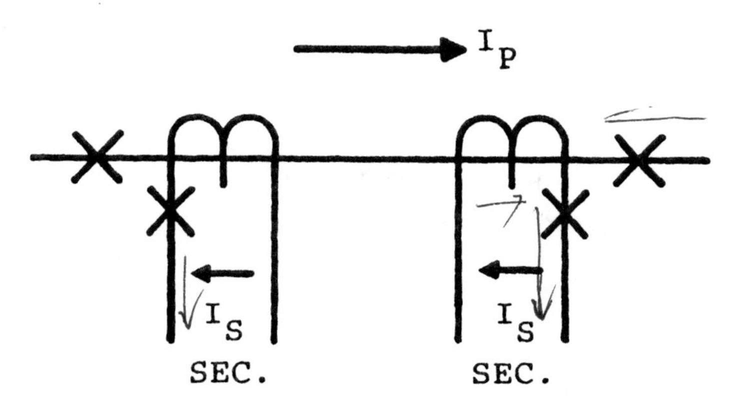

Figure 3.5 Polarity markings for current transformers.

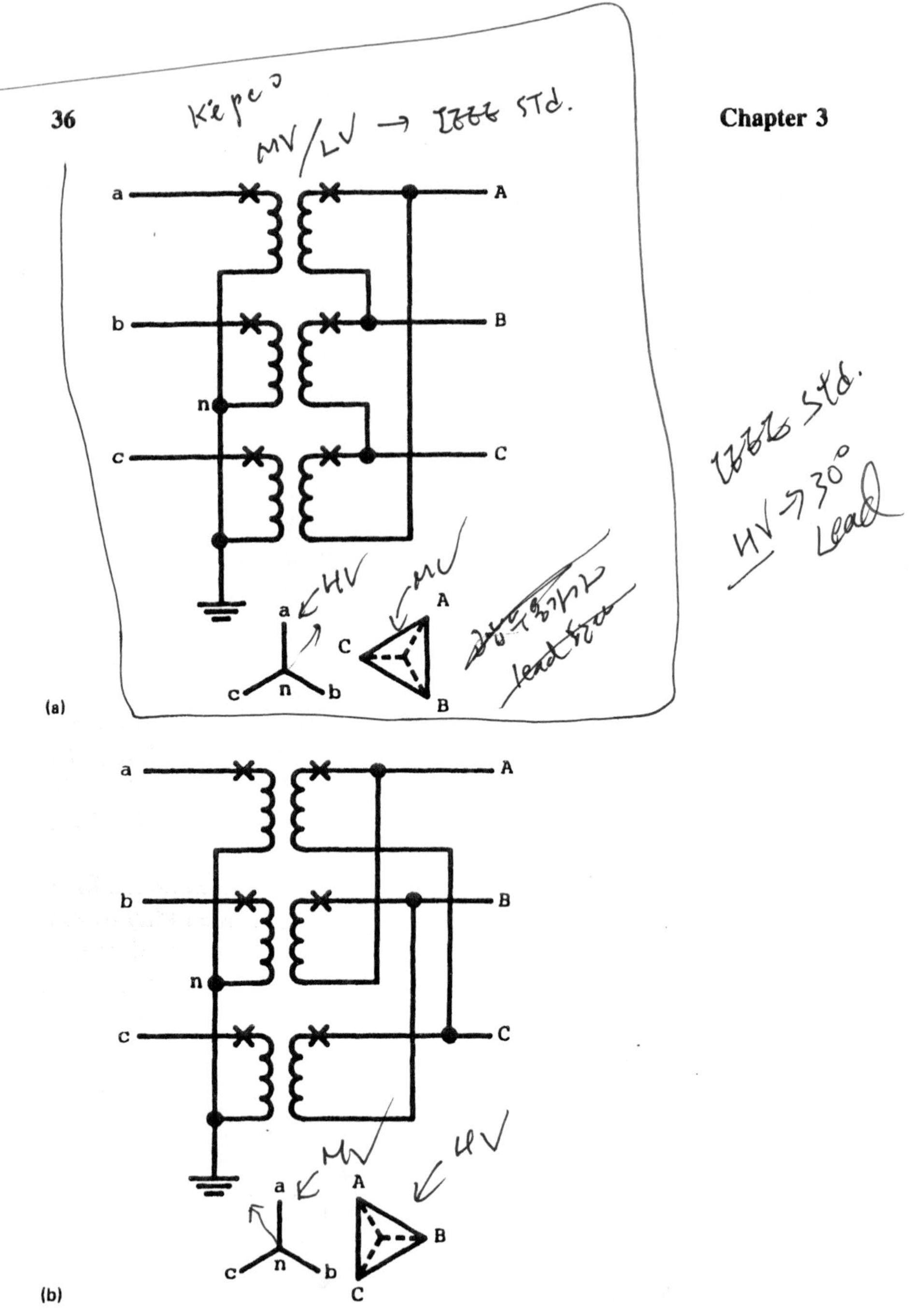

Figure 3.6 Voltage-drop polarity rule useful in checking or connecting wye–delta transformer banks: (a) wye-connected side leads the delta-connected side 30°; (b) delta-connected side leads the wye-connected side 30°.

C (polarity to nonpolarity) across the middle transformer, and c to n (polarity to nonpolarity) is in phase with C to A (polarity to nonpolarity) across the lower transformer. From this, by comparing the line-to-neutral voltages on the two sides, it is observed that phase-a-to-n voltage leads phase-A-to-neutral voltage. Thus the wye side would be the high-voltage side if this were an ANSI/IEEE standard transformer.

This same technique of applying voltage drops to Fig. 3.6b shows that for this three-phase bank connection the voltage drop polarity to nonpolarity or phase a to n is in phase with the voltage drop polarity to nonpolarity or phase A to phase C. Similarly, voltage drop phase b to n is in phase with voltage drop phase B to phase A, and voltage drop phase c to n is in phase with voltage drop phase C to phase B. By comparing similar voltages on the two sides of the transformer, phase-A-to-neutral voltage drop leads the phase-a-to-n voltage drop by 30°, so the delta winding would be the high-voltage side if this is an ANSI/IEEE standard transformer bank. This technique is very useful in making the proper three-phase transformer connections from a desired or known voltage diagram or phase-shift requirement. It is a very powerful tool, simple and straightforward to use.

Since the ANSI/IEEE standards have been in existence for a number of years, most transformer banks in service today follow this standard except where it is not possible because of preexisting system conditions. Many years ago, in the absence of a standard, a great many different connections were used. Some of the older references and textbooks reflect this.

For example, Wagner and Evans, *Symmetrical Components* (see the Bibliography), was published before the present-day standard. They used a 90° relation with the positive-sequence delta voltage (E_{A1}) leading the positive-sequence wye voltage (E_{a1}).

3.6 POWER SYSTEM HARMONICS

The method of symmetrical components is concerned basically with the fundamental frequency of the power system. However, the circulation of system harmonic frequencies is related to the three sequences as follows:

1. Balanced fundamental frequency has positive phase sequence.
2. Third harmonic frequency ($3 \times 120°$) is in time phase and so flows like zero phase sequence quantities. Hence they can flow only when (a) the neutral is grounded and (b) a fourth wire wye is used. They also circulate inside a delta similar to zero sequence.
3. Fifth harmonic frequency ($5 \times 120°$) is similar to negative sequence.
4. Seventh harmonic frequency ($7 \times 120°$) is similar to positive sequence.
5. Ninth harmonic frequency ($9 \times 120°$) is similar to zero sequence.
6. Eleventh harmonic frequency ($11 \times 120°$) is similar to negative sequence.
7. Thirteenth harmonic frequency ($13 \times 120°$) is similar to positive sequence.
8. Fifteenth harmonic frequency ($15 \times 120°$) is similar to zero sequence.

The higher the frequency, the higher the inductive reactance and the lower the capacitive reactance. In summary:

Three-phase wye systems. The fundamental, 5th, 7th, 11th, 13th, 17th, etc. harmonics can all flow in the three-phase system *without* a neutral conductor or ground return. The 3rd, 9th, 15th, etc. harmonics can flow *only* when a ground or neutral return exists.

Delta systems. The 5th, 7th, 11th, 13th, etc. harmonics can flow in the lines and in the separate phases of the delta. The 3rd, 9th and 15th, etc. harmonics can flow inside the delta and not in the system.

4

Basic Fundamentals and the Sequence Networks

4.1 INTRODUCTION

For three-phase electric power systems there are three distinct sets of symmetrical components: positive, negative, and zero sequence for both currents and voltages. Throughout this discussion, the sequence quantities are *always* line to neutral or line to ground, as appropriate. This is an exception to the general power system practice of using line-to-line voltages. It is very important to remember that throughout the sequence networks, *all* voltages are line to neutral (to the neutral or the zero-potential buses).

Next, we define the three sequence quantities.

4.2 POSITIVE-SEQUENCE SET

This set consists of the balanced (symmetrical) three-phase currents and three-phase line-to-neutral voltages supplied by the power system generators. Thus they are *always* equal in magnitude and phase displaced by 120°. With the power system phase sequence of a, b, c, Fig. 4.1 shows a positive-sequence set of phase currents. A voltage set is similar except for line-to-neutral voltage of the three phases, again all equal in magnitude and displaced 120°. These are phasors rotating counterclockwise at the system frequency.

To document the angle displacement, it is convenient to use a unit phasor with an angle displacement of 120°. This is designated as a, so that

$$
\begin{aligned}
a &= 1\angle 120^\circ = -0.5 + j0.866 \\
a^2 &= 1\angle 240^\circ = -0.5 - j0.866 \\
a^3 &= 1\angle 360^\circ = 1\angle 0^\circ = 1.0 + j0
\end{aligned}
\tag{4.1}
$$

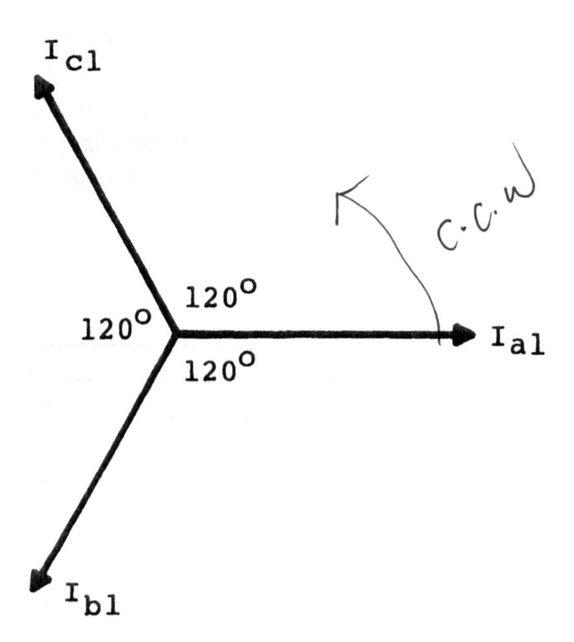

Figure 4.1 Positive-sequence current phasors.

Thus the positive-sequence set can be designated as

$$
\begin{aligned}
I_{a1} &= I_1 & V_{a1} &= V_1 \\
I_{b1} &= a^2 I_{a1} = a^2 I_1 = I_1\angle 240^\circ & V_{b1} &= a^2 V_1 = V_1\angle 240^\circ \qquad (4.2) \\
I_{c1} &= a I_{a1} = a I_1 = I_1\angle 120^\circ & V_{c1} &= a V_1 = V_1\angle 120^\circ
\end{aligned}
$$

It is most important to emphasize that the set of sequence currents or sequence voltages *always* exists as defined. I_{a1} or I_{b1} or I_{c1} can *never* exist alone or in pairs, *always* all three. Thus it is necessary to define only one of the phasors (any one), from which the other two will be as documented in Eqs. (4.2).

4.3 NOMENCLATURE CONVENIENCE

It will be noted that the designation subscript for phase a was dropped in the second expression for the currents and voltages in Eqs. (4.2) (and also in the following equations). This is a common shorthand notation used for convenience. Whenever the phase subscript does not appear, it can be assumed that the reference is to phase a. If phase b or phase c quantities are intended, the phase subscript must be correctly designated; otherwise, it is assumed to be phase a. This shortcut will be used throughout the book and is common in practice.

4.4 NEGATIVE-SEQUENCE SET

This set is also balanced with three equal magnitude quantities at 120° apart, but with the phase rotation or sequence reversed as illustrated in Fig. 4.2. Thus if positive sequence is a, b, c, negative will be a, c, b. Where positive sequence is a, c, b, as in some power systems, negative sequence is a, b, c. The negative-sequence set can be designated as

$$
\begin{aligned}
I_{a2} &= I_2 & V_{a2} &= V_2 \\
I_{b2} &= a I_{a2} = a I_2 = I_2\angle 120^\circ & V_{b2} &= a V_2 = V_2\angle 120^\circ \qquad (4.3) \\
I_{c2} &= a^2 I_{a2} = a^2 I_2 = I_2\angle 240^\circ & V_{c2} &= a^2 V_2 = V_2\angle 240^\circ
\end{aligned}
$$

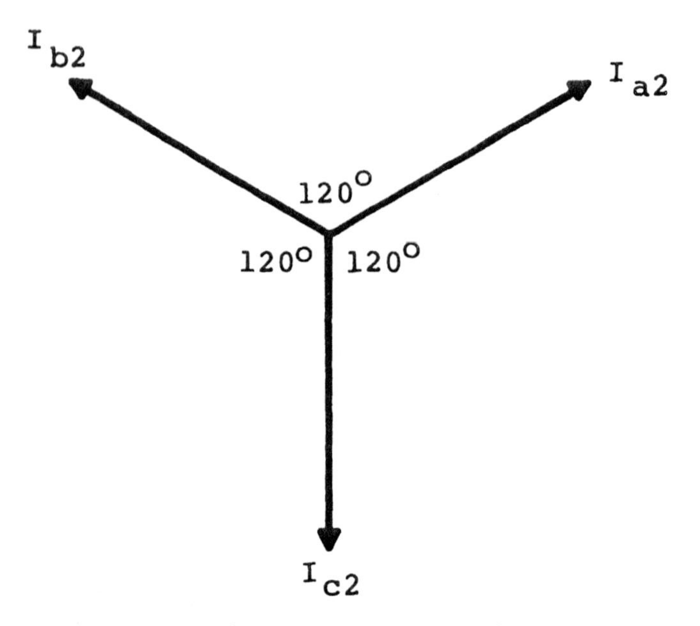

Figure 4.2 Negative-sequence current phasors.

Again, negative sequence always exists as a set of current or voltage as defined above or in Fig. 4.2. I_{a2} or I_{b2} or I_{c2} can *never* exist alone. When one current or voltage phasor is known, the other two of the set can be defined as above.

4.5 ZERO-SEQUENCE SET

The members of this set of rotating phasors are always equal in magnitude and always in phase (Fig. 4.3).

$$I_{a0} = I_{b0} = I_{c0} = I_0 \qquad V_{a0} = V_{b0} = V_{c0} = V_0 \tag{4.4}$$

Again, I_0 or V_0, if it exists, exists equally in all three phases, *never* alone in one phase.

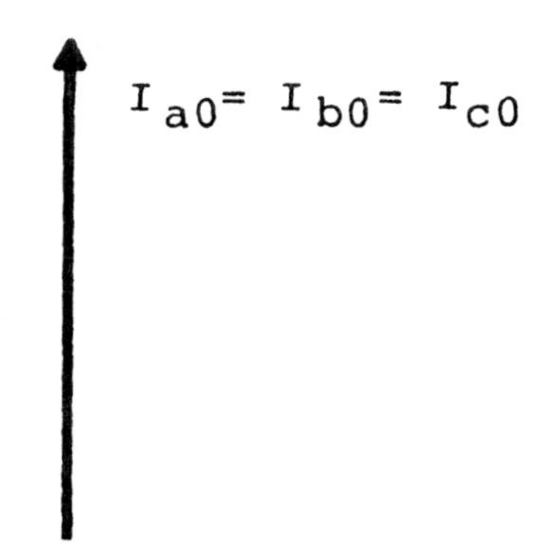

Figure 4.3 Zero-sequence current phasors.

4.6 GENERAL EQUATIONS

Any unbalanced current or voltage can be determined from the sequence components from the following fundamental equations:

$$I_a = I_1 + I_2 + I_0 \qquad V_a = V_1 + V_2 + V_0 \tag{4.5}$$

$$I_b = a^2 I_1 + a I_2 + I_0 \qquad V_b = a^2 V_1 + a V_2 + V_0 \tag{4.6}$$

$$I_c = a I_1 + a^2 I_2 + I_0 \qquad V_c = a V_1 + a^2 V_2 + V_0 \tag{4.7}$$

where I_a, I_b, and I_c or V_a, V_b, and V_c are general unbalanced line-to-neutral phasors. From these, equations defining the sequence quantities from a three-phase unbalanced set can be determined:

$$I_0 = \tfrac{1}{3}(I_a + I_b + I_c) \qquad V_0 = \tfrac{1}{3}(V_a + V_b + V_c) \tag{4.8}$$

$$I_1 = \tfrac{1}{3}(I_a + a I_b + a^2 I_c) \qquad V_1 = \tfrac{1}{3}(V_a + a V_b + a^2 V_c) \tag{4.9}$$

$$I_2 = \tfrac{1}{3}(I_a + a^2 I_b + a I_c) \qquad V_2 = \tfrac{1}{3}(V_a + a^2 V_b + a V_c) \tag{4.10}$$

These last three fundamental equations are the basis for determining if the sequence quantities exist in any given set of unbalanced three-phase currents or voltages.

Zero- and negative-sequence current and voltage are widely used in protective relaying. Connecting the three-phase current transformers in parallel provides zero-sequence current ($3I_0$), and connecting the phase voltage transformers in "broken delta" provides zero-sequence voltage ($3V_0$) for ground relaying. Both are the physical solution of Eqs. (4.8). More details on this appear in Blackburn's *Protective Relaying: Principles and Applications* (see the Bibliography).

Networks operating from CTs or VTs are used to provide an output proportional to I_2 or V_2 and are based on physical solutions of Eq. (4.10). This can be accomplished with resistors, transformers, reactors, or with operational amplifier components (see Westinghouse Electric Corp., *Applied Protective Relaying*).

4.7 SEQUENCE INDEPENDENCE

The *key* that makes dividing unbalanced three-phase quantities into the sequence components practical is the independence of the components in a balanced system network. For all practical purposes electric power systems are balanced or symmetrical from the generators to the point of single-phase loading except in an area of a fault or unbalance such as an open conductor. In this essentially balanced area, the following conditions exist:

1. Positive-sequence currents flowing in the symmetrical or balanced network produce *only* positive-sequence voltage drops, *no* negative- or zero-sequence drops.
2. Negative-sequence currents flowing in the balanced network produce *only* negative-sequence voltage drops, *no* positive- or zero-sequence voltage drops.
3. Zero-sequence currents flowing in the balanced network produce *only* zero-sequence voltage drops, *no* positive- or negative-sequence voltage drops.

This is not true for any unbalanced or nonsymmetrical point or area such as an unsymmetrical fault, open phase, and so on. In these:

4. Positive-sequence current flowing in an unbalanced system produces positive-, negative-, and possibly zero-sequence voltage drops.
5. Negative-sequence current flowing in an unbalanced system produces positive-, negative-, and possibly zero-sequence voltage drops.
6. Zero-sequence current flowing in an unbalanced system produces all three: positive-, negative-, and zero-sequence voltage drops.

This important fundamental permits setting up three independent networks, one for each of the three sequences, which can be interconnected only at the point or area of unbalance.

4.8 SEQUENCE NETWORKS

These represent one of the three-phase-to-neutral or ground circuits of the balanced three-phase power system, and documents how that sequence currents will flow if they can exist. These networks are best explained by an example, so consider the section of a power system of Fig. 4.4.

Reactance values only have been shown for the generator and the transformers. Theoretically, impedance values should be used, but the resistances of these units are very small and generally negligible for fault studies.

It is very important that all values be specified with a base [voltage if ohms are used, or MVA (kVA) and kV if per unit or percent impedances are used]. Before applying these to the sequence networks, all values must be changed to one common base. In most cases per unit (percent) values are used and a common base in practice is 100 MVA at the particular system kV.

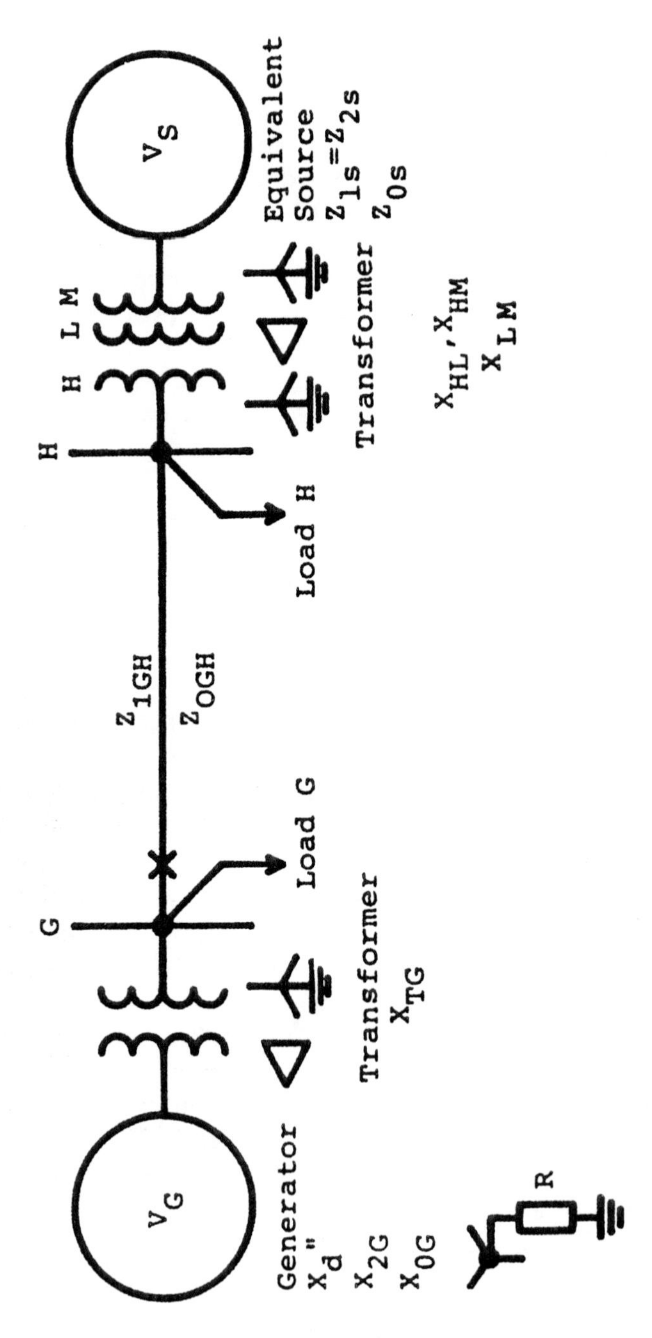

Figure 4.4 Single-line diagram of a section of a power system.

4.9 POSITIVE-SEQUENCE NETWORK

This is the usual line-to-neutral system diagram for any one of the three symmetrical phases modified for fault conditions. The positive-sequence network for the system of Fig. 4.4 is shown in Fig. 4.5. V_S and V_G are the system line-to-neutral voltages. V_G is the voltage behind the generator subtransient direct-axis reactance X''_d (see Chapter 10) and V_S is the voltage behind the system equivalent impedance Z_{1S}. Generators and synchronous motors that can continue to contribute power to the unbalance must be included in the positive (and negative)-sequence networks.

The U.S. ANSI Standards require that induction motors 50 hp or larger be considered as sources in determining circuit breaker ratings. This is a mandatory requirement. In many cases the contribution of induction motors to faults is quite small relative to the total fault current and is of very short duration. As a result, its impact on protective relay applications and operation is very small and generally negligible. Thus from a practical standpoint induction motors are normally not included as sources in the positive (and negative)-sequence networks for protective relaying fault studies.

X_{TG} is the transformer leakage reactance for the bank at bus G (Fig. 4.9a) and X_H is the leakage reactance for the bank at bus H between the H and M windings (Fig. 4.10b). The delta winding L of this three-winding transformer bank is not involved in the positive-sequence network unless a generator or synchronous motor is connected to this delta, or unless a fault is to be considered in the L delta system.

For the line between buses G and H, Z_{1GH} is the line-to-neutral impedance of three-phase circuit (see Chapter 11 if overhead lines, Chapter 14 if cable). For open-wire transmission lines an approximate estimating value is 0.8 Ω/mi at 60 Hz for a single conductor or 0.6 Ω/mi at 60 Hz for bundled conductors.

Normally, capacitance is neglected, as it is very high in relation to all other impedances involved in fault calculations. Typical estimating values for shunt capacitance on overhead lines is

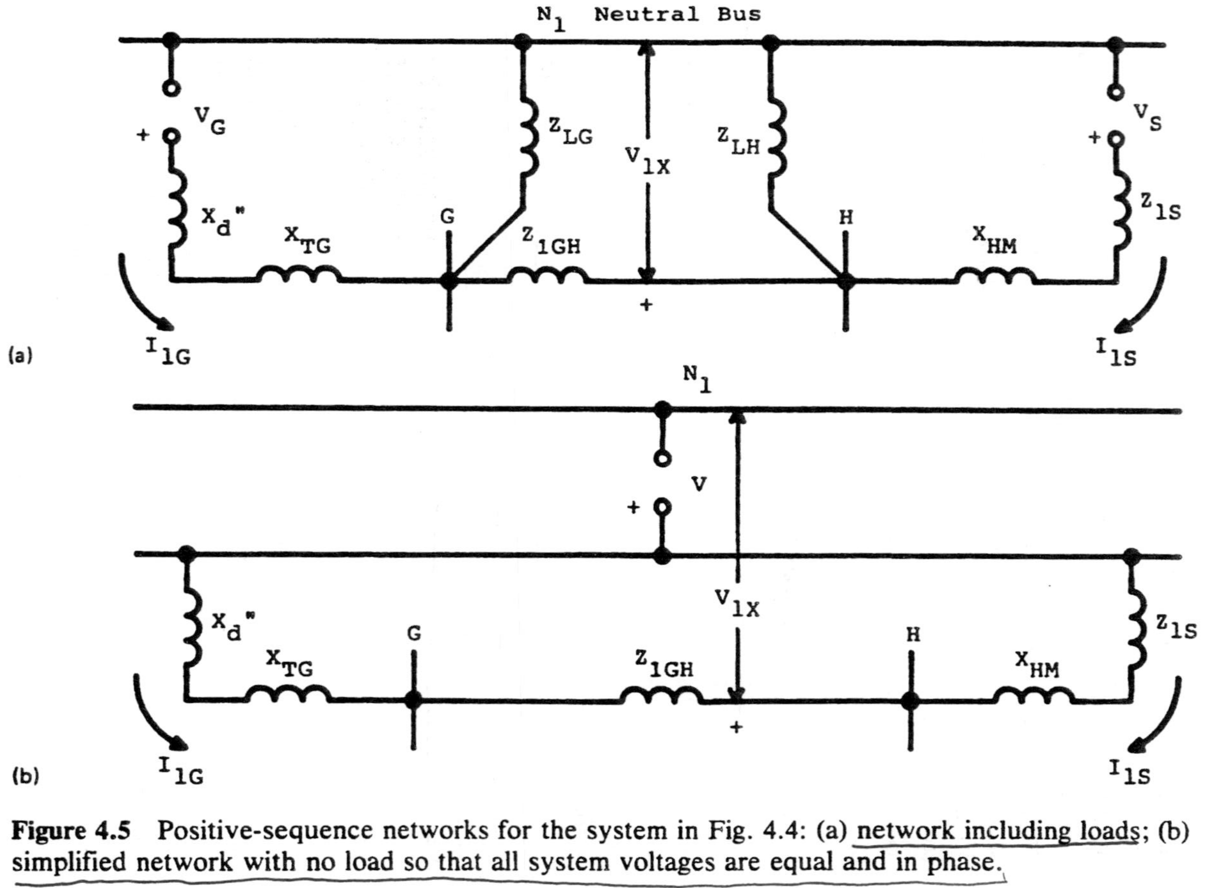

Figure 4.5 Positive-sequence networks for the system in Fig. 4.4: (a) network including loads; (b) simplified network with no load so that all system voltages are equal and in phase.

0.2 MΩ/mi at 60 Hz for single conductors and 0.14 MΩ/mi at 60 Hz for bundled conductors. On long lines on the order of 100 mi or more, the long-line constants should be considered.

The impedance angle of lines can vary quite widely depending on the voltage and whether cable or open wire is used. In computer fault programs the angles are considered and included, but for hand calculation it is practical in most cases to simplify calculations by assuming that all the equipment involved in the fault calculation is at 90° or use reactance values only. Sometimes it may be preferred to use the line impedance values and treat them as reactances. Unless the network consists of a large proportion of low-angle circuits, the error of using all values as 90° will not be too significant.

Load is shown connected at buses G and H. Normally, this would be specified as kVA or MVA and can be converted into impedance:

$$I_{\text{load}} = \frac{1000\,\text{MVA}_{\text{load}}}{\sqrt{3}\,\text{kV}} \quad \text{and} \quad V_{LN} = \frac{1000\,\text{kV}}{\sqrt{3}}$$
$$Z_{\text{load}} = \frac{V_{LN}}{I_{\text{load}}} = \frac{\text{kV}^2}{\text{MVA}_{\text{load}}} = \text{ohms at kV} \tag{4.11}$$

Equation (4.11) is a line-to-neutral value and would be used for Z_{LG} and Z_{LH}, representing the loads at G and H in Fig. 4.5a. If load is represented, the voltages V_G and V_S will be different in magnitude and angle, varying as the system load varies.

The value of load impedance is usually quite large compared to the system impedances, so that load has a negligible effect on the faulted phase current. Thus it becomes practical and simplifies calculations to neglect load for shunt faults. With no load, Z_{LG} and Z_{LH} are infinite, V_G and V_S are equal and in phase and so are replaced by a common voltage V as in Fig. 4.5b. Normally, V is considered as 1 pu, the system-rated line-to-neutral voltages.

4.10 NEGATIVE-SEQUENCE NETWORK

This network defines the flow of negative-sequence currents when they exist. The system generators do not generate negative sequence, but negative-sequence current can flow through their windings. Thus these generators and sources are represented by an impedance without voltage, as shown in Fig. 4.6. In the transformers, lines, and so on, the phase sequence of the current does not change the impedance encountered, so the same values are used as in the positive-sequence network.

A rotating machine can be visualized as a transformer with one stationary and one rotating winding. Thus dc in the field produces positive sequence in the stator. Similarly, the dc offset in the stator ac current produces an ac component in the field. In this relative motion model with the one winding rotating at synchronous speed, negative sequence in the stator results in a double-frequency component in the field. Thus the negative-sequence flux component in the air gap is alternately between and under the poles at this double frequency. One common expression for the negative-sequence impedance of a synchronous machine is

$$X_2 = \tfrac{1}{2}(X''_d + X''_q) \tag{4.12}$$

or the average of the direct and quadrature axes subtransient reactances. For a round-rotor machine, $X''_d = X''_q$, so that $X_2 = X''_d$. For salient-pole machines, X_2 will be different, but this is frequently neglected unless calculating a fault very near the machine terminals. Where normally $X_2 = X''_d$, the negative-sequence network is equivalent to the positive-sequence network except for the omission of voltages.

Loads can be shown as in Fig. 4.6 and will be the same impedance as for positive sequence if they are static loads. Rotating loads such as those of induction motors have quite different positive- and negative-sequence impedances when running. This is discussed further in Sec. 10.9. Again with load normally neglected, the network is as shown in Fig. 4.6b and is the same as the positive-sequence network (Fig. 4.5b), except that there is no voltage.

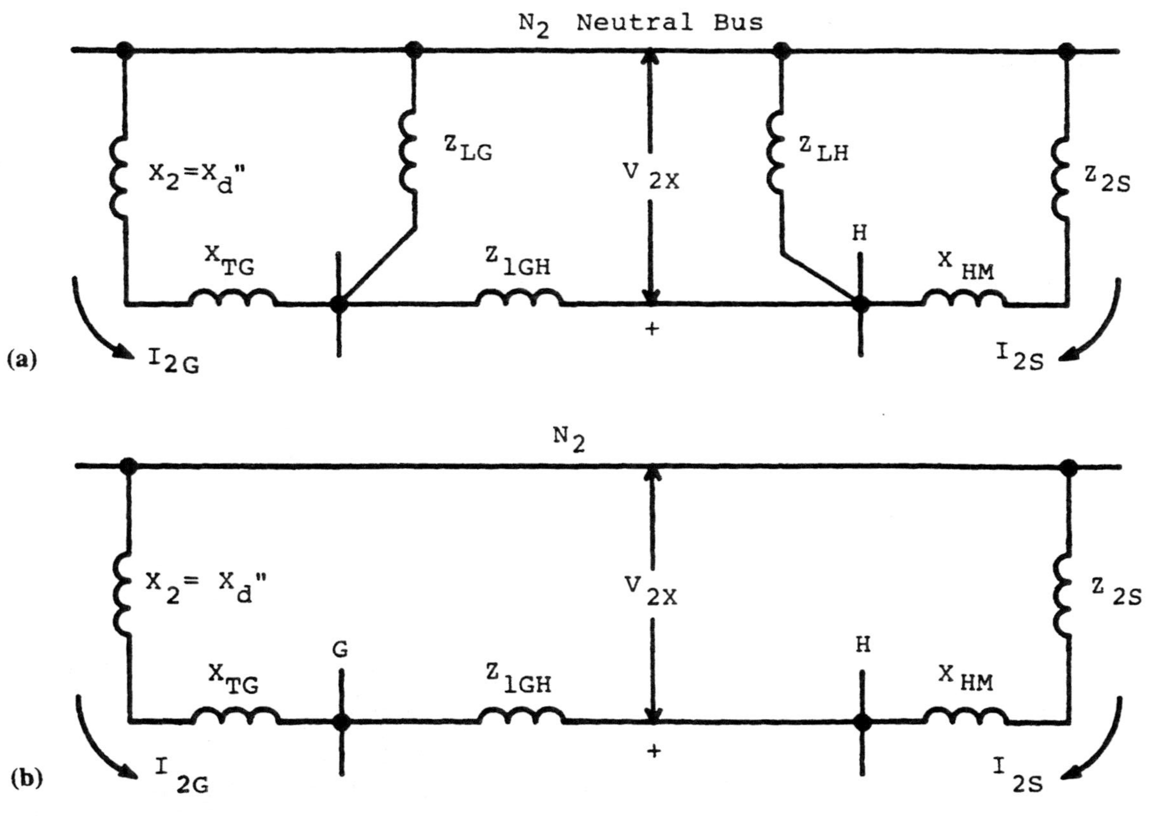

Figure 4.6 Negative-sequence networks for the system in Fig. 4.4: (a) network including loads; (b) network neglecting loads.

4.11 ZERO-SEQUENCE NETWORK

The zero-sequence network is always different. It must satisfy the flow of equal and in-phase currents in the three phases. If the connections for this network are not apparent, or there are questions or doubts, these can be resolved by drawing the three-phase system to see how the equal in-phase zero-sequence currents can flow. For the example of Fig. 4.4, a three-phase diagram is shown in Fig. 4.7. The convention is that current always

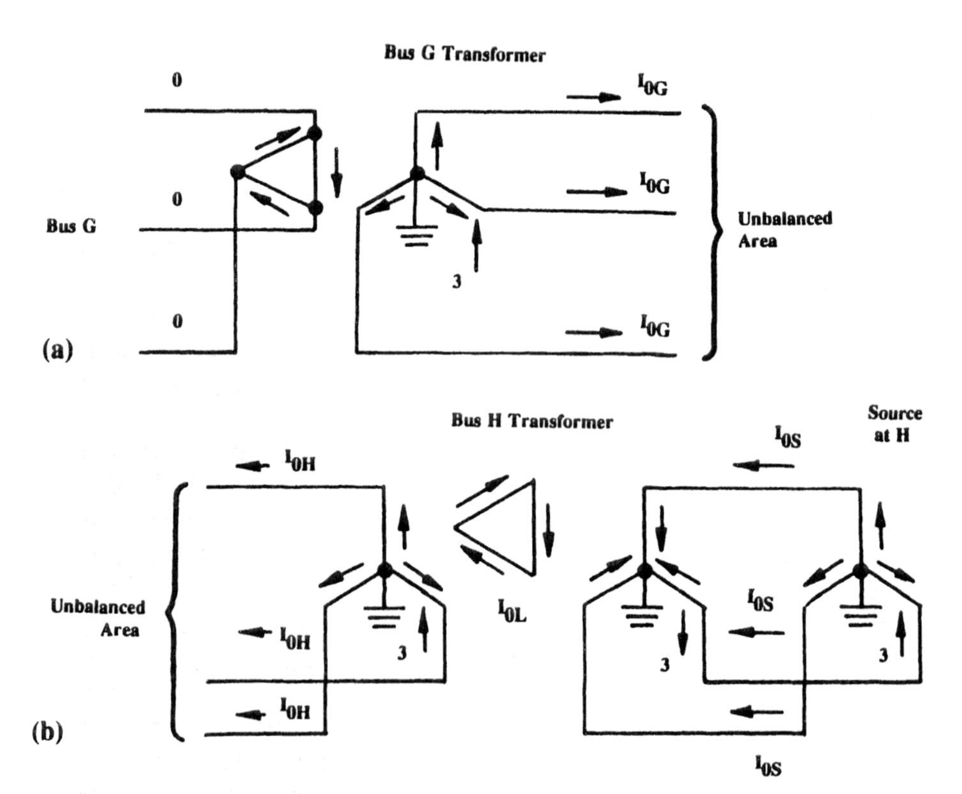

Figure 4.7 Diagrams illustrating the flow of zero-sequence current as an aid in drawing the zero-sequence network. Arrows indicate current directions only, not relative magnitudes. (a) Current flow to unbalanced area from left part of the system. (b) Current flows to unbalanced area from right part of the system.

flows to the unbalance, so assume an unbalance between buses G and H. Diagram (a) shows I_{0G} flowing from the transformer at bus G. Zero sequence can flow in the grounded wye and to the fault since there is a path for it to flow in the delta. Thus X_{TG} is connected between the zero-potential bus and bus G as shown in Fig. 4.8. This connection for the grounded wye–delta transformer bank is also shown in Fig. 4.9a.

Zero-sequence impedance for transformer banks is equal to the positive and negative sequence and is the transformer leakage impedance. The exception to this is for three-phase core-type transformers, where the construction does not provide an iron flux path for zero sequence. For these the zero-sequence flux must pass from the core to the tank and return. Hence for these types X_0 usually is 0.85 to $0.9X_1$, and when known the specific value should be used.

Figure 4.7b is for the system connected to bus H (Fig. 4.4). Currents out of the three-winding transformer will flow as shown in the L and M windings. The three currents can flow in the M-grounded wye since the equivalent source is shown grounded with Z_{0S} given. Thus the three-winding equivalent circuit is connected in the zero-sequence network (Fig. 4.8) as shown, which follows the connections documented in Fig. 4.10b.

Note that in the right-hand part of Fig. 4.7, if any of the wye connections were not grounded, the connections would be different. If the equivalent system, or the M winding were ungrounded, the network would be open between Z_M and Z_{0S}, as zero-sequence currents could not flow as shown. Loads, if desired, would be shown in the zero-sequence network only if they were wye grounded. Delta loads would not pass zero sequence.

Zero-sequence line impedance is always different, as it is a loop impedance: the impedance of the line plus a return path either in the earth, or in a parallel combination of the earth and ground wire, cable sheath, and so on. The positive-sequence impedance is a one-way impedance: from one end to the other end. As a result, zero sequence varies from 2 to 6 times X_1 for lines. For estimating open wire lines, a value of $X_0 = 3$ or $3.5X_1$ is commonly used (see Chapter 11).

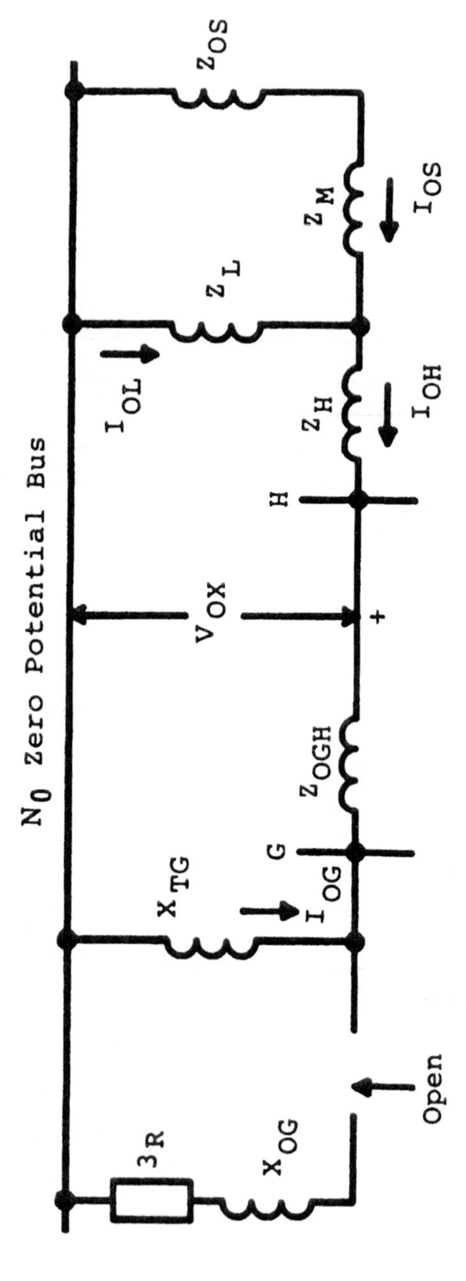

Figure 4.8 Zero-sequence network for the system of Fig. 4.4.

The zero-sequence impedance of generators is low and variable, depending on the winding design. Except for very low voltage units, generators are never solidly grounded. This is discussed in Chapter 10. In Fig. 4.4, the generator G is shown grounded through a resistor R. Faults on bus G and in the system to the right do not involve the generator as far as zero sequence since the transformer delta blocks the flow of zero-sequence current as shown.

However, for ground faults in the generator area of Fig. 4.4, the zero-sequence network (Fig. 4.8) shows the generator grounding resistor multiplied by 3. This is correct because the zero-sequence network represents one of the three phases where one zero-sequence unit current flows. With equal current in all three phases, zero-sequence current adds to three units in the neutral resistor R and ground. Thus the actual drop is $3I_0(R)$, which is equivalent to $3R(I_0)$ in the zero-sequence network. This means that the impedance of any equipment connected between the power system neutral and ground return is *always* multiplied by 3.

4.12 IMPEDANCE AND SEQUENCE CONNECTIONS FOR TRANSFORMER BANKS

4.12.1 Two-Winding Transformer Banks

Typical two-winding transformer banks with their positive-, negative-, and zero-sequence circuits are shown in Fig. 4.9. H is the high-voltage winding and L the low-voltage winding. These designations can be interchanged as required.

Z_T is the transformer leakage impedance between the two windings of each phase. Normally, it is designated in per unit or percent by the manufacturer, stamped on the nameplate of the transformer, and included with the technical information supplied. Unless otherwise specified, this value is on the self-cooled kVA or MVA rating at the rated voltages. As discussed in Section 4.11, impedance connected in the neutral is multiplied by 3 as shown. The positive- and negative-sequence impedances are

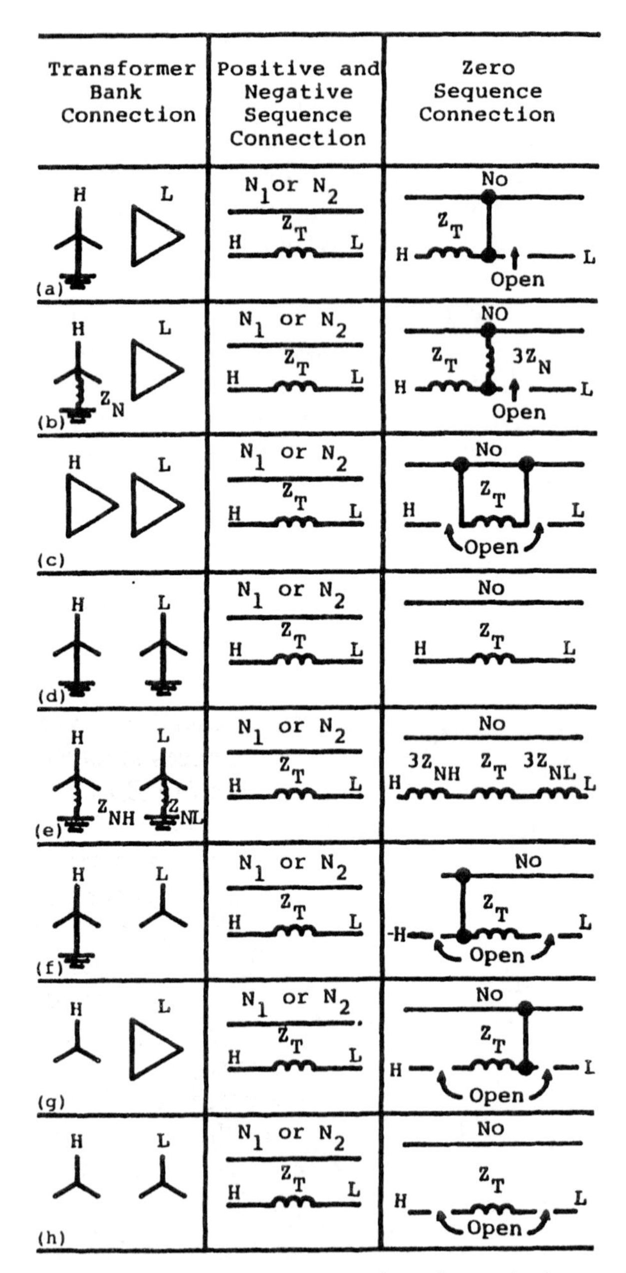

Figure 4.9 Sequence connections for typical two-winding transformer banks.

always equal and are the same for all connections, as noted in Fig. 4.9.

4.12.2 Three-Winding and Autotransformer Banks

Typical banks are shown in Fig. 4.10. H, M, and L are the high-, medium-, and low-voltage windings. These designations can be interchanged as required. Normally, the manufacturer provides the leakage impedances between pairs of windings as Z_{HM}, Z_{HL}, and Z_{ML}, usually on different kVA or MVA ratings of the winding pairs and at the rated voltages. First, they must be converted to a common base.

To use these impedances in the sequence networks, they must be converted to an equivalent wye type network as shown. This conversion is

$$\begin{aligned} Z_H &= \tfrac{1}{2}(Z_{HM} + Z_{HL} - Z_{ML}) \\ Z_M &= \tfrac{1}{2}(Z_{HM} + Z_{ML} - Z_{HL}) \\ Z_L &= \tfrac{1}{2}(Z_{HL} + Z_{ML} - Z_{HM}) \end{aligned} \tag{4.13}$$

It is easy to remember this conversion, as the equivalent wye value is always half the sum of the leakage impedances involved less the one not involved. For example, Z_H is half of Z_{HM} and Z_{HL}, both involving H, minus Z_{ML} that does not involve H.

After determining Z_H, Z_M, and Z_L, a good check is to see if they add up as $Z_H + Z_M = Z_{HM}$, etc. If these values are not available, they can be measured as described in Chapter 9. For the three-winding or autotransformers: Z_{HM} is the impedance looking into H winding with M shorted, L open; Z_{HL} is the impedance looking into H winding with L shorted, M open; Z_{ML} is the impedance looking into M winding with L shorted, H open.

This equivalent wye is a mathematical network representation valid for determining currents and voltages at the transformer terminals or in the associated system. The wye point has no physical meaning. Quite often, one of the values will be negative and should be used as such in the network. It does not represent a capacitor.

The positive- and negative-sequence connections are all the

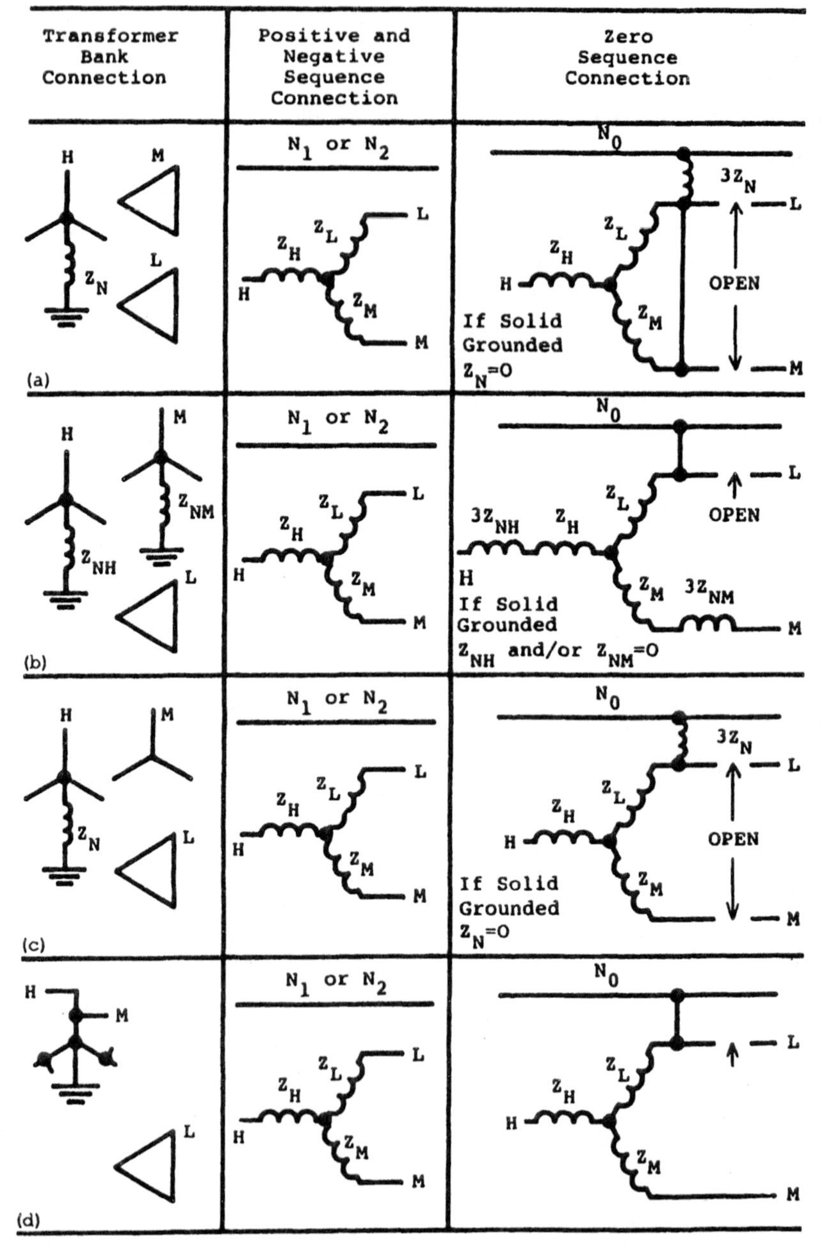

Figure 4.10 Sequence connections for typical three-winding and autotransformer banks.

same and independent of the actual bank connections. However, the connections for the zero-sequence network are all different and depend on the transformer bank connections. If the neutrals are solidly grounded, the Z_N and $3Z_N$ components shown are shorted out in the system and sequence circuits.

4.13 SEQUENCE PHASE SHIFTS THROUGH WYE–DELTA TRANSFORMER BANKS

The phase shift of the positive- and negative-sequence quantities through wye–delta or delta–wye transformer banks with reference to Fig. 4.11 is as follows: The analysis is based on the ANSI standard connections where the high-voltage quantities lead the low-voltage quantities by 30°. All quantities are phase-to-neutral values in amperes or volts. For per unit, $N = 1$ and $n = 1/\sqrt{3}$.

4.13.1 High-Voltage Wye, Low-Voltage Delta

See Fig. 4.11, left side. From the connections;

$$I_a = n(I_A - I_C) \quad \text{and} \quad V_A = n(V_a - V_b) \tag{4.14}$$

With positive-sequence quantities from Eqs. (4.2):

$$\begin{aligned}
I_{a1} &= n(I_{A1} - aI_{A1}) & V_{A1} &= n(V_{a1} - a^2V_{a1}) \\
I_{a1} &= n(1 - a)I_{A1} & V_{A1} &= n(1 - a^2)V_{a1} \\
I_{a1} &= n\sqrt{3}\,I_{A1}\underline{/-30^\circ} & V_{A1} &= n\sqrt{3}\,V_{a1}\underline{/+30^\circ} \\
I_{a1} &= NI_{A1}\underline{/-30^\circ} & V_{A1} &= NV_{a1}\underline{/+30^\circ}
\end{aligned} \tag{4.15}$$

or

$$I_{A1} = \frac{I_{a1}\underline{/+30^\circ}}{N} \tag{4.16}$$

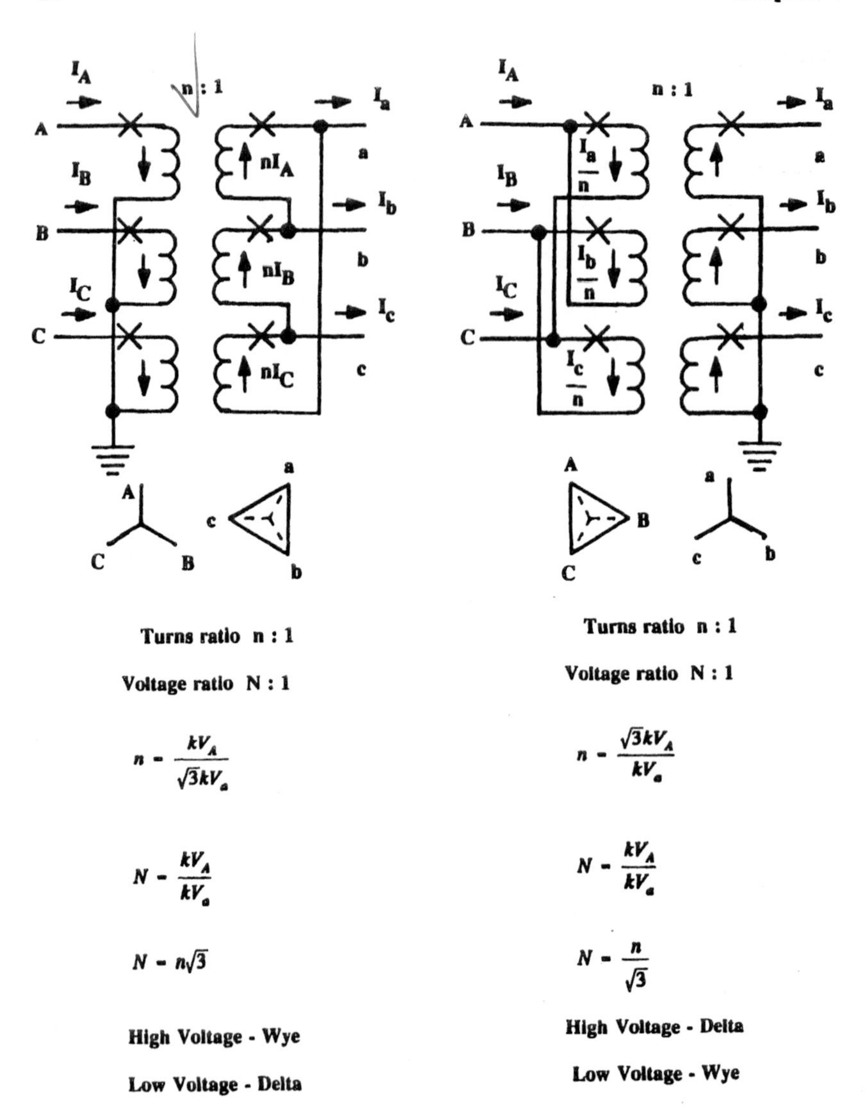

Figure 4.11 ANSI connected wye–delta transformer banks. The high line-to-neutral voltage leads the low line-to-neutral voltage by 30° for either connection.

With negative-sequence quantities from Eqs. (4.3):

$$
\begin{aligned}
I_{a2} &= n(I_{A2} - a^2 I_{A2}) & V_{A2} &= n(V_{a2} - aV_{a2}) \\
I_{a2} &= n(1 - a^2)I_{A2} & V_{A2} &= n(1 - a)V_{a2} \\
I_{a2} &= n\sqrt{3}\, I_{A2}\underline{/+30^\circ} & V_{A2} &= n\sqrt{3}\, V_{a2}\underline{/-30^\circ} \\
I_{a2} &= N I_{A2}\underline{/+30^\circ} & V_{A2} &= N V_{a2}\underline{/-30^\circ}
\end{aligned}
\tag{4.17}
$$

or

$$
I_{A2} = \frac{I_{a2}\underline{/-30^\circ}}{N} \tag{4.18}
$$

4.13.2 High-Voltage Delta, Low-Voltage Wye

See Fig. 4.11, right side. From the connections,

$$
I_A = \frac{I_a - I_b}{n} \qquad V_a = \frac{V_A - V_C}{n} \tag{4.19}
$$

With positive-sequence quantities from Eqs. (4.2),

$$
\begin{aligned}
I_{A1} &= \frac{I_{a1} - a^2 I_{a1}}{n} & V_{a1} &= \frac{V_{A1} - aV_{A1}}{n} \\
I_{A1} &= \frac{(1 - a^2)I_{a1}}{n} & V_{a1} &= \frac{(1 - a)V_{A1}}{n} \\
I_{A1} &= \frac{\sqrt{3}\, I_{a1}\underline{/+30^\circ}}{n} & V_{a1} &= \frac{\sqrt{3}\, V_{A1}\underline{/-30^\circ}}{n} \\
I_{A1} &= \frac{I_{a1}\underline{/+30^\circ}}{N} & V_{a1} &= \frac{V_{A1}\underline{/-30^\circ}}{N}
\end{aligned}
\tag{4.20}
$$

$$
V_{A1} = N V_{a1}\underline{/+30^\circ} \tag{4.21}
$$

With negative-sequence quantities from Eqs. (4.3),

$$I_{A2} = \frac{I_{a2} - aI_{a2}}{n} \qquad V_{a2} = \frac{V_{A2} - a^2V_{A2}}{n}$$

$$I_{A2} = \frac{(1 - a)I_{a2}}{n} \qquad V_{a2} = \frac{(1 - a^2)V_{A2}}{n} \tag{4.22}$$

$$I_{A2} = \frac{\sqrt{3}\, I_{a2}\underline{/-30^\circ}}{n} \qquad V_{a2} = \frac{\sqrt{3}\, V_{A2}\underline{/+30^\circ}}{n}$$

$$I_{A2} = \frac{I_{a2}\underline{/-30^\circ}}{N} \qquad V_{a2} = \frac{V_{A2}\underline{/+30^\circ}}{N}$$

$$V_{A2} = NV_{a2}\underline{/-30^\circ} \tag{4.23}$$

4.13.3 Summary

An examination of the preceding equations shows that for ANSI standard connected wye–delta transformer banks: (1) if both the positive-sequence current and voltage on one side lead the positive-sequence current and voltage on the other side by 30°, the negative-sequence current and voltage correspondingly will both lag by 30°; and (2) similarly, if the positive-sequence quantities lag in passing through the bank, the negative-sequence quantities correspondingly will lead 30°. This fundamental is useful in transferring currents and voltages through these banks.

The effect of these shifts on the various faults is illustrated in Fig. 4.12. For the three-phase (a) and phase-to-phase (b) faults, the per unit currents shown are equivalent. However, 1.0 pu phase-to-ground in (c) is not equivalent to 1.0 pu in (a) and (b) since the phase-to-ground fault involves zero sequence X_0.

It will be noted that (1) the three-phase fault is unaffected by the phase shift through the transformer. (2) A phase-to-phase fault on the wye side has somewhat the appearance of a phase-to-ground fault on the delta. There is one large current relative to two smaller currents, which is characteristic of phase-to-ground faults, but there is no zero sequence current. (3) A phase-to-ground fault on the wye side appears as phase-to-phase fault

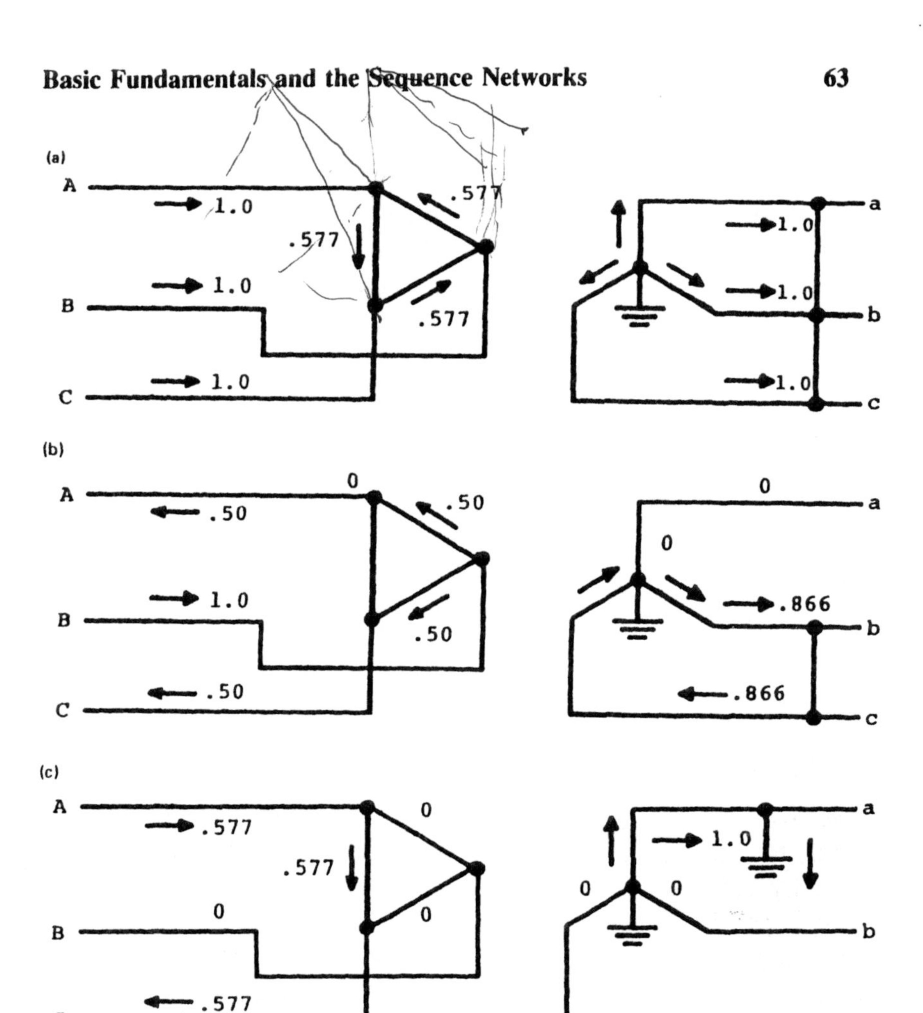

Figure 4.12 Review of faults through delta-wye transformer banks (currents shown in per unit): (a) three-phase faults; (b) phase-to-phase faults; (c) phase-to-ground faults where $X_1 = X_2 = X_0$.

on the delta side. This reduced current on the delta side can be a problem in protective relaying.

An example calculating these current and voltage across a wye-delta transformer is given in Sec. 6.17.

Zero sequence is not phase shifted *if* it can pass through and flow in the transformer bank. The zero-sequence circuits for various transformer banks are shown in Figs. 4.9 and 4.10.

4.14 SEQUENCE NETWORK VOLTAGES

The voltage drops in the three-sequence networks as indicated previously are *always* line-to-neutral or line-to-ground voltage drops. Conventional current flow in each network is assumed to be from the neutral or zero potential bus to the area of the unbalance.

4.14.1 Positive-Sequence Voltages

In the positive-sequence network, the positive-sequence voltage drop V_{1x} at any point x in the network is always

$$V_{1x} = V - \sum I_1 Z_1 \tag{4.24}$$

V is the source voltage (V, V_G, or V_H in Fig. 4.5) and $\sum I_1 Z_1$ is the sum of the drops along any path from the neutral bus (N_1) to the point of measurement (x).

4.14.2 Negative-Sequence Voltages

In the negative-sequence network, the negative-sequence voltage drop V_{2x} at any point x in the negative-sequence network is always

$$V_{2x} = 0 - \sum I_2 Z_2 \tag{4.25}$$

where $\sum I_2 Z_2$ is the sum of the drops along any path from the neutral bus N_2 to the point of measurement (x). See Fig. 4.6.

4.14.3 Zero-Sequence Voltages

In the zero-sequence network, the zero-sequence voltage drop V_{0X} at any point x in the network is always

$$V_{0X} = 0 - \sum I_0 Z_0 \tag{4.26}$$

where $\sum I_0 Z_0$ is the sum of the drops along any path from the neutral or zero-potential bus (N_0) to the point of measurement (x). See Fig. 4.8.

4.15 SEQUENCE NETWORK REDUCTION

For shunt fault calculations (three-phase, phase-to-phase, two-phase-to-ground and single-phase-to-ground faults), each of the three independent sequence networks can be reduced to a single equivalent impedance. These are commonly designated as Z_1 or X_1, Z_2 or X_2, Z_0 or X_0 for the respective sequence networks. Each represents the total impedance (reactance) from their neutral or zero potential bus to the fault location. These values will be different for each fault location and operating condition. In the positive-sequence network this is the Thévenin theorem equivalent impedance and is associated with the Thévenin voltage (see Sec. 6.14).

For the system of Fig. 4.4, and neglecting loads as is common in many calculations, the positive-sequence network, Fig. 4.5b, is reduced for a fault at bus H by paralleling the impedances on either side of the fault point:

$$Z_1 = \frac{(X''_d + X_{TG} + Z_{1GH})(Z_{1S} + X_{HM})}{X''_d + X_{TG} + Z_{1GH} + Z_{1S} + X_{HM}} \tag{4.27}$$

Each term in parentheses in the numerator, divided by the denominator, provides a per unit value to define the portion of current flowing in the two parts of the network. These are known as distribution factors and are necessary to determine the fault currents in various parts of the system. Thus the per unit current through bus G is

$$I_{1G} = \frac{Z_{1S} + X_{HM}}{X''_d + X_{TG} + Z_{1GH} + Z_{1S} + X_{HM}} \quad \text{pu} \tag{4.28}$$

and the current through bus H is

$$I_{1S} = \frac{X''_d + X_{TG} + Z_{1GH}}{X''_d + X_{TG} + Z_{1GH} + Z_{1S} + X_{HM}} \quad \text{pu} \tag{4.29}$$

The negative- and zero-sequence networks can be reduced in a manner similar to a single impedance or reactance value to a fault point and with appropriate distribution factors. These three independent equivalent networks are shown in Fig. 4.13 with I_1, I_2, and I_0 representing the respective sequence currents in the fault, and V_1, V_2, and V_0 representing the respective sequence voltages at the fault point.

Fault and unbalance studies with computers use various techniques to reduce complex power system networks and determine fault currents and voltages. A discussion of these is beyond the scope of this book. Where a computer or program may not be available, convenient, or practical to use, especially in small networks or additions, the following can be helpful.

4.16 THÉVENIN THEOREM IN NETWORK REDUCTION

The reduction of the positive-sequence network with load as illustrated in Fig. 4.5a requires determining the load current flow throughout the network before the fault. With load current flowing before the fault, determining the voltage at the fault point. This is the Thévenin voltage and is the equivalent of an open-circuit voltage at the fault since it is prefault. Then with all system

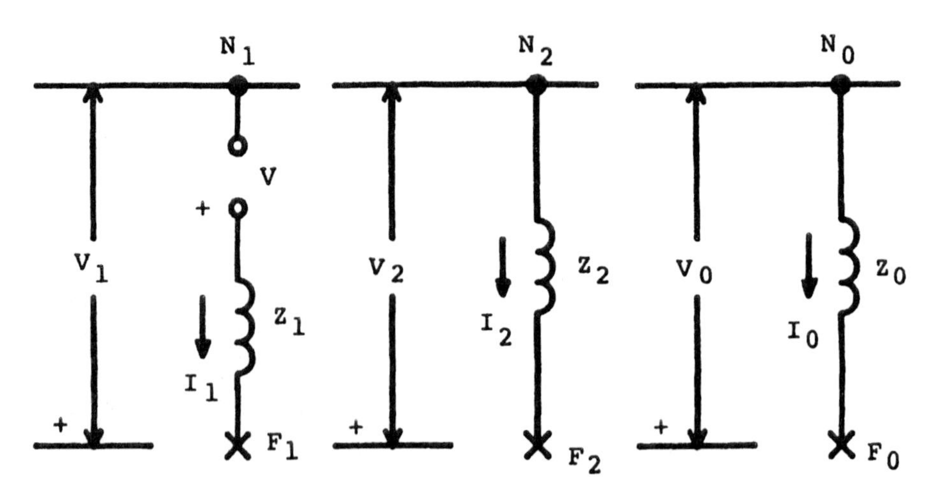

Figure 4.13 Reduced sequence networks where Z_1, Z_2, and Z_0 are the equivalent impedances of the network to the fault point.

voltages shorted out, determine the equivalent impedance looking into the network. This is the Thévenin impedance. These Thévenin quantities are the V and Z_1 in the reduced positive network of Fig. 4.13. After the fault currents are determined, the total currents in the network are the phasor sum of the prefault load and the fault currents.

Since there are no voltages in the negative- or zero-sequence networks, there is no Thévenin voltage and the normal reduction to Z_2 and Z_0 as in Fig. 4.13 are equivalent to the Thévenin impedances.

4.17 WYE–DELTA NETWORK TRANSFORMATIONS

In reducing loop-type sequence networks, the wye–delta transformations are very useful.

4.17.1 Delta to Wye

In Fig. 4.14 the delta impedances Z_R, Z_S, and Z_T between points X, Y, and Z can be replaced by Z_X, Z_Y, and Z_Z, where

$$\begin{aligned} Z_X &= \frac{Z_S Z_R}{Z_R + Z_S + Z_T} \\ Z_Y &= \frac{Z_R Z_T}{Z_R + Z_S + Z_T} \\ Z_Z &= \frac{Z_S Z_T}{Z_R + Z_S + Z_T} \end{aligned} \tag{4.30}$$

This is easy to remember; the wye branch is the product of the two impedances of the delta branch on either side of the wye equivalent divided by the sum of the three delta impedances.

For the balanced case where $Z_R = Z_S = Z_T$, the equivalent wye branches are

$$Z_{\text{wye}} = \frac{Z_{\text{delta}}}{3} \tag{4.31}$$

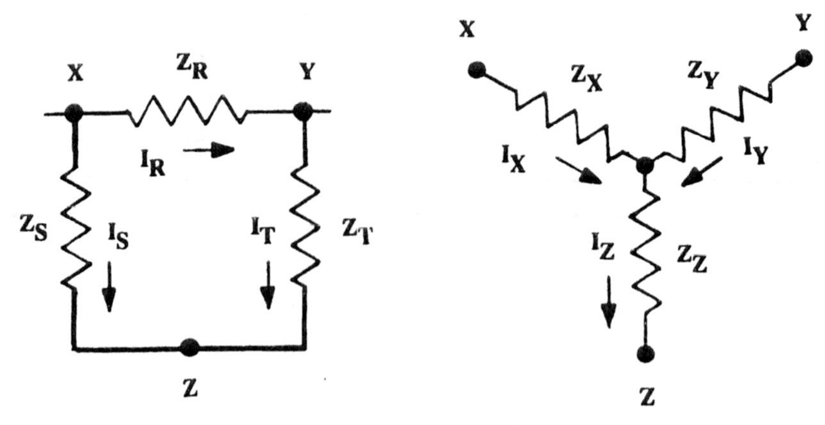

Figure 4.14 Wye–delta equivalent networks.

4.17.2 Wye to Delta

In Fig. 4.14, the wye impedances Z_X, Z_Y, and Z_Z between points X, Y, and Z can be replaced by Z_R, Z_S, and Z_T, where

$$
\begin{aligned}
Z_R &= \frac{Z_X Z_Y + Z_Y Z_Z + Z_Z Z_X}{Z_Z} \\
Z_S &= \frac{Z_X Z_Y + Z_Y Z_Z + Z_Z Z_X}{Z_Y} \\
Z_T &= \frac{Z_X Z_Y + Z_Y Z_Z + Z_Z Z_X}{Z_X}
\end{aligned}
\tag{4.32}
$$

In determining the current distribution factors in a delta node that had previously been converted to an equivalent wye, voltage drops provide a useful method. Thus in Fig. 4.14 with I_X, I_Y, and I_Z flowing in the wye branch, the currents in the delta equivalent I_S, I_R, and I_T can be obtained from the voltage drops. Thus the voltage drop between points X and Y must be the same in the wye or delta. With I_X and I_Y flowing toward the neutral in the wye,

$$V_{XY} = I_X Z_X - I_Y Z_Y = I_R Z_R$$

$$I_R = \frac{I_X Z_X - I_Y Z_Y}{Z_R}$$

Similarly,

$$I_S = \frac{I_X Z_X + I_Z Z_Z}{Z_S}$$
$$I_T = \frac{I_Y Z_Y + I_Z Z_Z}{Z_T} \qquad (4.33)$$

4.18 SHORT-CIRCUIT MVA AND EQUIVALENT IMPEDANCE

Where a new circuit is to be added to an existing bus in a complex power system, short-circuit MVA (or kVA) data provide the equivalent impedance of the power system up to that bus. One value is for three-phase faults, another is for single-phase-to-ground faults. The derivation of this and conversion into system impedance are as follows.

4.18.1 Three-Phase Short-Circuit MVA

$$\text{MVA}_{SC} = \text{three-phase short-circuit MVA}$$
$$= \frac{\sqrt{3}\, I_{3\phi}\, \text{kV}}{1000} \qquad (4.34)$$

where $I_{3\phi}$ is the total three-phase fault current in amperes and kV is the system phase-to-phase voltage in kilovolts. Normally, this is the rated system voltage. From this,

$$I_{3\phi} = \frac{1000\ \text{MVA}_{SC}}{\sqrt{3}\ \text{kV}} \qquad (4.35)$$

$$Z_{\Omega} = \frac{V_{LN}}{I_{3\phi}} = \frac{1000\ \text{kV}}{\sqrt{3}\ I_{3\phi}} = \frac{\text{kV}^2}{\text{MVA}_{SC}} \qquad (4.36)$$

Substituting Eq. (2.15), which is

$$Z_{pu} = \frac{MVA_{base} Z_{\Omega}}{kV^2} \quad (2.15)$$

the positive-sequence impedance to the fault location is

$$Z_1 = \frac{MVA_{base}}{MVA_{SC}} \quad \text{pu} \quad (4.37)$$

$Z_1 = Z_2$ for all practical cases. Z_1 can be assumed to be X_1 unless X/R data are provided to determine an angle.

4.18.2 Single-Phase-to-Ground Faults

$$MVA_{\phi GSC} = \phi\text{G-fault short-circuit MVA}$$
$$= \frac{\sqrt{3}\, I_{\phi G}\, kV}{1000} \quad (4.38)$$

where $I_{\phi G}$ is the total single-line-to-ground fault current in amperes and kV is the system line-to-line voltage in kilovolts.

$$I_{\phi G} = \frac{1000\, MVA_{\phi GSC}}{\sqrt{3}\, kV} \quad (4.39)$$

However,

$$I_{\phi G} = I_1 + I_2 + I_0 = \frac{3\, V_{LN}}{Z_1 + Z_2 + Z_0} = \frac{3\, V_{LN}}{Z_g} \quad (4.40)$$

where $Z_g = Z_1 + Z_2 + Z_0$. From Eqs. (4.36) and (4.40),

$$Z_g = \frac{3\, kV^2}{MVA_{\phi GSC}} \quad \Omega \quad (4.41)$$

$$Z_g = \frac{3\, MVA_{base}}{MVA_{\phi GSC}} \quad \text{pu} \quad (4.42)$$

Then $Z_0 = Z_g - Z_1 - Z_2$, or in most practical cases, $X_0 = X_g - X_1 - X_2$, since the resistance is usually very small in relation to the reactance.

4.18.3 Example

A short-circuit study indicates that at bus X in the 69-kV system,

$$\mathrm{MVA}_{SC} = 594 \text{ MVA}$$

$$\mathrm{MVA}_{\phi GSC} = 631 \text{ MVA}$$

on a 100-MVA base. Thus the total reactance to the fault is

$$X_1 = X_2 = \frac{100}{594} = 0.1684 \text{ pu}$$

$$X_g = \frac{300}{631} = 0.4754 \text{ pu}$$

$X_0 = 0.4754 - 0.1684 - 0.1684 = 0.1386$ pu, all values on a 100-MVA 69-kV base.

4.19 EQUIVALENT NETWORK FROM A PREVIOUS FAULT STUDY

It may be desirable to obtain an equivalent circuit within a network system for additional study without modifying an existing program. Typical examples can be the addition of circuits, tapping an existing line, inserting a new station in the system, and so on.

This equivalent assumes a no-load condition where the system multiple source voltages are all equal, as is common in fault studies. It is in the form of a wye or delta, as shown in Fig. 4.15. The required parameters are the existing circuit impedance Z_{1GH} and Z_{0GH} between buses G and H, and from previous studies the following fault currents (per unit or amperes):

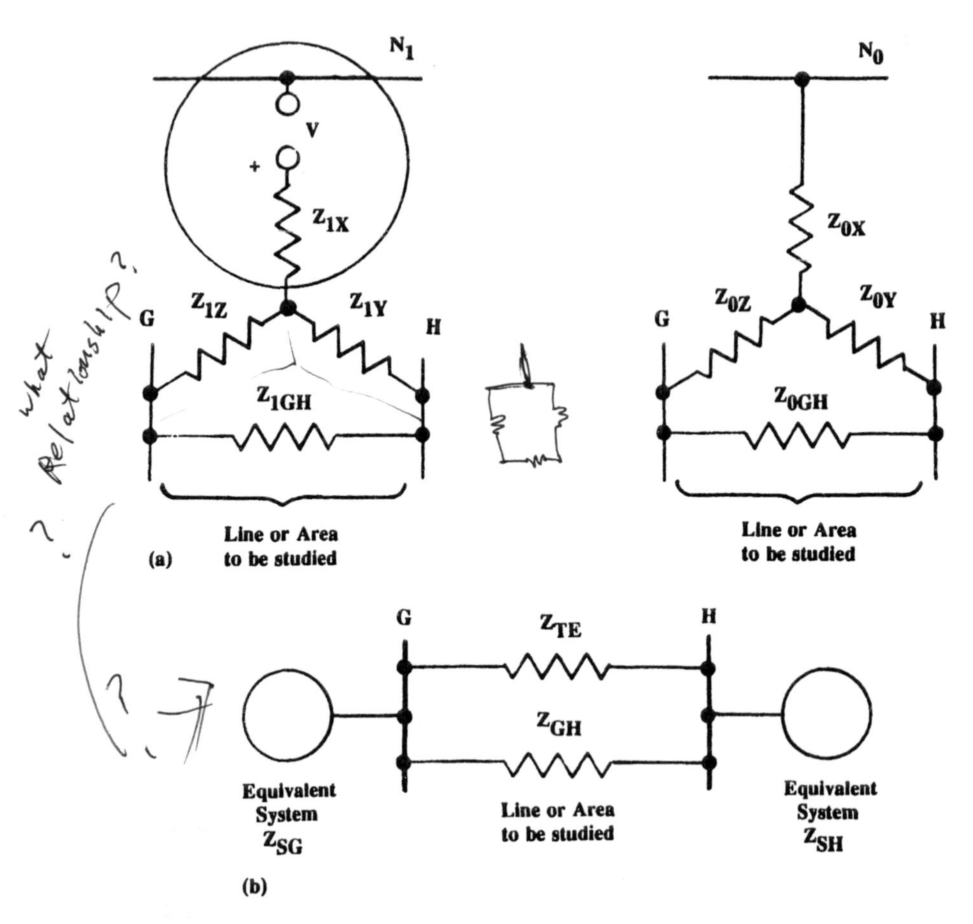

Figure 4.15 Equivalent networks between two buses G and H with known Z_{GH} values.

For the positive- and negative-sequence equivalents,

$I_{3\phi G}$ = total fault current for a three-phase fault at bus G

$I_{3\phi GL}$ = fault current in line GH for $I_{3\phi G}$ bus G fault

$I_{3\phi H}$ = total fault current for a three-phase fault at bus H

$I_{3\phi HL}$ = fault current in line GH for $I_{3\phi H}$ bus H fault

For the zero-sequence equivalents,

I_{0G} = total zero-sequence fault current for a single-phase-to-ground fault at bus G

I_{0GL} = zero-sequence fault current in line GH for I_{0G} bus G fault

I_{0H} = total zero-sequence fault current for a single-phase-to-ground fault at bus H

I_{0HL} = zero-sequence fault current in line GH for I_{0H} bus H fault

The above are zero-sequence components or one-third of the $3I_0$ values normally available.

The Z_{1X}, Z_{1Y}, and Z_{1Z} positive- and negative-sequence equivalents, and Z_{0X}, Z_{0Y}, and Z_{0Z} zero-sequence equivalents, are derived for the wye network of Fig. 4.15a. These can be changed into Fig. 4.15b delta equivalent values using the wye–delta transformations (Section 4.17.2).

4.19.1 Positive-Sequence Equivalents

For a three-phase fault at bus G in Fig. 4.15a, the total positive-sequence impedance is

$$Z_{1G} = Z_{1X} + \frac{(Z_{1Z})(Z_{1Y} + Z_{1GH})}{Z_{1Y} + Z_{1Z} + Z_{1GH}} = \frac{V}{I_{3\phi G}} \tag{4.43}$$

$$I_{3\phi GL} = I_{3\phi G}\frac{Z_{1Z}}{Z_{1Y} + Z_{1Z} + Z_{1GH}} \tag{4.44}$$

$I_{3\phi GL}$ is the current in line GH flowing from H to G.

For a three-phase fault at bus H in Fig. 4.15a, the total positive-sequence impedance is (note that these are separate faults, not simultaneous):

$$Z_{1H} = Z_{1X} + \frac{(Z_{1Y})(Z_{1Z} + Z_{1GH})}{Z_{1Y} + Z_{1Z} + Z_{1GH}} = \frac{V}{I_{3\phi H}} \tag{4.45}$$

$$I_{3\phi HL} = I_{3\phi H}\frac{Z_{1Y}}{Z_{1Y} + Z_{1Z} + Z_{1GH}} \tag{4.46}$$

$I_{3\phi HL}$ is the current flowing in line GH from G to H.
Ratio equations (4.44) and (4.46):

$$\frac{I_{3\phi GL}}{I_{3\phi HL}} = \frac{I_{3\phi G} Z_{1Z}}{I_{3\phi H} Z_{1Y}} \quad \text{or} \quad Z_{1Z} = Z_{1Y} \frac{I_{3\phi H} I_{3\phi GL}}{I_{3\phi G} I_{3\phi HL}} \tag{4.47}$$

Substitute Eq. (4.47) in Eq. (4.46):

$$I_{3\phi HL} = (I_{3\phi H}) \frac{Z_{1Y}}{Z_{1Y} + (Z_{1Y})(I_{3\phi H} I_{3\phi GL}/I_{3\phi G} I_{3\phi HL}) + Z_{1GH}} \tag{4.48}$$

Solve for Z_{1Y}:

$$Z_{1Y} = \frac{I_{3\phi G} I_{3\phi HL} Z_{1GH}}{I_{3\phi G} I_{3\phi H} - I_{3\phi H} I_{3\phi GL} - I_{3\phi G} I_{3\phi HL}} \tag{4.49}$$

Substitute Eq. (4.49) in Eq. (4.47):

$$Z_{1Z} = \frac{I_{3\phi H} I_{3\phi GL} Z_{1GH}}{I_{3\phi G} I_{3\phi H} - I_{3\phi H} I_{3\phi GL} - I_{3\phi G} I_{3\phi HL}} \tag{4.50}$$

From Eq. (4.43),

$$Z_{1X} = \frac{V}{I_{3\phi G}} - \frac{(Z_{1Z})(Z_{1Y} + Z_{1GH})}{Z_{1Y} + Z_{1Z} + Z_{1GH}} \tag{4.51}$$

Add and subtract $(Z_{1Z})^2$ in the second part of the right part of Eq. (4.51), and reduce:

$$\begin{aligned} Z_{1X} &= \frac{V}{I_{3\phi G}} - \frac{Z_{1Z}^2 + Z_{1Z} Z_{1Y} + Z_{1Z} Z_{1GH} - Z_{1Z}^2}{Z_{1Y} + Z_{1Z} + Z_{1GH}} \\ &= \frac{V}{I_{3\phi G}} - Z_{1Z} + \frac{Z_{1Z}^2}{Z_{1Z} + Z_{1Y} + Z_{1GH}} \end{aligned} \tag{4.52}$$

Substitute Eq. (4.44):

$$Z_{1X} = \frac{V}{I_{3\phi G}} - \left(1 - \frac{I_{3\phi GL}}{I_{3\phi G}}\right)(Z_{1Z}) \tag{4.53}$$

Thus Eqs. (4.49), (4.50), and (4.53) provide the equivalents of Fig. 4.15a. These can be converted into Z_{1SG}, Z_{1SH} and Z_{1TE} as in Section 4.17, Eqs. (4.32).

4.19.2 Zero-Sequence Equivalents

A similar analysis can be made for the zero-sequence equivalent impedances using single-phase-to-ground-fault data assuming that $Z_1 = Z_2$ in the networks. For a single-phase-to-ground fault at bus G, the total zero-sequence impedance is

$$Z_{0G} = \frac{(Z_{0Z})(Z_{0Y} + Z_{0GH})}{Z_{0Y} + Z_{0Z} + Z_{0GH}} + Z_{0X} \tag{4.54}$$

$$I_{1G} = I_{2G} = I_{0G} = \frac{V}{Z_{1G} + Z_{2G} + Z_{0G}} \tag{4.55}$$

From Fig. 4.15a:

$$I_{0GL} = I_{0G} \frac{Z_{0Z}}{Z_{0Y} + Z_{0Z} + Z_{0GH}} \tag{4.56}$$

where I_{0GL} is the current flowing from H to G.

For a single-phase-to-ground fault at bus H, the total zero-sequence impedance is

$$Z_{0H} = Z_{0X} + \frac{(Z_{0Y})(Z_{0Z} + Z_{0GH})}{Z_{0Y} + Z_{0Z} + Z_{0GH}} \tag{4.57}$$

$$I_{1H} = I_{2H} = I_{0H} = \frac{V}{Z_{1H} + Z_{2H} + Z_{0H}} \tag{4.58}$$

From Fig. 4.15a:

$$I_{0HL} = I_{0H} \frac{Z_{0Y}}{Z_{0Y} + Z_{0Z} + Z_{0GH}} \tag{4.59}$$

I_{0HL} is the current flowing in line *GH* from *G* to *H*.
Ratio equations (4.56) and (4.59):

$$\frac{I_{0GL}}{I_{0HL}} = \frac{I_{0G}Z_{0Z}}{I_{0H}Z_{0Y}} \quad \text{or} \quad Z_{0Z} = Z_{0Y}\frac{I_{0GL}I_{0H}}{I_{0HL}I_{0L}} \tag{4.60}$$

Substitute Eq. (4.60) in Eq. (4.59):

$$I_{0HL} = (I_{0H}) \frac{Z_{0Y}}{Z_{0Y} + (I_{0GL}I_{0H}/I_{0HL}I_{0G})(Z_{0Y}) + Z_{0GH}} \tag{4.61}$$

Solve for Z_{0Y}:

$$Z_{0Y} = \frac{I_{0G}I_{0HL}Z_{0GH}}{I_{0G}I_{0H} - I_{0H}I_{0GL} - I_{0G}I_{0HL}} \tag{4.62}$$

Substitute Eq. (4.62) in Eq. (4.60):

$$Z_{0Z} = \frac{I_{0H}I_{0GL}Z_{0GH}}{I_{0G}I_{0H} - I_{0H}I_{0GL} - I_{0G}I_{0HL}} \tag{4.63}$$

From Eq. (4.55), and substituting Eqs. (4.43) ($Z_{1G} = Z_{2G}$), and (4.54), we obtain

$$I_{0G} = \frac{V}{2V/I_{3\phi G} + Z_{0X} + Z_{0Z}(Z_{0Y} + Z_{0GH})/(Z_{0Y} + Z_{0Z} + Z_{0GH})} \tag{4.64}$$

From Eq. (4.56),

$$\frac{Z_{0Z}}{Z_{0Y} + Z_{0Z} + Z_{0GH}} = \frac{I_{0GL}}{I_{0G}} \tag{4.56}$$

Substitute Eq. (4.56) in Eq. (4.64) and solve for Z_{0X}:

$$Z_{0X} = \frac{V}{I_{0G}} - \frac{2V}{I_{3\phi G}} - \frac{I_{0GL}}{I_{0G}}(Z_{0Y} + Z_{0GH}) \tag{4.65}$$

Thus Eqs. (4.65), (4.62), and (4.63) provide the zero-sequence equivalents of Fig. 4.15a. These can be converted into Z_{0SG}, Z_{0SH}, and Z_{0TE} according to Eqs. (4.32). Normally, the current values for single-phase-to-ground faults are $3I_0$ values, so these values must be divided by 3 for Eqs. (4.65), (4.62), and (4.63).

4.20 EXAMPLE: DETERMINING AN EQUIVALENT NETWORK FROM A PREVIOUS FAULT STUDY

Determine the equivalent network at the 69-kV busses G and H that has a single 69-kV transmission line between them. The line constants on a 100-MVA base are

$$Z_1 = Z_2 = Z_{1GH} = 0.189\angle 90^\circ \text{ pu}$$

$$Z_0 = Z_{0GH} = 0.598\angle 90^\circ \text{ pu}$$

From a previous study the following data were available at 100 MVA:

At bus G: $MVA_{3\phi SC} = 2355$ with 291.7 MVA supplied over line GH from bus H. For a ground fault the total $3I_0 = 3.0$ per unit with 0.222 per unit $3I_0$ supplied over line GH from bus H.

At bus H: $MVA_{3\phi SC} = 1653$ with 362.5 MVA supplied over line GH from bus G. For a ground fault the total $3I_0 = 3.0$ per unit with 0.549 per unit $3I_0$ supplied over line GH from bus G.

The equivalents are determined as follows:

From Eqs. (2.11) and (4.35), $I_{pu} = MVA_{SC}/100$ at 100 MVA base. Thus, the per unit currents at 100 MVA are

MVA are

$$I_{3\phi G} = 23.55 \qquad I_{0G} = 1.0$$

$$I_{3\phi GL} = 2.917 \qquad I_{0GL} = 0.074$$

$$I_{3\phi H} = 16.53 \qquad I_{0H} = 1.0$$

$$I_{3\phi HL} = 3.625 \qquad I_{0HL} = 0.183$$

The positive (and negative) sequence equivalents from Eqs. (4.49), (4.50), and (4.53) are

$$Z_{1y} = \frac{(23.55)(3.625)(0.189\underline{/90^\circ})}{(23.55)(16.53) - (16.53)(2.917) - (23.55)(3.625)}$$

$$= \frac{16.135\underline{/90^\circ}}{255.695} = 0.063\underline{/90^\circ}\,\text{pu} \tag{4.68}$$

$$Z_{1z} = \frac{(16.53)(2.917)(0.189\underline{/90^\circ})}{255.695}$$

$$= \frac{9.113\underline{/90^\circ}}{255.695} = 0.0356\underline{/90^\circ}\,\text{pu} \tag{4.69}$$

$$Z_{1x} = \frac{\underline{/90^\circ}}{23.55} - \left(1 - \frac{2.917}{23.55}\right)(0.0356\underline{/90^\circ})$$

$$= 0.0425\underline{/90^\circ} - 0.0312\underline{/90^\circ} \tag{4.70}$$

$$= 0.0113\,\underline{/90^\circ}\,\text{pu}$$

$$= 0.0463\underline{/90^\circ}\,\text{pu}$$

The positive sequence equivalent wye per Fig. 4.15a is shown in Fig. 4.16.

Similarly, for the zero sequence equivalent per Eqs. (4.62),

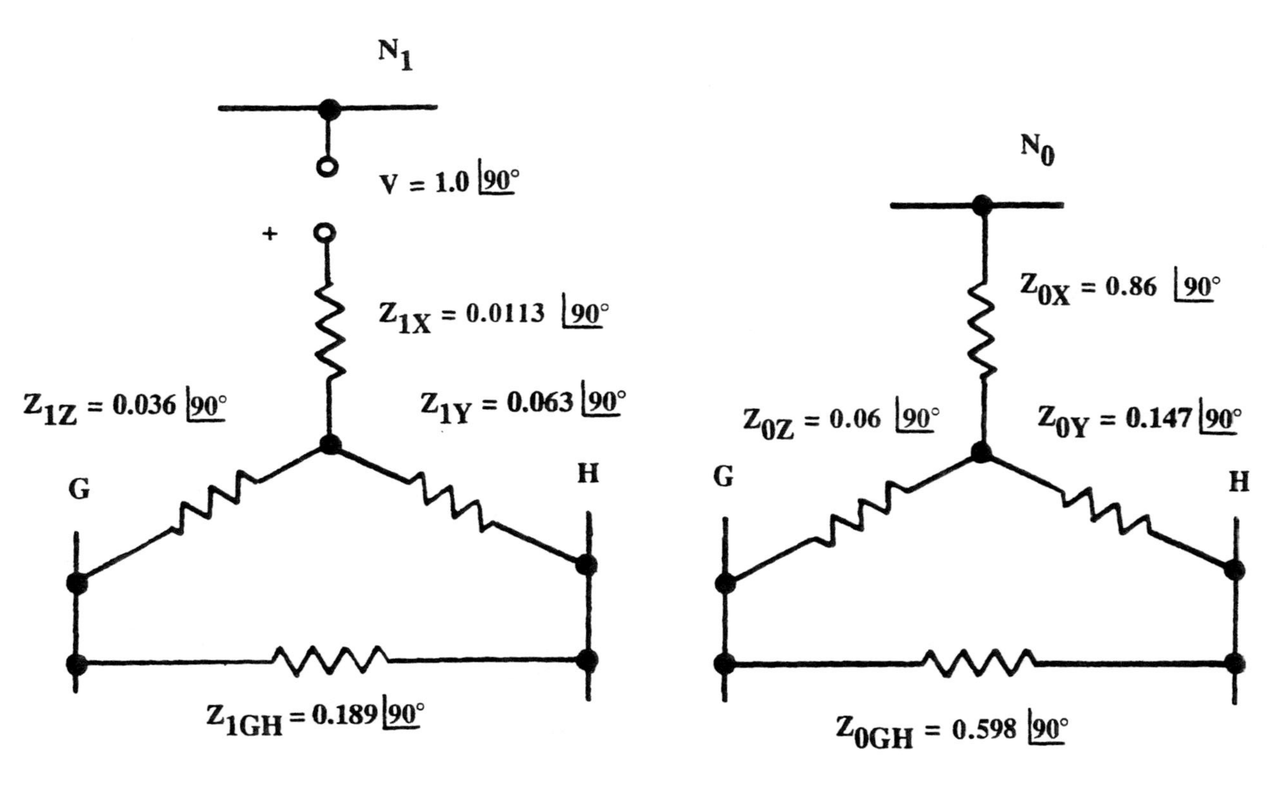

Figure 4.16 Equivalent wye-type networks for the example of Section 4.20. Values are in per unit at 100 MVA, 69 kV.

(4.63), and (4.65),

$$Z_{0y} = \frac{(1)(0.183)(0.598\angle 90^\circ)}{(1)(1) - (1)(0.074) - (1)(0.183)}$$

$$= \frac{0.1094\angle 90^\circ}{0.743} = 0.147\angle 90^\circ \text{ pu} \tag{4.71}$$

$$Z_{0z} = \frac{(1)(0.074)(0.598\angle 90^\circ)}{0.743}$$

$$= \frac{0.044\angle 90^\circ}{0.743} = 0.060\angle 90^\circ \text{ pu} \tag{4.72}$$

$$Z_{0x} = \frac{1\angle 90^\circ}{1} - \frac{2\angle 90^\circ}{23.55} - \frac{0.074}{1}(0.147\angle 90^\circ + 0.598\angle 90^\circ)$$

$$= (1 - 0.0849 - 0.0551)\angle 90^\circ = 0.86\angle 90^\circ \text{ pu} \tag{4.73}$$

The equivalent zero sequence wye per Fig. 4.15a is shown in Fig. 4.16.

With Eqs. (4.32), the positive (negative) and zero sequence equivalents can be determined as in Fig. 4.15b. The equivalents are shown in Fig. 4.17, where, from Figs. 4.14 and 4.15b,

Z_{SG} of Fig. 4.15b = Z_S of Fig. 4.14

Z_{SH} of Fig. 4.15b = Z_R of Fig. 4.14

Z_{TE} of Fig. 4.15b = Z_T of Fig. 4.14

From Eqs. (4.32) for the positive and negative sequence:

$$Z_{1x}Z_{1y} + Z_{1y}Z_{1z} + Z_{1z}Z_{1x} = [(0.0113)(0.063) + (0.063)(0.036) + (0.036)(0.0113)]\angle 180^\circ$$

$$= 0.00339\angle 180^\circ \text{ pu}$$

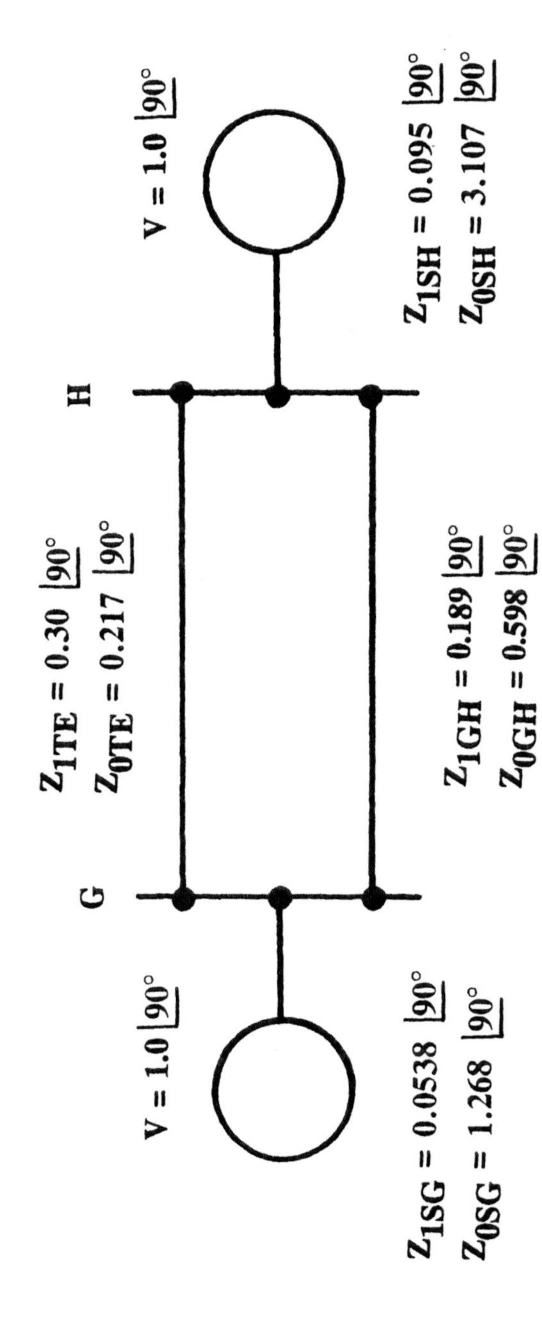

Figure 4.17 Equivalent delta-type network for the example of Section 4.20. Values are in per unit at 100 MVA, 69 kV.

Then,

$$Z_{1SG} = \frac{0.00339}{0.063} \angle 90^\circ = 0.0538\angle 90^\circ \text{ pu} \tag{4.74}$$

$$Z_{1SH} = \frac{0.00339}{0.0356} \angle 90^\circ = 0.0951\angle 90^\circ \text{ pu} \tag{4.75}$$

$$Z_{1TE} = \frac{0.00339}{0.0113} \angle 90^\circ = 0.030\angle 90^\circ \text{ pu} \tag{4.76}$$

In a similar manner, from Eqs. (4.32) for the zero sequence:

$$\begin{aligned} Z_{0x}Z_{0y} + Z_{0y}Z_{0z} + Z_{0z}Z_{0x} &= [(0.86)(0.147) + (0.147)(0.060) \\ &\quad + (0.060)(0.86)]\angle 180^\circ \\ &= (0.126 + 0.0088 + 0.0516)\angle 180^\circ \\ &= 0.1864\angle 180^\circ \end{aligned}$$

Then,

$$Z_{0SG} = \frac{0.1864}{0.147} \angle 90^\circ = 1.268\angle 90^\circ \tag{4.77}$$

$$Z_{0SH} = \frac{0.1864}{0.06} \angle 90^\circ = 3.107\angle 90^\circ \tag{4.78}$$

$$Z_{TE} = \frac{0.1864}{0.86} \angle 90^\circ = 0.217\angle 90^\circ \tag{4.79}$$

These quantities are shown in Fig. 4.17.

Other parallel lines or taps now can be added at or between buses G and H, and either Fig. 4.16 or Fig. 4.17 can be used to calculate faults on these circuits.

4.21 NETWORK REDUCTION BY SIMULTANEOUS EQUATIONS

The reduction of interconnected sequence networks, especially where series unbalances and load are involved, often requires the solution of three simultaneous equations with three unknown currents. The solution by formula is as follows.

Kirchhoff's voltage drops can be written around the positive- and negative-sequence networks, the positive- and zero-sequence networks and the negative- and zero-sequence networks. Any two of these plus other relations apparent from the interconnections will provide the three independent equations with three unknown currents. They must be arranged in the following order;

$$
\begin{aligned}
a_1X + b_1Y + c_1Z &= d_1 \\
a_2X + b_2Y + c_2Z &= d_2 \\
a_3X + b_3Y + c_3Z &= d_3
\end{aligned}
\tag{4.66}
$$

As long as

$$a_1b_2c_3 + a_2b_3c_1 + a_3b_1c_2 - a_3b_2c_1 - a_1b_3c_2 - a_2b_1c_3 \neq 0$$

the values of X, Y, and Z are

$$
\begin{aligned}
X &= \frac{d_1b_2c_3 + d_2b_3c_1 + d_3b_1c_2 - d_3b_2c_1 - d_1b_3c_2 - d_2b_1c_3}{a_1b_2c_3 + a_2b_3c_1 + a_3b_1c_2 - a_3b_2c_1 - a_1b_3c_2 - a_2b_1c_3} \\
Y &= \frac{a_1d_2c_3 + a_2d_3c_1 + a_3d_1c_2 - a_3d_2c_1 - a_1d_3c_2 - a_2d_1c_3}{a_1b_2c_3 + a_2b_3c_1 + a_3b_1c_2 - a_3b_2c_1 - a_1b_3c_2 - a_2b_1c_3} \\
Z &= \frac{a_1b_2d_3 + a_2b_3d_1 + a_3b_1d_2 - a_3b_2d_1 - a_1b_3d_2 - a_2b_1d_3}{a_1b_2c_3 + a_2b_3c_1 + a_3b_1c_2 - a_3b_2c_1 - a_1b_3c_2 - a_2b_1c_3}
\end{aligned}
\tag{4.67}
$$

Application of the above is illustrated in Chapter 7.

4.22 OTHER NETWORK REDUCTION TECHNIQUES

There are other techniques of reducing networks, many of which are the basis of computer programs. These are not essential to the understanding and application of symmetrical components and are seldom necessary in calculations for systems where computer and/or program may not be available or practical.

5
Shunt Unbalance Sequence Network Interconnections

5.1 INTRODUCTION

The principal abnormal shunt unbalances on a power system are commonly called *faults*. These are three-phase, phase-to-phase, two-phase-to-ground, and (single) phase-to-ground. The most common fault is the phase-to-ground, or line-to-ground as it is often called. This results primarily from lightning (induced voltage or direct strike), causing the high voltage to flash over the insulators, and wind, especially in the lower voltages, causing tree contacts. Other causes are ice, earthquakes, fire, explosions, falling trees, flying objects, physical contact by animals, and contamination. Faults result from accidents such as vehicles hitting line poles or live equipment, people contact, digging into underground cables, and so on.

The percentage or frequency of fault incidents on a power system varies with time and with many other factors, such as climate, physical location, construction, and so on. In very general terms 70 to 85% of the faults are phase-to-ground, 8 to 15%

phase-to-phase, 4 to 10% double-phase-to-ground, and 3 to 5% three-phase. Faults can also evolve from one type to another type, especially where the protective equipment is slow in dissipating or isolating the fault. Thus a phase-to-ground fault may develop into a double-phase-to-ground fault or a three-phase fault. A phase-to-phase fault may become a double-phase-to-ground or three-phase fault. With moderate to high-speed relay protection, this generally does not happen.

A fault study on a power system usually provides data only for three-phase and single-phase-to-ground faults. Both provide information for equipment selection, with the first data for phase relaying, and the second, for ground relays. The other two types of faults are rarely necessary in practice, as will be shown later.

5.2 GENERAL REPRESENTATION OF POWER SYSTEMS AND SEQUENCE NETWORKS

A useful technique as a convenient "shorthand" was developed by E. L. Harder to represent any complex or simple symmetrical power system up to an area or point of unbalance as a box. Then three other boxes provide a representation of the positive-, negative-, and zero-sequence networks for that system. This is shown in Fig. 5.1. The left set is for shunt faults and unbalances discussed in this chapter. The right set is for series unbalances in Chapter 7.

For shunt faults and unbalances, it is necessary to identify only the three phases and neutral or ground points at the area or point of the fault or unbalance. These are the a, b, c, and n points in the top box. In the sequence networks P, N, and O the neutral or zero-potential bus (n) and the fault area or point (x) are required for documenting sequence interconnections to represent various faults or unbalances. The (x) point is equivalent to the fault point (F) commonly used.

For a series unbalance, both sides of the unbalanced area or point must be identified. Hence two sets of a, b, c are required for the phases and neutral (or ground) on each side of the unbalance in the top right system box, and (x) and (y) in the sequence networks for the left and right sides of the unbalance.

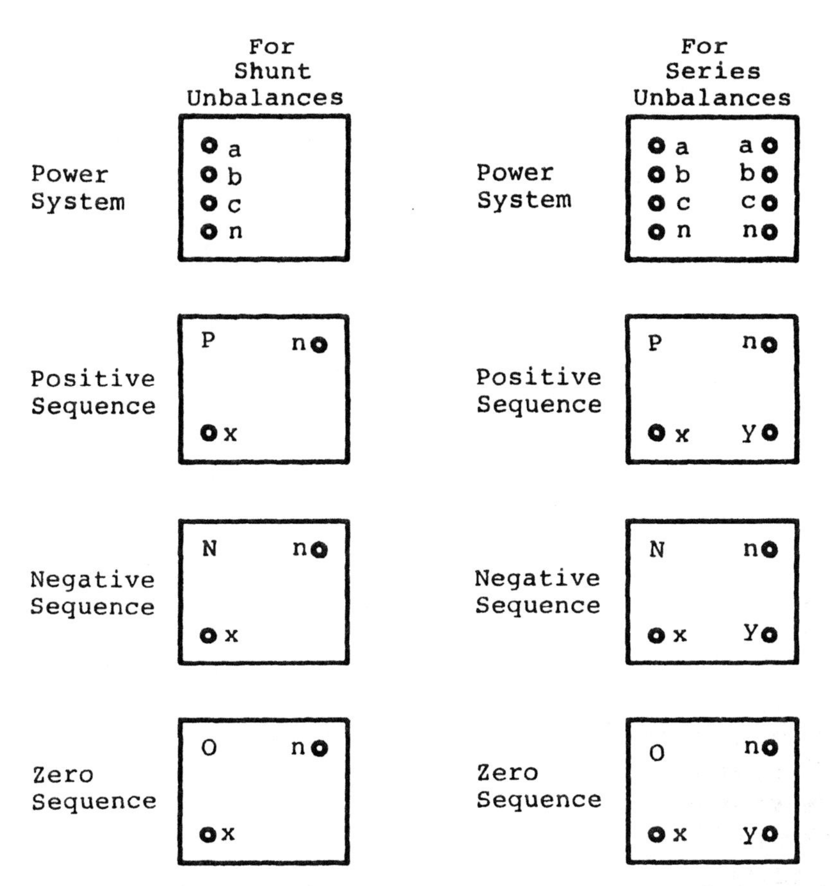

Figure 5.1 Box symbols used to represent the symmetrical or balanced power system and their sequence networks up to a point or area of unbalance.

5.3 SEQUENCE NETWORK INTERCONNECTIONS FOR THREE-PHASE FAULTS

Three-phase faults are assumed to be symmetrical; hence symmetrical components analysis is not necessary for their calculation but is convenient. The three-phase fault with optional fault resistance dashed is shown in Fig. 5.2a. The positive-sequence network reduced to a single impedance to the fault Z_{1F} (Z_1 in

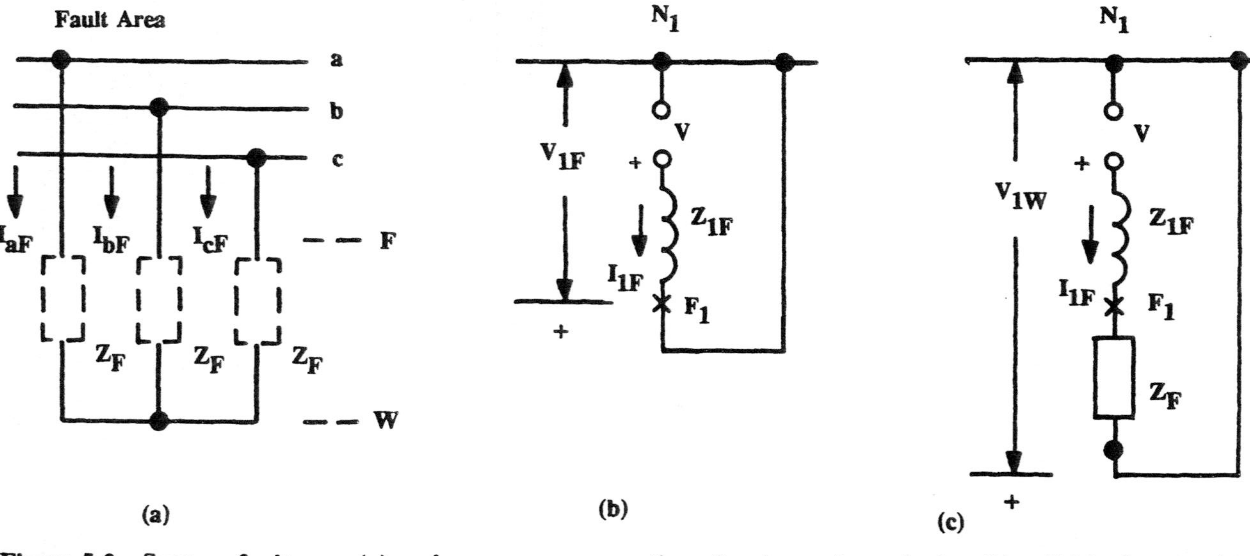

Figure 5.2 System fault area (a) and sequence connections for three-phase faults: (b) solid fault; (c) with fault impedance.

Fig. 4.13) is connected from the fault point back to the neutral bus to represent the three-phase fault as in Fig. 5.2b. A similar connection with fault resistance considered is Fig. 5.2c. From these,

$$I_{1F} = I_{aF} = \frac{V}{Z_{1F}} \quad \text{or} \quad I_{1F} = I_{aF} = \frac{V}{Z_{1F} + Z_F} \tag{5.1}$$

and from the basic equations (4.6) and (4.7),

$$I_{bF} = a^2 I_{1F} \quad \text{and} \quad I_{cF} = a I_{1F}$$

Using the box symbolism of Fig. 5.1, the sequence connections for symmetrical three-phase faults are shown in Fig. 5.3: parts (b) and (c) for solid faults, and part (a) for faults with impedance. It will be noted that these connections are equivalent to the connections of Fig. 5.2. The connections of Fig. 5.3b and c show that there is no difference between a three-phase fault involving neutral from that not involving neutral.

In Fig. 5.3a through c, the neutral bus n is shown connected to the fault point x in both the negative- and zero-sequence networks. This is academic and has no practical significance since these networks do not have any voltage sources and are not interconnected with the positive-sequence network.

5.4 SEQUENCE NETWORK INTERCONNECTIONS FOR PHASE-TO-GROUND FAULTS

The phase-a-to-ground fault is shown in Fig. 5.4a. I_{aF} is the fault current. $I_{bF} = I_{cF} = 0$ and for a solid fault, $V_{aF} = V_{aG} = 0$. Z_{1F}, Z_{2F}, and Z_{0F} are the respective sequence network impedances from the neutral bus to the fault (equivalent to Z_1, Z_2, Z_0 in Fig. 4.12). The parameters are satisfied by connecting the three sequence impedances (or networks) in series as in Fig. 5.4b or c. The equivalent connections are shown in Fig. 5.3f for a

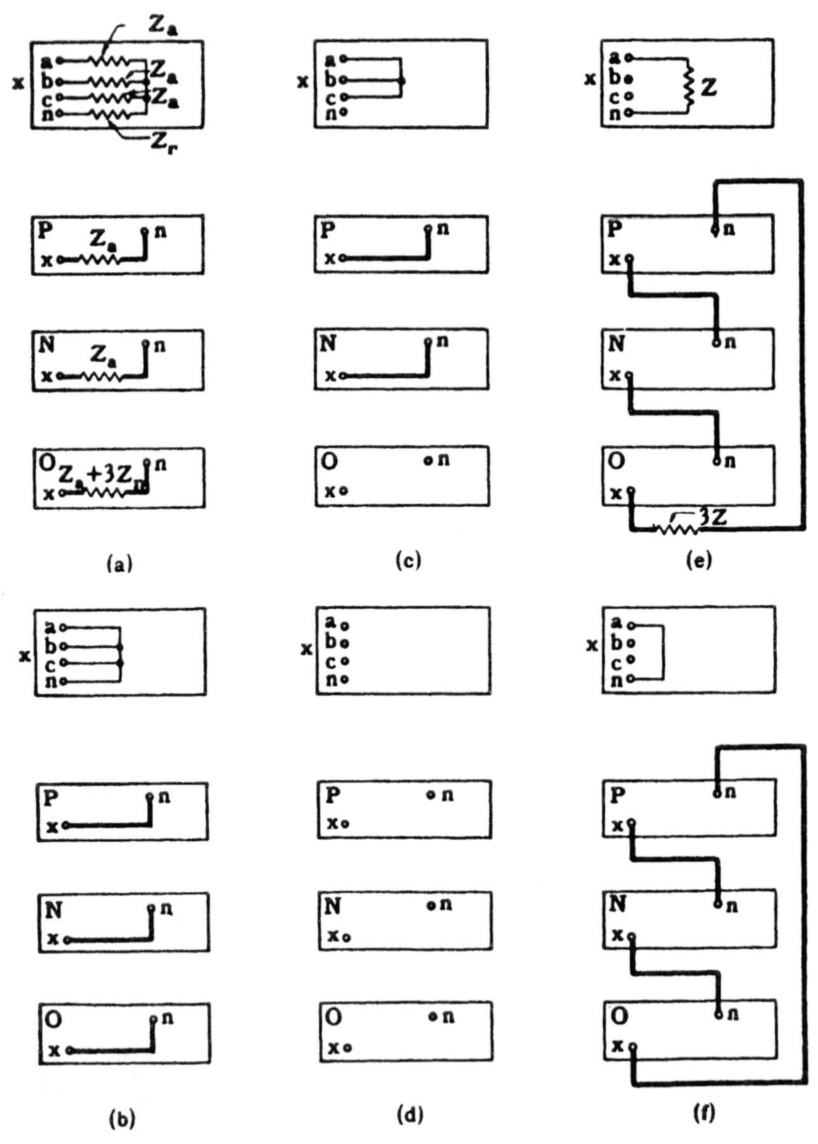

Figure 5.3 Box sequence connections for shunt balanced and unbalanced conditions: (a) balanced load or three-phase-to-ground fault with impedances; (b) three-phase-to-ground fault; (c) three-phase fault; (d) shunt circuit open; (e) phase-to-ground fault through an impedance; (f) phase-to-ground fault; (g) phase-to-phase fault through impedance; (h) phase-to-phase fault; (i) two-phase-to-ground fault through impedance;

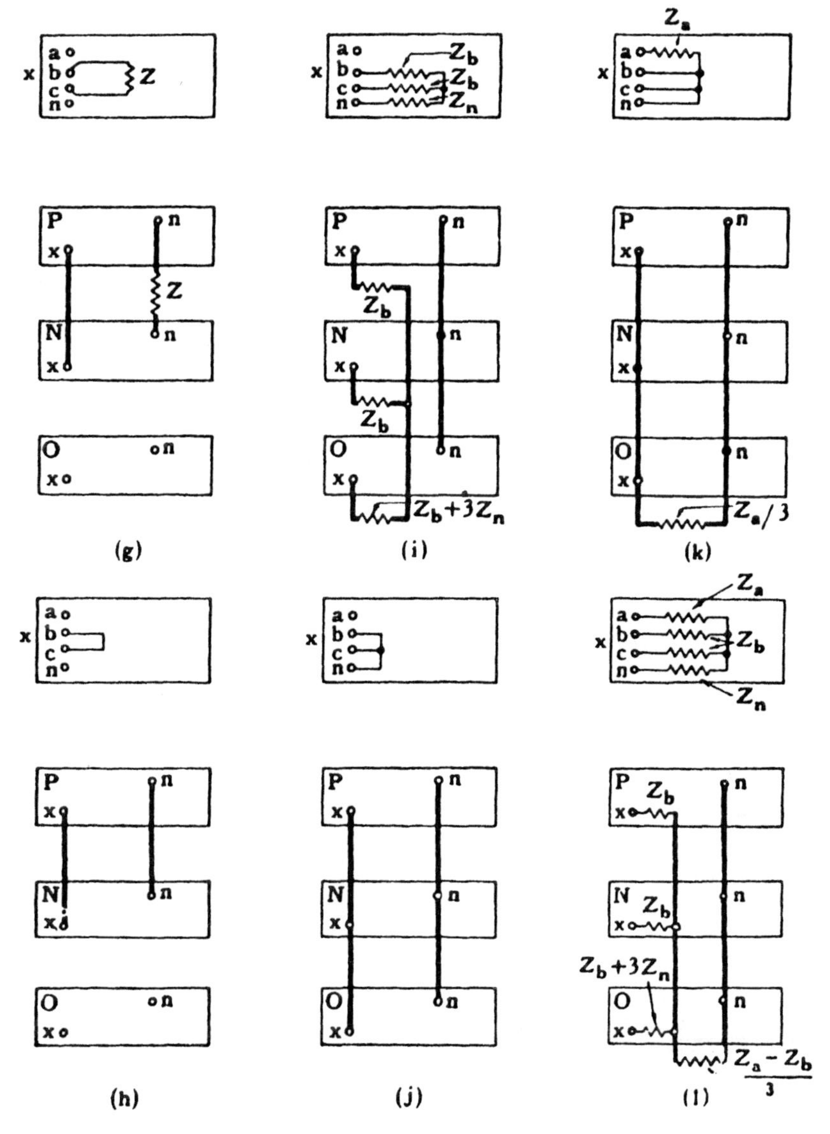

(j) two-phase-to-ground fault; (k) three-phase-to-ground fault with impedance in phase *a*; (l) unbalanced load or three-phase-to-ground fault with impedance. (From E. L. Harder, Sequence Network Connections for Unbalanced Load and Fault Conditions, *The Electrical Journal*, December 1937.)

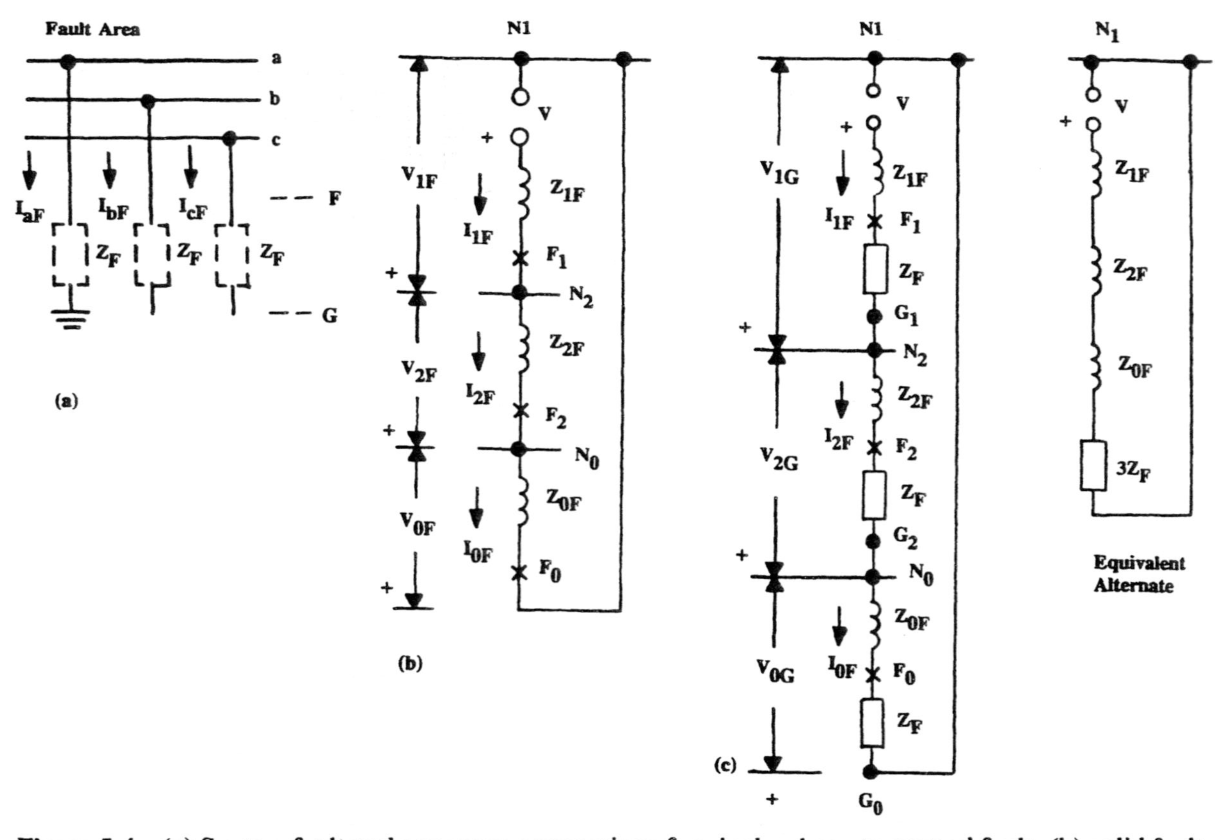

Figure 5.4 (a) System fault and sequence connections for single-phase-to-ground fault: (b) solid fault; (c) with fault impedance.

solid fault and Fig. 5.3e with fault impedance. From these,

$$I_{1F} = I_{2F} = I_{0F} = \frac{V}{Z_{1F} + Z_{2F} + Z_{0F}} \tag{5.2}$$

or

$$I_{1F} = I_{2F} = I_{0F} = \frac{V}{Z_{1F} + Z_{2F} + Z_{0F} + 3Z_F} \tag{5.3}$$

$$I_{aF} = I_{1F} + I_{2F} + I_{0F} = 3I_{1F} = 3I_{2F} = 3I_{0F} \tag{5.4}$$

From the basic equations (4.6) and (4.7) it can be seen that $I_{bF} = I_{cF} = 0$, which is correct in the fault. Also, $V_{aF} = 0$, which is supported by the sequence interconnections, where $V_{1F} + V_{2F} + V_{0F} = 0$.

5.5 SEQUENCE NETWORK INTERCONNECTIONS FOR PHASE-TO-PHASE FAULTS

It is convenient to consider a phase b-to-phase c fault as in Fig. 5.5a. Here $I_{aF} = 0$, $I_{bF} = -I_{cF}$, $V_{bW} = V_{cW}$, or for a solid fault, $V_{bF} = V_{cF}$. These parameters are satisfied by connecting the positive- and negative-sequence networks in parallel as shown in Fig. 5.5b for solid faults and in Fig. 5.5c for faults with impedance. Again Z_{1F} and Z_{2F} are the total respective sequence impedances from the neutral bus to the fault (as Z_1 and Z_2 in Fig. 4.13). These connections are equivalent to the connections in Fig. 5.3h and g, respectively. From these,

$$I_{1F} = -I_{2F} = \frac{V}{Z_{1F} + Z_{2F}} \quad \text{or}$$

$$I_{1F} = -I_{2F} = \frac{V}{Z_{1F} + Z_{2F} + Z_F} \tag{5.5}$$

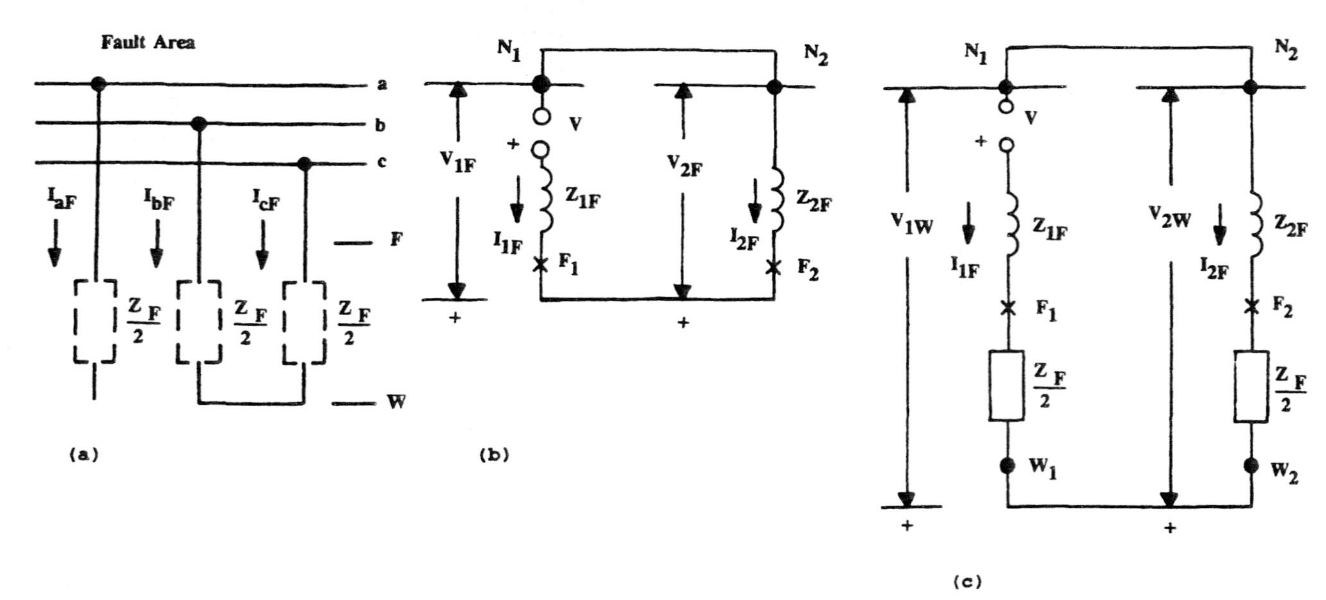

Figure 5.5 (a) System fault and sequence connections for phase-to-phase faults: (b) solid fault; (c) with fault impedance.

From the fundamental equations (4.5), (4.6), and (4.7), $I_{aF} = I_{1F} - I_{2F} = 0$, as it should be in the fault.

$$I_{bF} = a^2 I_{1F} - a I_{2F} = (a^2 + a) I_{1F} = -j\sqrt{3}\, I_{1F} \tag{5.6}$$

$$I_{cF} = a I_{1F} - a^2 I_{2F} = (a + a^2) I_{1F} = +j\sqrt{3}\, I_{1F} \tag{5.7}$$

As is common, $Z_1 = Z_2$. Then $I_{1F} = V/2Z_1$. Disregarding $+j$ and considering magnitude only, we obtain

$$I\phi\phi = \frac{\sqrt{3}\, V}{2Z_{1F}} = 0.866 \frac{V}{Z_{1F}} = 0.866 I_{3\phi} \tag{5.8}$$

Thus the solid phase-to-phase fault is 86.6% of the solid three-phase fault when $Z_1 = Z_2$.

5.6 SEQUENCE NETWORK INTERCONNECTIONS FOR TWO-PHASE-TO-GROUND FAULTS

A phase b-to-phase c-to-ground fault is shown in Fig. 5.6. Here $I_{aF} = 0$; $V_{bF} = V_{cF} = 0$ for the solid fault. This results in the three sequence networks connected in parallel as Fig. 5.6b or c with fault impedance. Z_{1F}, Z_{2F}, and Z_{0F} are the respective total impedances in the sequence networks from the neutral bus to the fault point (corresponding to Z_1, Z_2, and Z_0 in Fig. 4.13. The corresponding connections are shown in Fig. 5.3j and i, respectively. From these,

$$I_{1F} = \frac{V}{Z_{1F} + Z_{2F}Z_{0F}/(Z_{2F} + Z_{0F})} \tag{5.9}$$

or

$$I_{1F} = \frac{V}{Z_{1F} + \dfrac{Z_F}{2} + \dfrac{(Z_{2F} + Z_F/2)(Z_{0F} + Z_F/2 + 3Z_{FG})}{Z_{2F} + Z_{0F} + Z_F + 3Z_{FG}}} \tag{5.10}$$

$$I_{2F} = -I_{1F}\frac{Z_{0F}}{Z_{2F} + Z_{0F}} \quad \text{and} \quad I_{0F} = -I_{1F}\frac{Z_{2F}}{Z_{2F} + Z_{0F}} \tag{5.11}$$

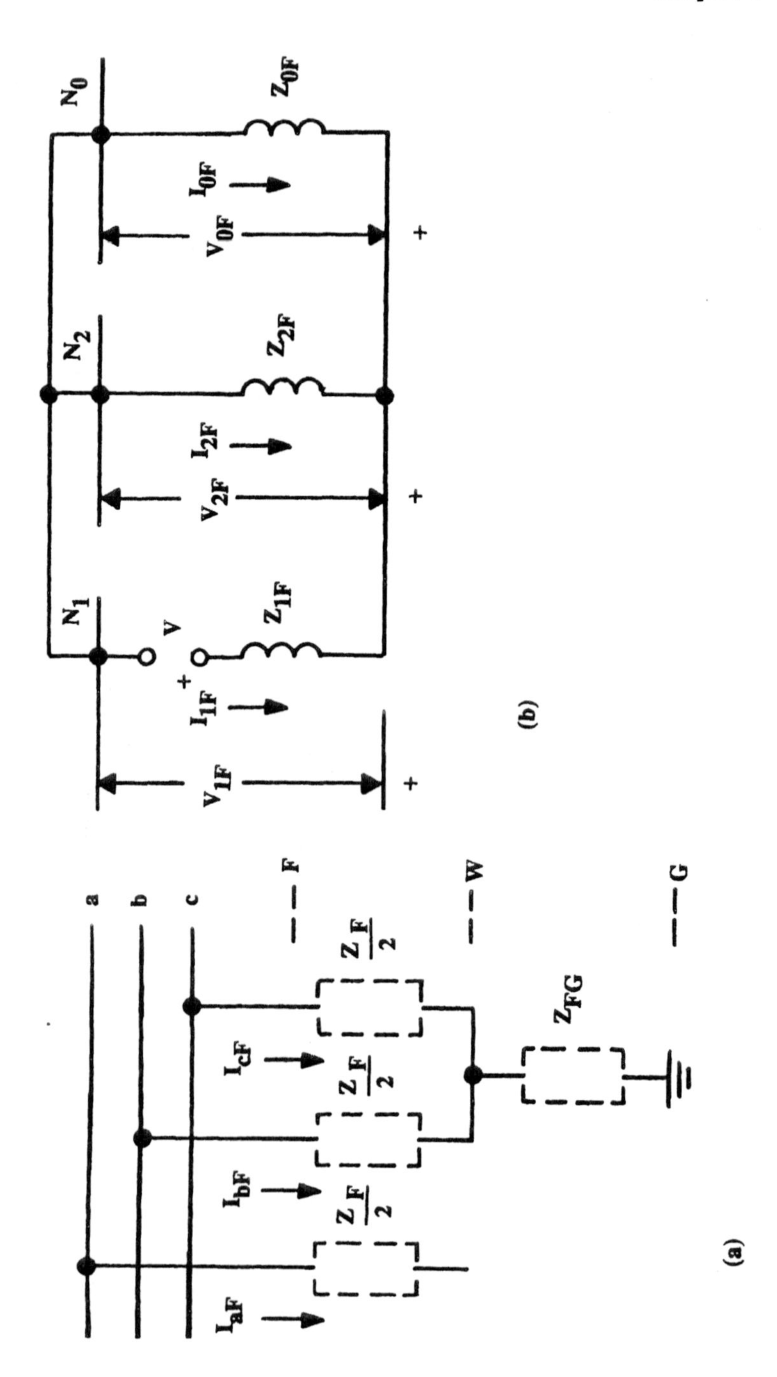
a
b
c
F
W
G
I_aF
I_bF
I_cF
Z_F/2
Z_F/2
Z_F/2
Z_FG
(a)
N_1
N_2
N_0
V
+
V_1F
I_1F
Z_1F
V_2F
I_2F
Z_2F
V_0F
I_0F
Z_0F
+
+
+
(b)

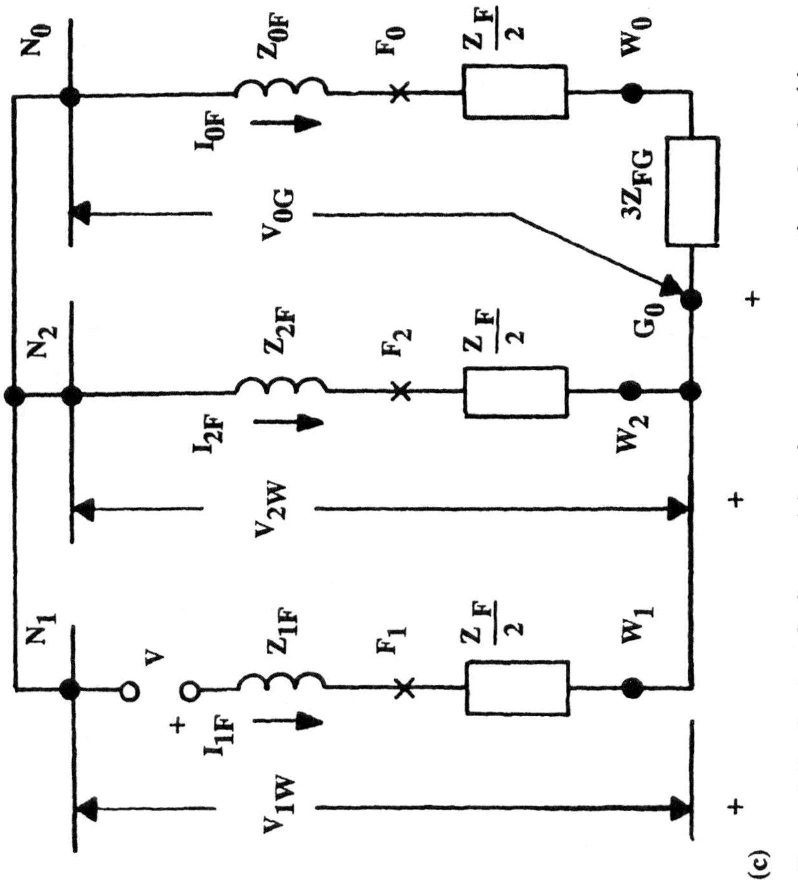

Figure 5.6 System fault area (a) and sequence connections for double-phase to ground faults: (b) solid fault; (c) with fault impedance.

or

$$I_{2F} = -I_{1F}\frac{Z_{0F} + Z_F/2 + 3Z_{FG}}{Z_{2F} + Z_{0F} + Z_F + 3Z_{FG}} \tag{5.12}$$

$$I_{0F} = I_{1F}\frac{Z_{2F} + Z_F/2}{Z_{2F} + Z_{0F} + Z_F + 3Z_{FG}} \tag{5.13}$$

5.7 OTHER SEQUENCE NETWORK INTERCONNECTIONS FOR SHUNT SYSTEM CONDITIONS

The impedances at the fault point in Figs. 5.2 through 5.6 were considered as resulting from the fault. However, they can also be considered as shunt loads, shunt reactor, and shunt capacitor connected, as indicated in the diagrams of Fig. 5.3. Thus the currents and voltages resulting from a single phase or phase-to-phase loads can be determined. Also It can be used to evaluate the effect of a shorted turn in a shunt reactor or the loss of a capacitor in a capacitor bank using the connections of Fig. 5.3l, where Z_a is made less than the normal Z_b to model the condition.

5.8 FAULT IMPEDANCE

Faults are seldom solid but involve varying amounts of resistance. However, it is generally assumed in protective relaying and most fault studies that the connection and/or contact with the ground involves very low and generally negligible impedance. For the higher voltages of transmission and subtransmission this is essentially true. In distribution systems (34.5 kV and below) very large to basically infinite impedance can exist. This is true particularly at the lower voltages. Many faults are tree contacts, which can be high impedance and intermittent and variable. Conductors lying on the ground may or may not result in significant fault current and again can be highly variable. Many tests have been conducted over the years on wet soil, dry soil, rocks, asphalt, concrete, and so on, with quite variable and sometimes unpredictable results. Thus in most fault studies, the practice is to assume zero ground mat and fault impedance for maximum fault values. Protective relays are set as sensitively as possible, yet respond properly to these maximum values.

Thus while arcs are quite variable, a commonly accepted value for currents between 70 and 20,000 A has been an arc drop of 440 V per phase, essentially independent of current magnitude. Therefore,

$$Z_{arc} = \frac{440l}{I} \quad \text{ohms} \tag{5.14}$$

where l is the arc length in feet and I the current in amperes. l/kV at 34.5 kV and higher is approximately 0.1 to 0.05. The arc essentially is resistance but can appear to protective relays as an impedance, with a significant reactive component resulting from out-of-phase contributions from remote sources. This is discussed in more detail in Chapter 12 of Blackburn, *Protective Relaying: Principles and Applications*. In low-voltage (480-V) switchboard-type enclosures, typical arc voltages of around 150 V can be experienced. This is relatively independent of current magnitude.

It appears that since arcs are variable, their resistances tend to start at a low value and continue at this value for an appreciable time, then build up exponentially. Upon reaching a high value, an arc breaks over to shorten its path and resistance.

Another difficult problem with fault arc can occur where distance-type protective relays are applied. Where the arc is fed by current from both sides of the fault, the arc resistance appears to the distance relays at either end as a much larger value and with load flowing over the circuit, as an impedance rather than resistance. This "in-feed" effect is discussed in detail in Blackburn's *Protective Relaying: Principles and Applications*.

5.9 SUBSTATION AND TOWER FOOTING IMPEDANCE

Another highly variable factor, which is difficult both to calculate and measure, is the resistance between a station ground mat or line pole or tower and ground. In recent years a number of technical papers have been written and computer programs developed in this area, but there are still many variables and assumptions. All this is beyond the scope of this book. As indicated above, the general practice is to neglect these in most fault studies and relay applications and settings.

5.10 GROUND FAULTS ON UNGROUNDED OR HIGH RESISTANCE GROUNDED SYSTEMS

The current magnitudes for these faults are very low, making it practically impossible to locate the ground fault in the system. They can be detected since $3V_0$ is approximately $3V_{LN}$ for a solid ground fault.

With low currents the equipment damage is minimal, so it is not necessarily essential that the fault area be isolated rapidly. This advantage is sometimes used in industrial plant systems where high continuity of service is important to minimize costly production process interruption, but use is limited to the lower voltages.

5.10.1 Ungrounded Systems

These systems have no intentional applied grounds but are grounded by the natural capacitance of the system as illustrated, in Fig. 5.7. While ungrounded systems exist, they are not recommended because of a high potential of restriking arcs, which can result in high, destructive transient overvoltages that can be a hazard to equipment and personnel.

Before the fault in Fig. 5.7 I_a flowing in phase *a* capacitance, I_b, and I_c are the charging currents of the system. When phase *a* is solidly grounded, the phasor currents and voltages shift essentially, as shown in Fig. 5.7. Thus the low fault current returns in the distributed shunt capacitance of the *b* and *c* phases. As shown, I_a is the fault current that shorts out the phase *a* capacitance but remote from the fault some current will also return in the phase *a* capacitance. The small currents flowing through the series phase impedances of the system will cause a very slight distortion of the voltage from that shown in Fig. 5.7. In other words, the solid fault shifts the voltage triangle from symmetry

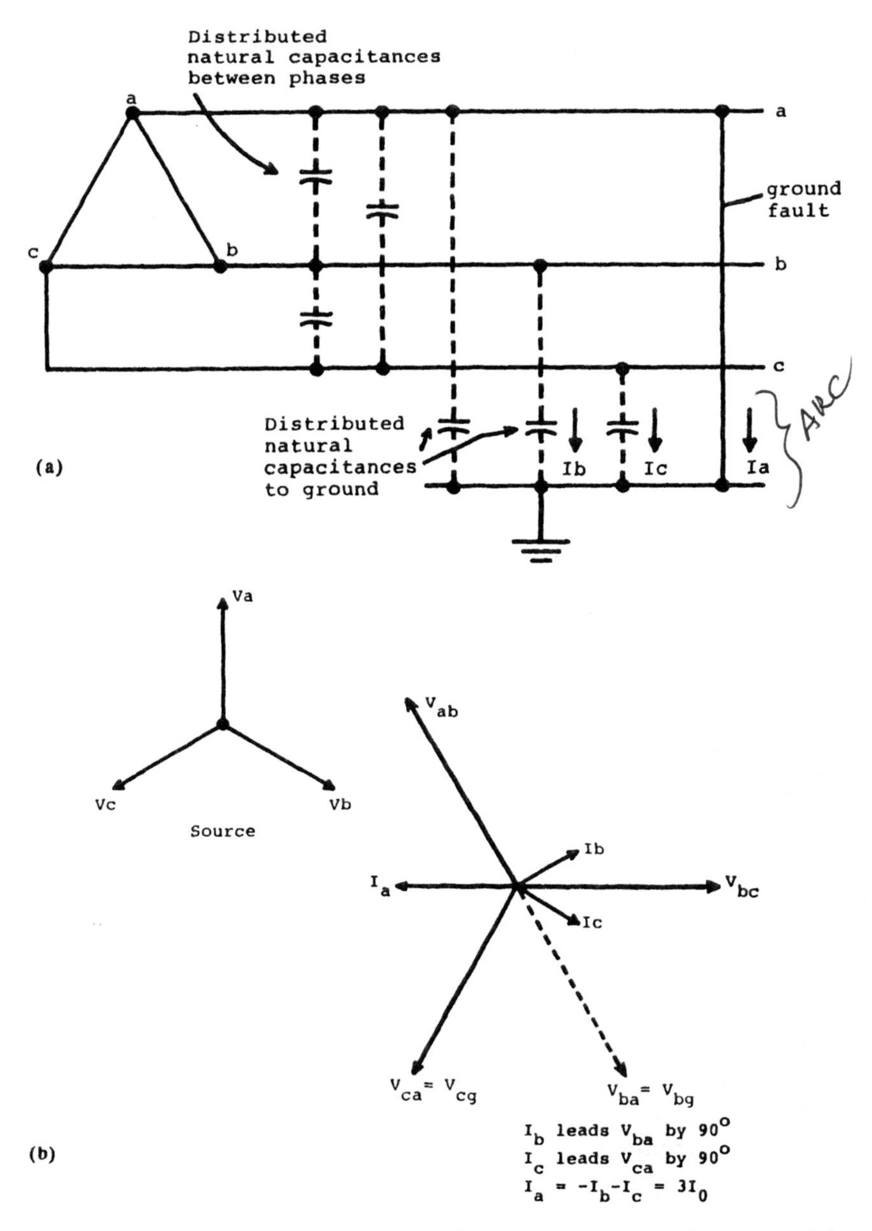

Figure 5.7 A phase-a-to-ground fault on an ungrounded system: (a) Ungrounded system, (b) phasor diagrams.

around the ground plane to that shown in Fig. 5.8b. Ideally, V_{bg} and V_{cg} are $\sqrt{3}$ times the prefault phase-to-ground voltages and at $-60°$ and $-120°$ angles. Because of the small currents in the phases these voltages are slightly different from those shown in magnitude and angle.

For a phase a-to-ground fault the sequence networks and their interconnection are shown in Fig. 5.9. The distributed capacitive reactances X_{1C}, X_{2C} and X_{0C} are very large, while the series reactance (or impedance) values X_{1S}, X_{2S}, X_T, X_{1L}, X_{0L}, and so on, are relatively very small. Thus, practically, X_{1C} is shorted out by $X_{1S} + X_T$ in the positive sequence network, and X_{2C} is shorted out by $X_{2S} + X_T$ in the negative sequence network. Since these series reactances (impedances) are very low, X_1 and X_2 approach zero relative to the large value of X_{0C}. Therefore, essentially,

$$I_1 = I_2 = I_0 = \frac{V_S}{Z_1 + Z_2 + X_{0C}} = \frac{V_S}{X_{0C}} \tag{5.15}$$

and

$$I_a = 3I_0 = \frac{3V_S}{X_{0C}} \tag{5.16}$$

This calculation can be made in per unit or amperes, remembering that V_S and all the reactances (impedances) are line-to-neutral quantities.

The unfaulted phase b and c currents will be zero when determined from the sequence currents of Eq. (5.15). This is correct for the fault itself. However, throughout the system the distributed capacitance X_{1C} and X_{2C} is actually parallel with the series reactances X_{1S}, X_T, and so on, so that in the system I_1 and I_2 are not quite equal to I_0. Thus I_b and I_c exist and are small but necessary as the return paths for I_a fault current. This is shown in Fig. 5.7. If $I_a = -1$ pu, then $I_b = 0.577\underline{/+30°}$ and $I_c = 0.577\underline{/-30°}$ pu.

In industrial applications where ungrounded systems might be used, the X_{0C} is equal practically to $X_{1C} = X_{2C}$ and is equivalent to the charging capacitance of the transformers, cables, motors, surge-suppression capacitors, local generators, and so on, in the ungrounded circuit area. Various reference sources provide tables and curves for typical charging capacitances per phase of the power system components. In an existing system the total capacitance can be determined by dividing the measured phase charging current into the line-to-neutral voltage.

Note that as faults occur in different parts of the ungrounded system, X_{0C} does not change significantly. Since the series impedances are quite small in comparison, the fault currents are the same practically and independent of the fault location. This makes impractical selective location of faults on these systems by the protective relaying.

When a phase-to-ground fault does occur, the unfaulted phase-to-ground voltages are increased essentially by $\sqrt{3}$, as shown in Figs. 5.7 and 5.8b. Thus these systems require line-to-line voltage insulation.

In the normal balanced system (Fig. 5.8a), $V_{an} = V_{ag}$, $V_{bn} = V_{bg}$, and $V_{cn} = V_{cg}$. When a ground fault occurs, the phase-to-neutral voltages and the phase-to-ground voltages are quite different. The neutral "*n*" or "*N*" is defined as the "point that has the same potential as the point of junction of a group (three for three-phase systems) of equal nonreactive resistances if connected at their free ends to the appropriate main terminals (phases of the power system)" (IEEE 100). This is the "*n*" shown in Fig. 5.8b.

From this figure, the voltage drop around the right-hand triangle is

$$V_{bg} - V_{bn} - V_{ng} = 0 \tag{5.17}$$

and around the left triangle,

$$V_{cg} - V_{cn} - V_{ng} = 0 \tag{5.18}$$

Also,

$$V_{ng} + V_{an} = 0 \tag{5.19}$$

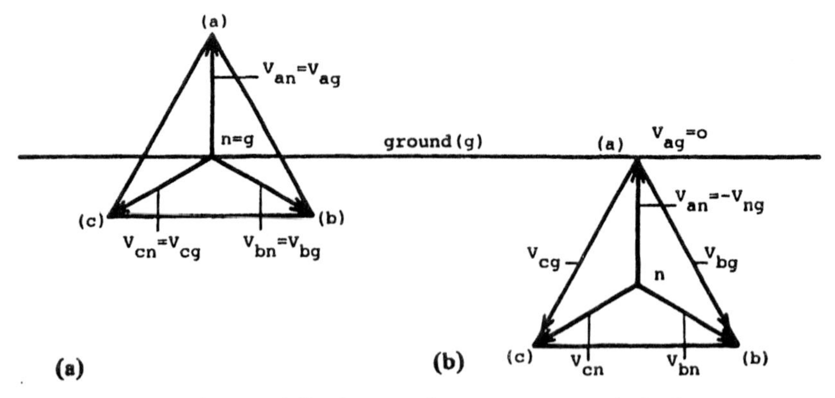

Figure 5.8 Voltage shift for a phase-a-to-ground fault on an ungrounded system: (a) Normal balanced system; (b) phase a solidly grounded.

From the basic equations,

$$V_{ag} + V_{bg} + V_{cg} = 3V_0 \tag{5.20}$$

$$V_{an} + V_{bn} + V_{cn} = 0 \tag{5.21}$$

Subtracting Eq. (5.20) from (5.21) substituting Eqs. (5.17) through (5.19), and with $V_{ag} = 0$:

$$V_{ag} - V_{an} + V_{bg} - V_{bn} + V_{cg} - V_{cn} = 3V_0$$

$$V_{ng} + V_{ng} + V_{ng} = 3V_0 \tag{5.22}$$

$$V_{ng} = V_0$$

Thus the neutral shift is the zero-sequence voltage. In the balanced system of Fig. 5.8a, $n = g$ and V_0 is zero and there is no neutral shift.

5.10.2 High-Resistance Grounded Systems

These systems are grounded through a resistor, and the accepted practice is use a value of resistance essentially equal to the total system capacitance to ground. This provides a low fault current to minimize damage and limits the potential transient overvoltages to not more than 2.5 times normal crest value to ground.

The grounding resistor may be connected in the neutral of a generator or power transformer or across the broken delta secondary of three phase-to-ground connected distribution transformers.

Resistor in the Neutral

This application is shown in Fig. 5.10 and is widely used to ground unit generators, and in industrial applications with a sin-

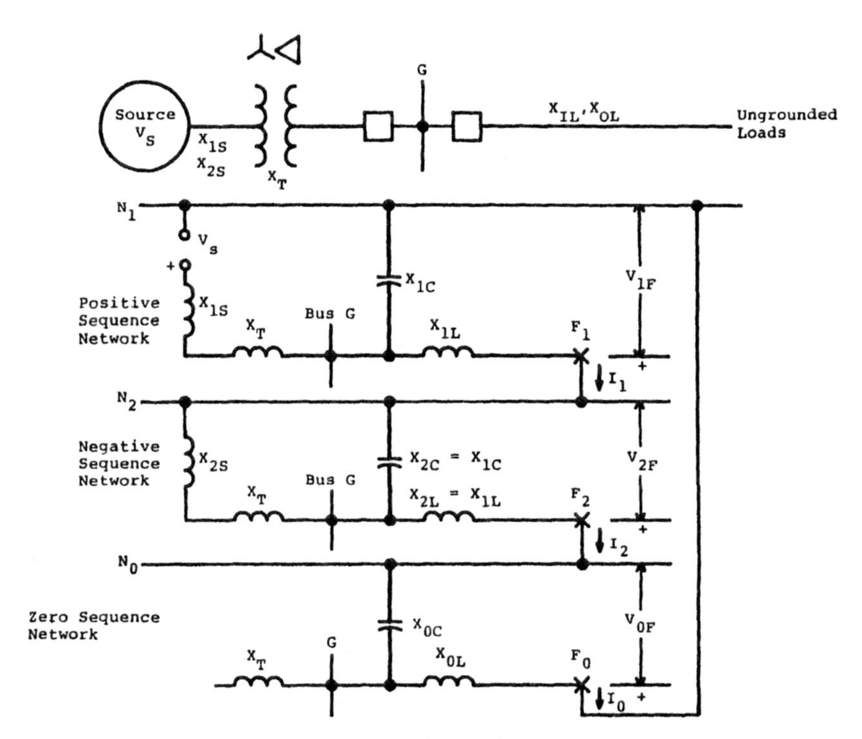

Figure 5.9 Sequence networks and interconnections for a phase-*a*-to-ground on an ungrounded system.

gle power transformer supply. With the resistor in the neutral, a solid ground fault can produce a maximum V_0 equivalent to the phase-to-neutral voltage (Eq. (5.22) and Fig. 5.8). Technically, the resistor could be inserted directly into the neutral but since a very high ohm resistor is required in most cases, eco-

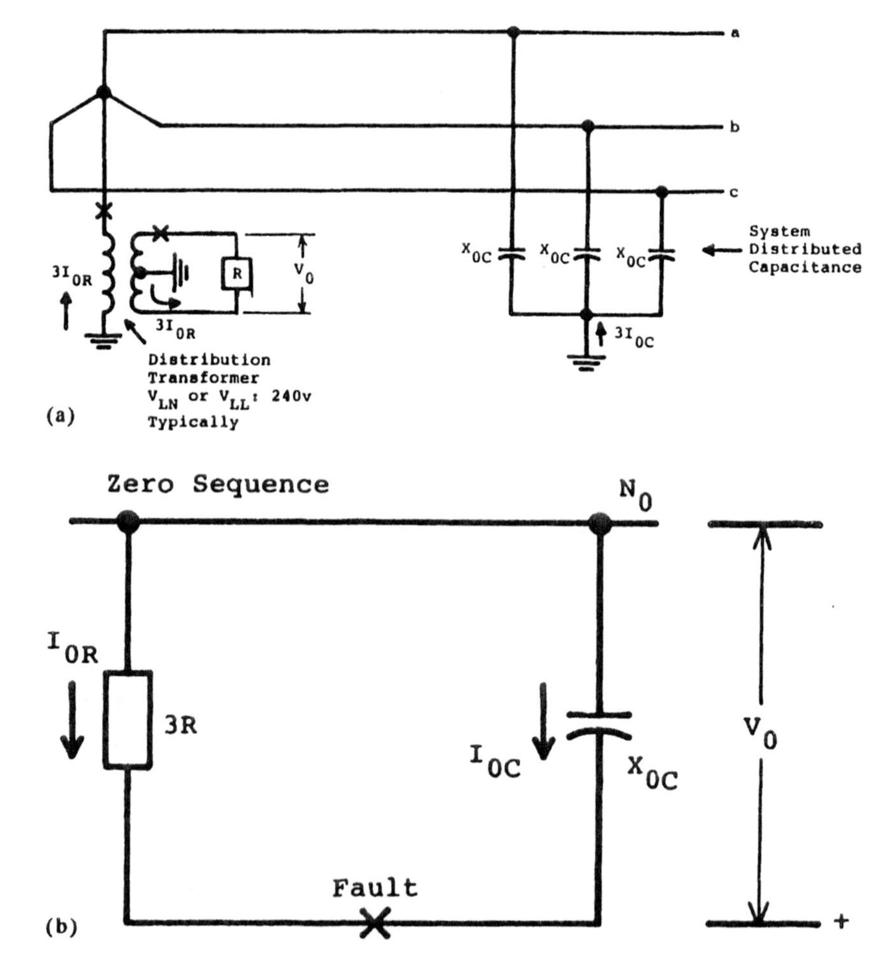

Figure 5.10 High resistance grounding with neutral resistor. (a) The power system: zero sequence current flow for a system ground fault. (b) The zero sequence network.

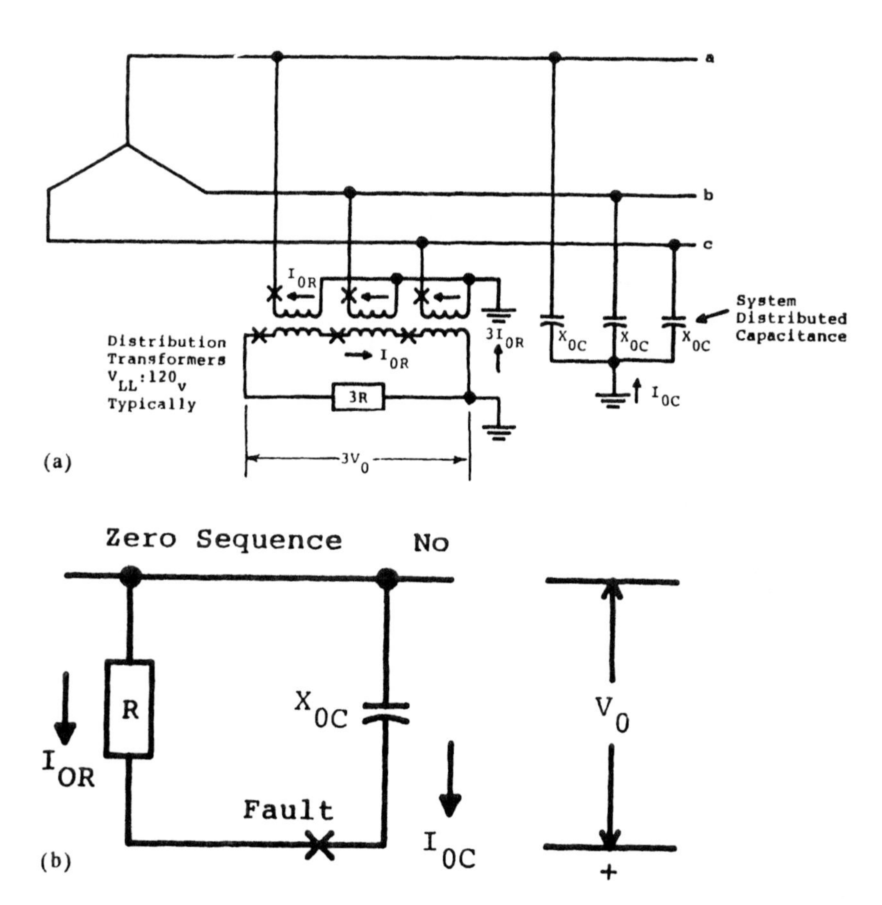

Figure 5.11 High-resistance grounding with resistor across three distribution transformer secondaries. (a) The power system: zero sequence current flow for a system ground fault; (b) The zero sequence network.

nomics dictates the use of a distribution transformer. A phase-to-neutral rated transformer can be applied but often a phase-to-phase rating is used.

The zero sequence network is shown in Fig. 5.10b. A value of 3R must be used, since only I_{0R} flows in this network in contrast to $3I_{0R}$ in the neutral.

Resistor Across the Broken Delta of a Wye-Grounded Distribution Transformer

This type of grounding is applied with multiple generators connected to a common bus or for systems with several power sources or multiple wye-connected transformers. The connections are illustrated in Fig. 5.11. Here R is used in the zero sequence network rather than $3R$, and $3R$ is connected across the broken delta. The logic of this can be seen by considering R as connected across each secondary winding.

Fault calculation for a typical example is given in Chapter 6.

6

Fault Calculation Examples for Shunt-Type Faults

6.1 INTRODUCTION

The application of the technology of Chapters 4 and 5 can best be demonstrated by several examples. The fault calculation process is the same for any other voltage and system arrangements, which would vary only the system constants and sequence network configurations. Faults at other locations in the system involve a repeat of the calculations.

6.2 FAULTS ON A LOOP-TYPE POWER SYSTEM

A part of a power system with four buses interconnected by lines is shown in Fig. 6.1. Arbitrarily, a 115-kV system was selected with one load bus and the other three having power sources. Three of the stations have solidly grounded wye–delta transformers and the other a delta–delta transformer.

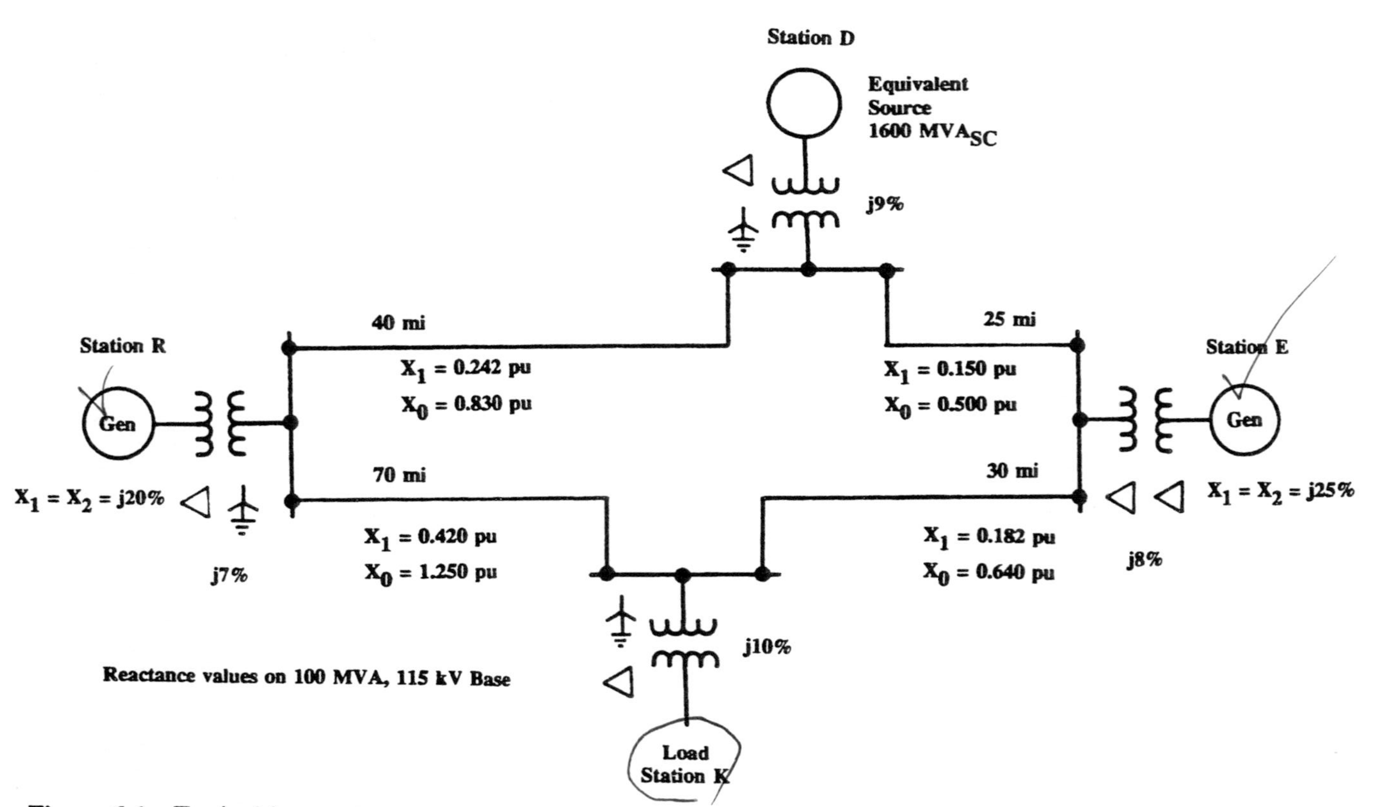

Figure 6.1 Typical loop-type system.

6.3 BASIC ASSUMPTIONS

The basic assumptions for this example are:

1. All generated voltages equal and in phase; hence no load is considered for the fault calculation. This considerably simplifies the calculations and so is a convenience. Generally, load has negligible impact on the faulted phase(s) and the associated protection operation. Load that can be quite variable increases the fault currents. Neglecting load provides fault values for relay applications at light or no load. Computer fault programs may or may not include load.
2. Neglect resistance; only reactance values are used. Except at the lower voltages the resistance component is relatively small compared to the reactance. At times it may be desirable to use impedance values and add them directly to the reactance. This would give slightly lower fault currents. All this a convenience to avoid tedious $r + jx$ calculations.
3. All shunt reactances neglected (loads, charging and magnetizing reactances).
4. All mutual reactances neglected. This is not valid where there are parallel lines. This aspect is discussed in a later chapter.

6.4 FAULT CALCULATION

Step 1: *Set up the positive-sequence network.* The reactance (impedance) values for all of the components must be obtained and using percent or per unit all reduced to a common base. This has been done in Fig. 6.1 using a base of 100 MVA, 115 kV. Figure 6.2 shows the positive-sequence network. Load is indicated at station K and in this example is assumed to be nonsynchronous. For protection application of the fault study any induction motors are neglected, as discussed in Section 4.9.

Step 2: *Reduce the positive-sequence network and determine*

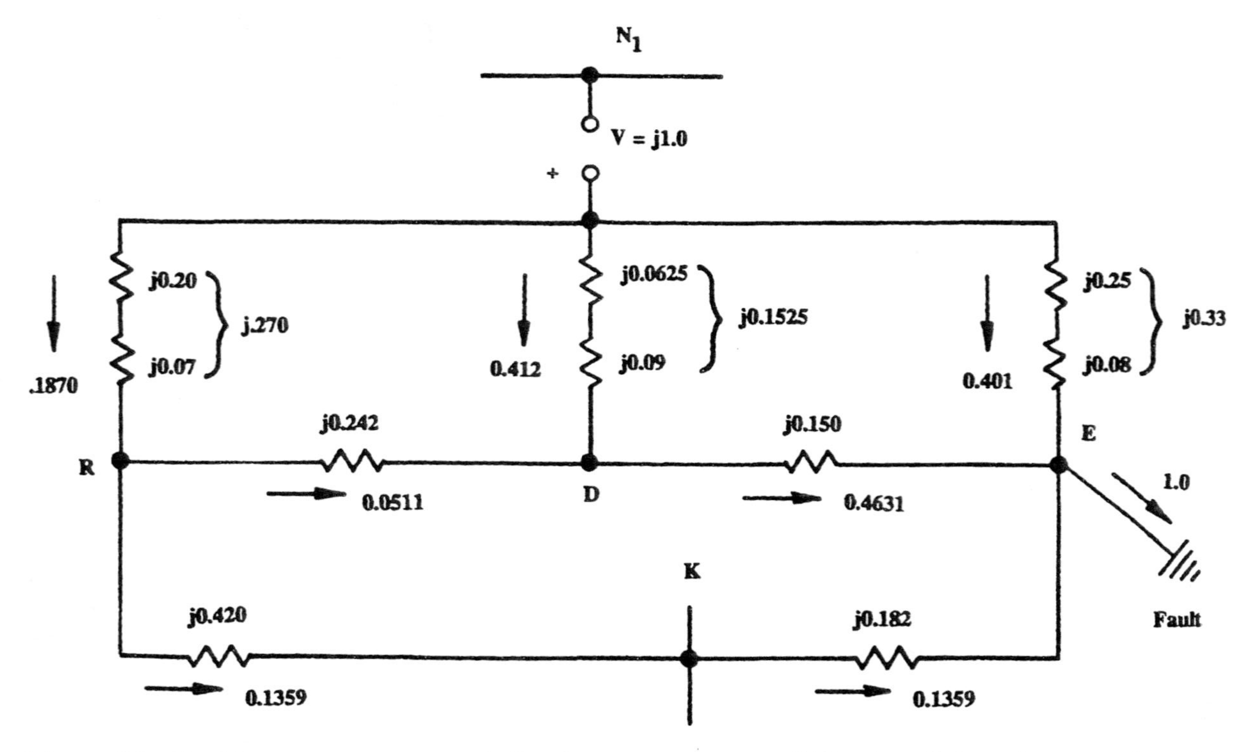

Figure 6.2 Positive-sequence network for the power system of Fig. 6.1. Values in per unit on 100-MVA, 115-kV base. Fault point at bus E.

the current distribution factors. A fault point for this is the 115-kV bus at station E. There are several delta nodes in Fig. 6.2. The upper left delta is selected and converted to an equivalent wye using Eqs. (4.32). Thus in Fig. 6.3:

$$X_x = \frac{0.270 \times 0.1525}{0.270 + 0.1525 + 0.242} = j0.0620 \tag{6.1}$$

$$X_y = \frac{0.1525 \times 0.242}{0.6645} = j0.0555 \tag{6.2}$$

$$X_z = \frac{0.270 \times 0.242}{0.6645} = j0.0983 \tag{6.3}$$

Now paralleling the lower two circuits:

$$\frac{(0.0555 + 0.150)(0.0983 + 0.602)}{\text{sum of the four values}} = \frac{\overset{0.2269\quad 0.7731}{(0.2055)(0.7003)}}{0.9058} = j0.1589 \tag{6.4}$$

The two numbers (0.2269 and 0.7731) are the ratio of 0.2055 and 0.9058 and of 0.7003 and 0.9058. They must add to 1.0 as a check and are useful in determining distribution factors.

Adding 0.1589 to 0.062 (X_x) and paralleling with 0.33 provides the final reduction for X_1, and assuming that the negative-sequence reactances are the same,

$$X_1 = X_2 = \frac{\overset{0.4010\ 0.5990}{(0.2209)(0.33)}}{0.5509} = j0.1323 \tag{6.5}$$

X_1 and X_2 are the equivalent impedance of the positive (and negative)-sequence networks from the neutral bus (N_1) to the fault point. The values above provide current distribution. The 0.401 value is the fraction of 1.0 pu current flowing in the 0.33 or station E generator and transformer. The 0.599 is the fraction of 1.0 pu current flowing through X_x. This divides using the fac-

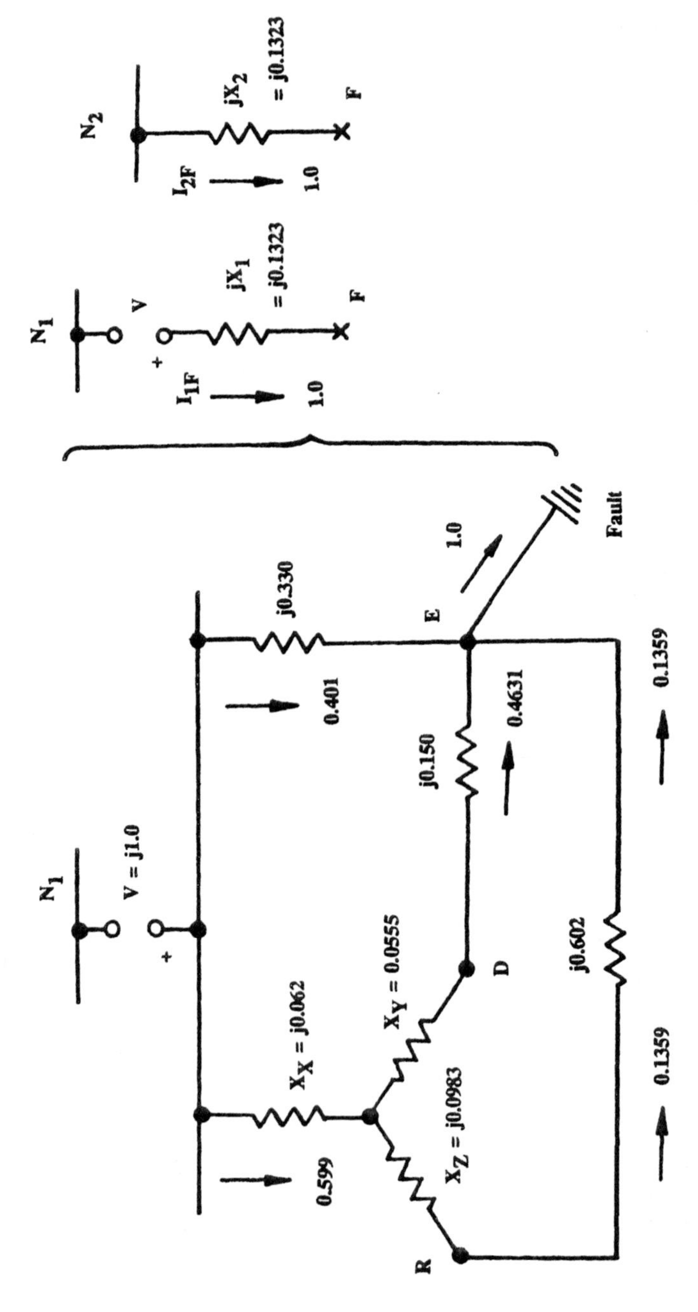

Figure 6.3 Reducing the positive-sequence network of Fig. 6.2. Values in per unit on 100-MVA, 115-kV base.

tors of Eq. (6.4):

$$0.599 \times 0.7731 = 0.4631 \quad \text{through the } ED \text{ path} \tag{6.6}$$

$$0.599 \times 0.2269 = 0.1359 \quad \text{through the } EKR \text{ path} \tag{6.7}$$

These current distributions are shown on the figures.

The current distributions for the line *RD* and generator–transformer of stations D and R can be determined by dividing the drop across $X_x + X_y$ by station D reactance (0.1525) and the drop across $X_x + X_z$ by station R reactance. The calculations are

$$\frac{(0.599)(0.062) + (0.4631)(0.0555)}{0.1525} = \frac{0.0371 + 0.0257}{0.1525} = j0.412 \tag{6.8}$$

$$\frac{(0.599)(0.062) + (0.1359)(0.0983)}{0.270} = \frac{0.0371 + 0.0134}{0.270} = j0.1870 \tag{6.9}$$

All of these distribution factors have been added to the positive-sequence network (Fig. 6.2). These multiplied by the positive-sequence (or negative-sequence) fault current provide the currents flowing in all parts of the system.

Step 3: Calculate the three-phase fault. For a three-phase fault at the station E 115-kV bus,

$$I_{1F} = I_{a3\phi} = \frac{j1}{j0.1323} = 7.56 \text{ pu} \tag{6.10}$$

$$= 7.56 \times \frac{100{,}000}{\sqrt{3}\ 115} = 3794.3 \text{ A at } 115 \text{ kV} \tag{6.11}$$

The fault current in any part of the system can be determined by multiplying the fault current of Eq. 6.10 [or (6.11)] by the proper distribution factor.

Step 4: Set up the negative-sequence network. Since the positive- and negative-sequence impedances are the same in all parts of this system, the negative-sequence network is the same as the positive-sequence network except for the voltage V.

Step 5: *Reduce the negative-sequence network and determine the current distribution factors*. This step is necessary only when the negative-sequence network is different. In this case repeat step 2 for the negative-sequence network.

Step 6: *Set up the zero-sequence network*. The reactance (impedance) values for all the system must be obtained and using percent (per unit) reduced to a common base. This has been done in Fig. 6.1. In the zero-sequence network only the grounded transformers are involved, as shown in Fig. 6.4.

Step 7: *Reduce the zero-sequence network and determine the current distribution factors*. The fault point is the 115-kV bus at station E. First reduce the top left delta to an equivalent wye using Eqs. (4.32). Thus in Fig. 6.5,

$$X_{0x} = \frac{0.07 \times 0.09}{0.07 + 0.09 + 0.83} = j0.00636 \tag{6.12}$$

$$X_{0y} = \frac{0.09 \times 0.830}{0.99} = j0.07545 \tag{6.13}$$

$$X_{0z} = \frac{0.07 \times 0.83}{0.99} = j0.05869 \tag{6.14}$$

Adding X_{0y} and 0.50, and X_{0z} and 1.25, then rearranging as in Fig. 6.6, points P, K, and E provide a delta that can be replaced with an equivalent wye:

$$X_{0r} = \frac{1.30869 \times 0.57545}{1.30869 + 0.57545 + 0.640} = j0.29835 \tag{6.15}$$

$$X_{0s} = \frac{1.30869 \times 0.640}{2.52414} = j0.33182 \tag{6.16}$$

$$X_{0t} = \frac{0.57545 \times 0.640}{2.52414} = j0.14591 \tag{6.17}$$

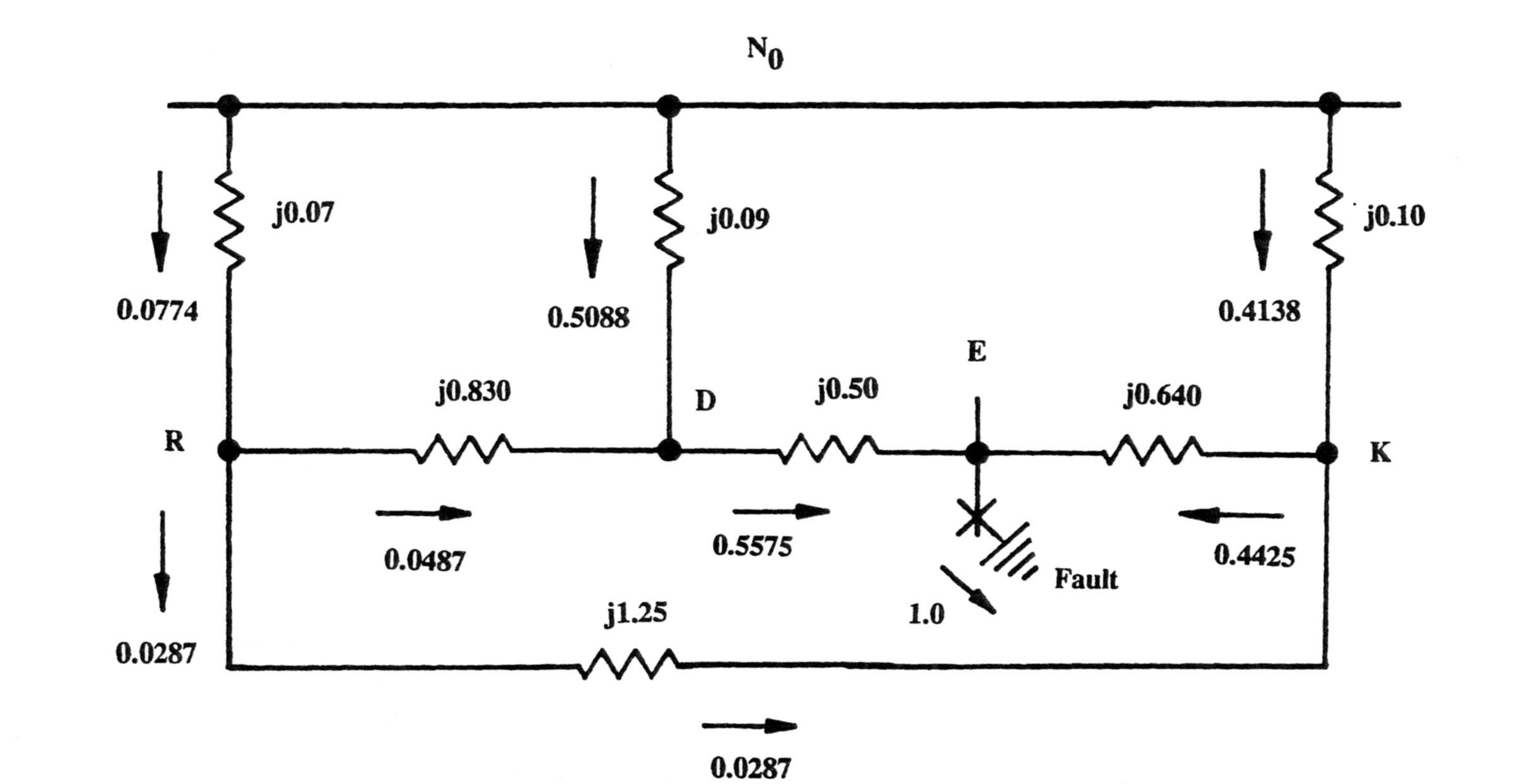

Figure 6.4 Zero-sequence network for the power system of Fig. 6.1. Values in per unit on 100-MVA, 115-kV base. Fault point at bus E.

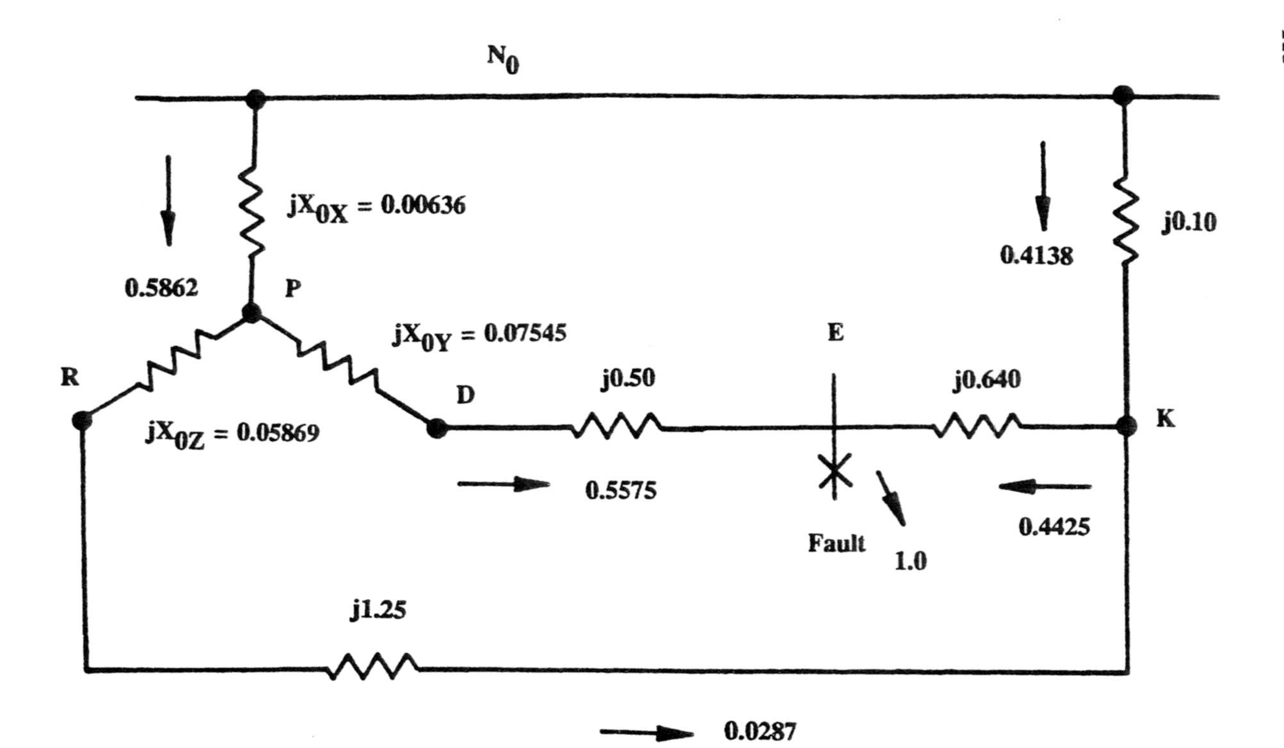

Figure 6.5 First reduction of the zero-sequence network of Fig. 6.4. Values in per unit at 100-MVA, 115-kV base.

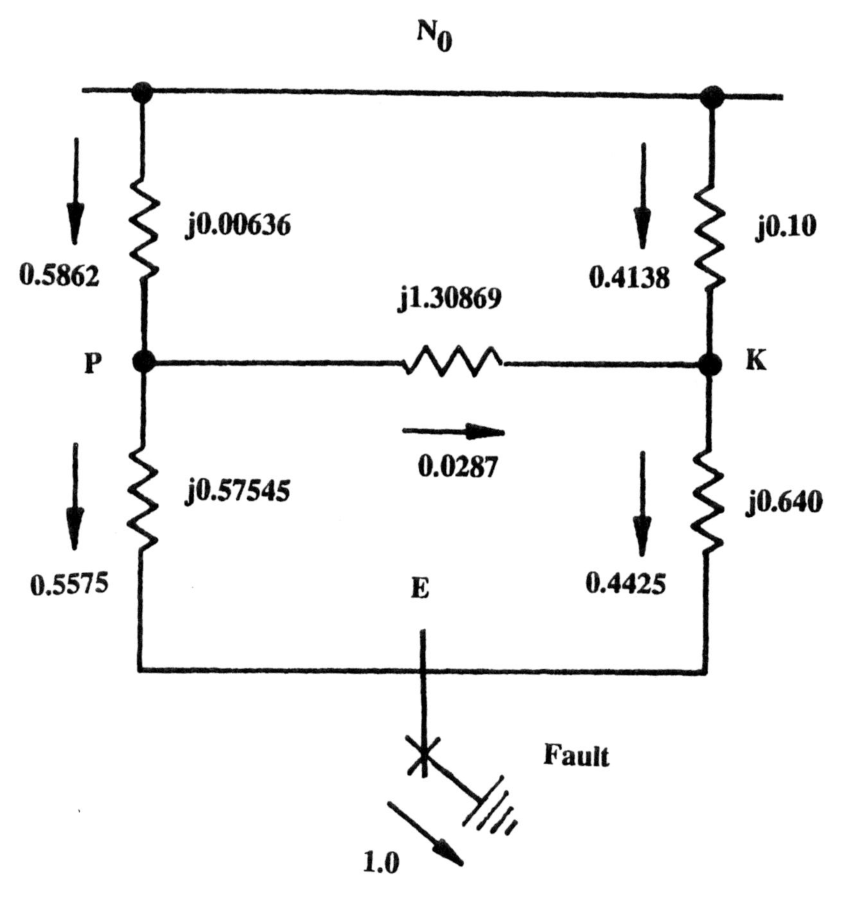

Figure 6.6 Zero-sequence network of Fig. 6.5 redrawn.

Finally,

$$X_0 = 0.14591 + \frac{(0.00636 + 0.29835)(0.10 + 0.33182)}{0.00636 + 0.29835 + 0.10 + 0.33182} \tag{6.18}$$

$$ \quad 0.4138 \quad 0.5862$$

$$X_0 = 0.14591 + \frac{(0.30471)(0.43182)}{0.73653} = j0.3246$$

Now determine the current distribution factors, beginning

with Fig. 6.7 and working back through Figs. 6.6 and 6.5 to 6.4. The current distribution in branch *PE*, Fig. 6.6, is from Fig. 6.7,

$$\frac{0.5862 \times 0.29835 + 1.0 \times 0.14591}{0.57545} = 0.5575 \tag{6.19}$$

In branch *PRK*,

$$\frac{0.5862 \times 0.29835 - 0.4138 \times 0.33182}{1.30869} = 0.0287 \tag{6.20}$$

Continuing the current distribution through N_0D is

$$\frac{0.5862 \times 0.00636 + 0.5575 \times 0.07545}{0.09} = 0.5088 \tag{6.21}$$

and through N_0R,

$$\frac{0.5862 \times 0.00636 + 0.0287 \times 0.05869}{0.07} = 0.0774 \tag{6.22}$$

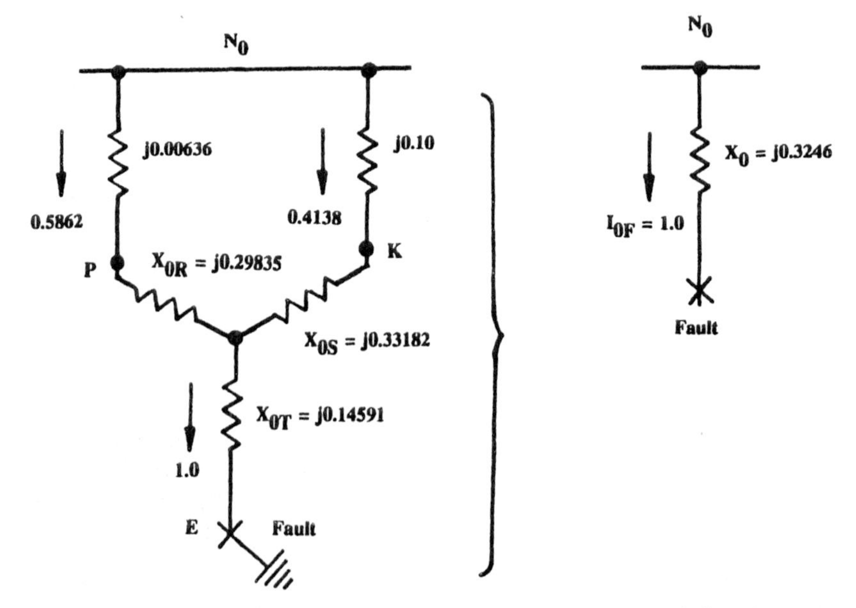

Figure 6.7 Final reduction of the zero-sequence network for the system of Fig. 6.1. Values in per unit on 100-MVA, 115-kV base.

All of these current distribution factors are shown on the figures as they are developed. The complete distribution is on the zero-sequence network (Fig. 6.4).

Step 8: *Calculate the single phase-to-ground fault.* For this fault at the station E 115-kV bus,

$$I_{1F} = I_{2F} = I_{0F} = \frac{j1.0}{j0.1323 + j0.1323 + j0.3246} = \frac{j1.0}{j0.5892} = 1.697 \text{ pu} \tag{6.23}$$

$$I_{a\phi G} = 3I_{0F} = 5.09 \text{ pu} = 5.09 \times \frac{100{,}000}{\sqrt{3}\ 115} = 2556.23 \text{ A at 115 kV} \tag{6.24}$$

Reactance values were used in these calculations so should all be prefixed by the operator j. By using the voltage as jV, all the currents will be at zero angle. Thus as a shortcut and simplification, the j operator can be omitted.

Step 9: *Calculate the fault current distribution in the power system.* In ground fault studies normally only the $3I_0$ currents around the system are documented. These are the currents that operate the ground protective relays. The phase currents have little to no value and so are seldom available. In the example both the phase and ground currents for the fault at station E are shown in Fig. 6.8 for comparative purposes.

6.5 SUMMARY OF FAULT CURRENT

With reference to Fig. 6.8, it will be noted that the phase *a* currents are different from the $3I_0$ currents, usually quite different, and can be either larger or smaller. This is a function of the current distribution factors, which are different in the positive (negative)-sequence and zero-sequence networks. Since ground relays operate on $3I_0$, it is necessary to have these zero-sequence fault currents.

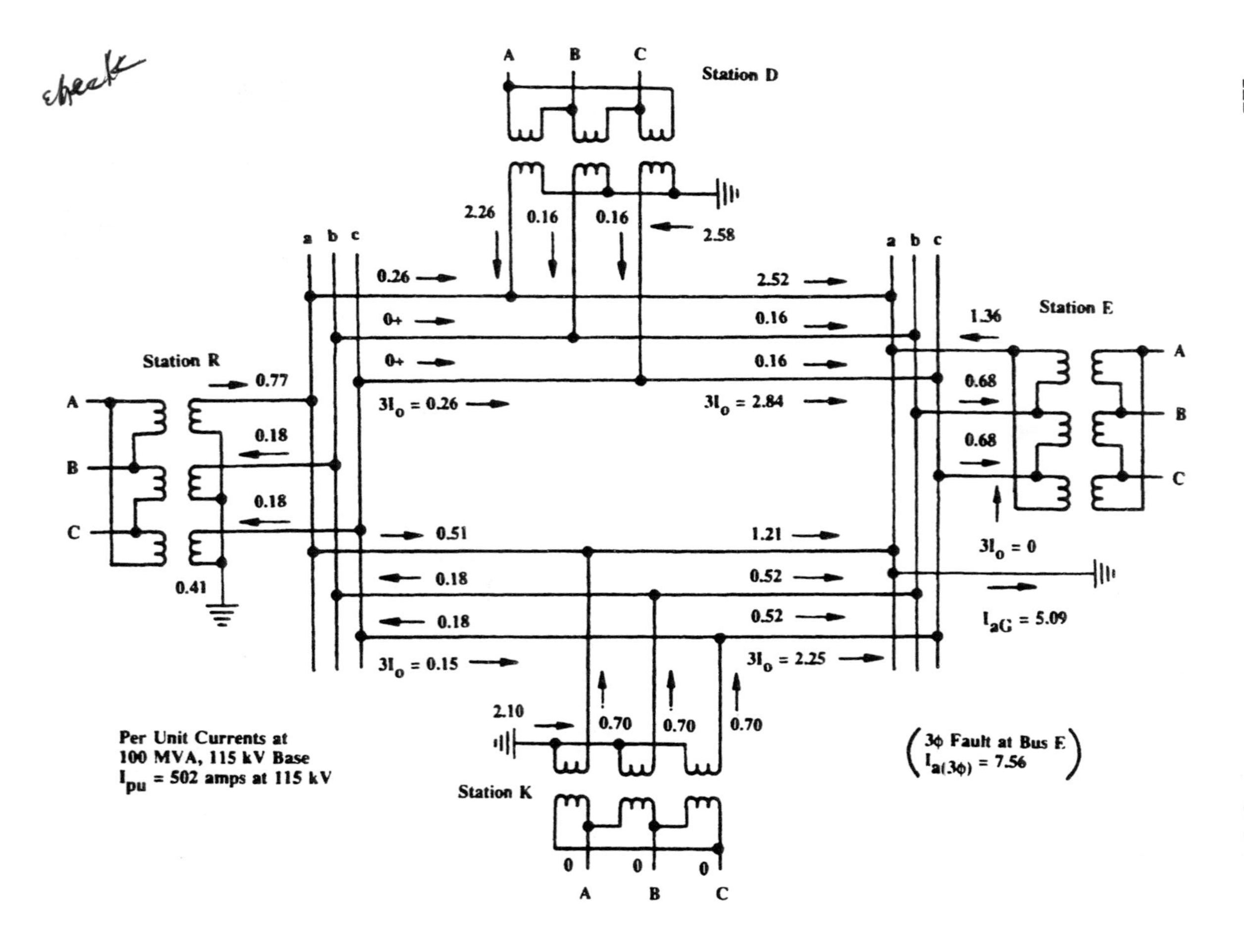
A
B
C
Station D
2.26
0.16
0.16
2.58
a b c
a b c
0.26
2.52
0+
0.16
0+
0.16
$3I_o = 0.26$
$3I_o = 2.84$
Station R
A
B
C
0.77
0.18
0.18
0.41
Station E
1.36
0.68
0.68
A
B
C
$3I_o = 0$
$I_{aG} = 5.09$
0.51
1.21
0.18
0.52
0.18
0.52
$3I_o = 0.15$
$3I_o = 2.25$
2.10
0.70
0.70
0.70
Station K
0
0
0
A
B
C
Per Unit Currents at
100 MVA, 115 kV Base
I_{pu} = 502 amps at 115 kV
(3φ Fault at Bus E
$I_{a(3\phi)} = 7.56$)

To satisfy Kirchhoff's law, all the currents throughout the system add or subtract correctly. Thus the currents up the several grounded transformer neutrals must equal the fault current to ground.

In this example the calculations were made with zero load current. With a phase-a-to-ground fault, phase a currents would be expected, but note that there are also currents in the unfaulted b and c phases. This results because the zero-sequence network is *always* different from the positive- and negative-sequence networks. From the basic equations (4.5) through (4.7), and with $Z_1 = Z_2$,

$$I_a = I_1 + I_2 + I_0 \tag{4.5}$$

$$I_b = a^2 I_1 + aI_2 + I_0 = (a^2 + a)I_1 = -I_1 + I_0 \tag{6.25}$$

$$I_c = aI_1 + a^2 I_2 + I_0 = (a + a^2)I_1 = -I_1 + I_0 \tag{6.26}$$

Thus, with the common fact or assumption that the positive and sequence impedances are the same, phase b and phase c currents are the difference between the positive- and zero-sequence components. In any loop-type system as in the example, unequal current distribution factors in the positive (negative)- and zero-sequence networks will result in different I_1 and I_0 components, hence phase b and phase c current.

In a radial-type system, typical of many distribution and industrial systems, the power source and grounded transformer are at the same end without a power or zero-sequence source at the other end. In these cases, although the positive (negative)- and zero-sequence impedances are different, the current distribution factors are the same, or 1.0 pu, in both networks. Then from Eqs. (6.25) and (6.26), $I_b = I_c = 0$, and $I_a = 3I_0$ throughout the radial system. Again this assumes no-load current flow.

In these radial-type system fuse protection commonly is used

Figure 6.8 Phase and ground fault current distribution for a phase-a-to-ground fault at station E on the power system of Fig. 6.1.

with ground relays. Fuses operate on the phase currents, while ground relays operate on $3I_0$. When these currents are the same or essentially equal, fuses and relays are easier to coordinate. In a loop-type system such as Fig. 6.8, coordination would be very difficult to impossible. Fortunately, fuses are rarely used on these types of systems.

For illustrative and comparison purposes, Fig. 6.9 documents the currents for a fault at the station R 115-kV bus. In both Figs. 6.8 and 6.9 the three-phase total fault current (given as a reference) is roughly the same, but the total ground fault current is quite different. The ground current is lower (67% of the three-phase fault current) for the fault at station E since the power transformer at that station is delta–delta (no ground source). With a grounded transformer bank at station R, the total ground fault current at that bus is 122% higher than the three-phase fault current.

6.6 VOLTAGES DURING FAULTS

The positive sequence and the phase voltages during faults generally are not of too much practical use in protection. However, the negative- and zero-sequence voltages are used in ground fault protection and it is highly recommended that all fault studies for relaying include these data in their output.

The per unit voltages can be determined from the sequence networks from the fundamental equations (4.24) through (4.26), and the phase voltages from the fundamental equations (4.5) through (4.7).

For the example of Fig. 6.1 and the sequence networks for phase-a-to-ground fault at the 115-kV bus at station E, the voltages are as follows: The current distribution factors in Figs. 6.2 and 6.4 are based on 1.0 fault current, so they must be multiplied by the actual sequence fault current (1.697) as determined in Eq. (6.23).

Figure 6.9 Phase and ground fault current distribution for a phase-a-to-ground fault at station R on the power system of Fig. 6.1.

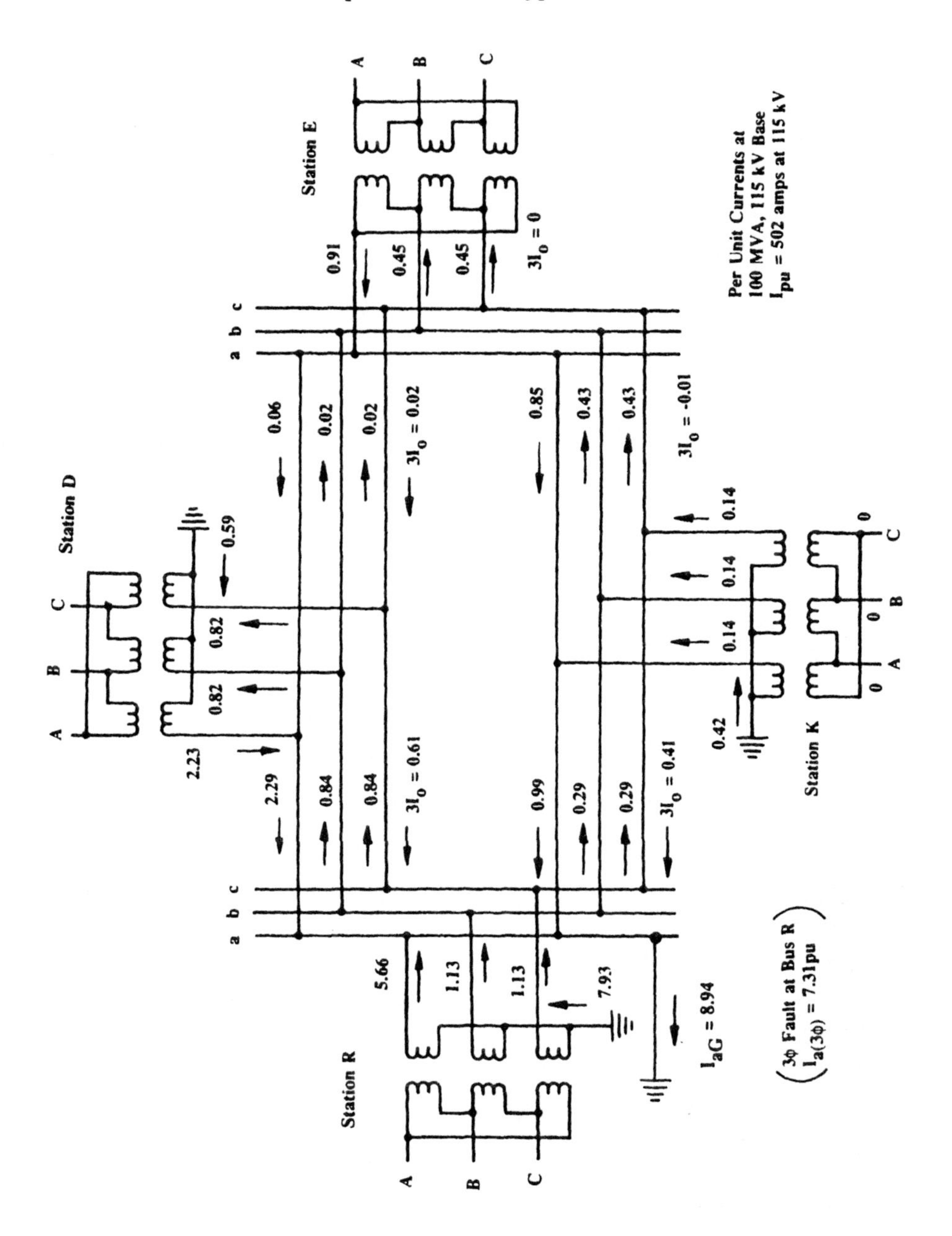
Station D
Station E
Station R
Station K
A
B
C
a b c
0.91
0.45
0.45
$3I_0 = 0$
0.06
0.02
0.02
$3I_0 = 0.02$
0.85
0.43
0.43
$3I_0 = -0.01$
0.59
0.82
0.82
2.23
2.29
0.84
0.84
$3I_0 = 0.61$
0.99
0.29
0.29
$3I_0 = 0.41$
0.14
0.14
0.14
0.42
0
0
0
5.66
1.13
1.13
7.93
$I_{aG} = 8.94$
Per Unit Currents at
100 MVA, 115 kV Base
I_{pu} = 502 amps at 115 kV
(3φ Fault at Bus R
$I_{a(3\phi)}$ = 7.31pu)

At the station E 115-kV bus,

$$V_{1E} = j1.0 - 1.697 \times 0.401 \times j0.330 = j0.7754\,\text{pu} \tag{6.27}$$

$$V_{2E} = 0 - 1.697 \times 0.401 \times j0.330 = -j0.2246\,\text{pu} \tag{6.28}$$

$$V_{0E} = 0 - 1.697(0.4138 \times j0.10 + 0.4425 \times j0.640) = -j0.5509\,\text{pu} \tag{6.29}$$

From these,

$$V_{aE} = j(0.7754 - 0.2246 - 0.5509) = 0 \tag{6.30}$$

$$\begin{aligned} V_{bE} &= a^2(j0.7754) + a(-j0.2246) - j0.5509 \\ &= 0.7754\angle 330^\circ + 0.2246\angle 30^\circ - j0.5509 \\ &= 0.866 - j0.8263 = 1.197\angle -43.67^\circ\ \text{pu} \end{aligned} \tag{6.31}$$

$$\begin{aligned} V_{cE} &= a(j0.7754) + a^2(-j0.2246) - 10.5509 \\ &= 0.7754\angle 210^\circ + 0.2246\angle 150^\circ - j0.5509 \\ &= -0.866 - j0.8263 = 1.197\angle -136.33^\circ\ \text{pu} \end{aligned} \tag{6.32}$$

At the station D 115-kV bus,

$$\begin{aligned} V_{1D} &= j1.0 - 1.697 \times 0.412 \times j0.1525 \\ &= j0.8934\ \text{pu} \end{aligned} \tag{6.33}$$

$$\begin{aligned} V_{2D} &= 0 - 1.697 \times 0.412 \times j0.1525 \\ &= -j0.1066\ \text{pu} \end{aligned} \tag{6.34}$$

$$\begin{aligned} V_{0D} &= 0 - 1.697 \times 0.5088 \times j0.09 \\ &= -j0.0777 \end{aligned} \tag{6.35}$$

From these the phase voltages are

$$V_{aD} = j0.7091\ \text{pu} \tag{6.36}$$

$$V_{bD} = j0.866 - j0.4711 = 0.9858\angle -28.55^\circ\ \text{pu} \tag{6.37}$$

$$V_{cD} = -j0.866 - j0.4711 = 0.9858\angle -151.45^\circ\ \text{pu} \tag{6.38}$$

At the station K 115-kV bus,

$$V_{1K} = j1.0 - 1.697(0.187 \times j0.27 + 0.1359 \times j0.42) = j0.8174 \text{ pu} \tag{6.39}$$

$$V_{2K} = 0 - 1.697(0.187 \times j0.27 + 0.1359 \times j0.42) = -j0.1826 \text{ pu} \tag{6.40}$$

$$V_{0K} = 0 - 1.697 \times 0.4138 \times j0.10 = -j0.0702 \text{ pu} \tag{6.41}$$

From these the phase voltages are

$$V_{aK} = j0.5646 \text{ pu} \tag{6.42}$$

$$V_{bK} = 0.8666 - j0.3876 = 0.9488\angle -24.11^\circ \text{ pu} \tag{6.43}$$

$$V_{cK} = -0.866 - j0.3876 = 0.9488\angle -155.89^\circ \text{ pu} \tag{6.44}$$

At the station R 115-kV bus,

$$V_{1R} = 1.0 - 1.697 \times 0.187 \times j0.27 = j0.9143 \text{ pu} \tag{6.45}$$

$$V_{2R} = 0 - 1.697 \times 0.187 \times j0.27 = -j0.0857 \text{ pu} \tag{6.46}$$

$$V_{0R} = 0 - 1.697 \times 0.0774 \times j0.07 = -j0.0092 \text{ pu} \tag{6.47}$$

From these the phase voltages are

$$V_{aR} = j0.8194 \text{ pu} \tag{6.48}$$

$$V_{bR} = 0.866 - j0.4235 = 0.9640\angle -26.06^\circ \text{ pu} \tag{6.49}$$

$$V_{cR} = -0.866 - j0.4235 = 0.9640\angle -153.94^\circ \text{ pu} \tag{6.50}$$

6.7 SUMMARY OF FAULT VOLTAGES

With the normal phase a voltage of $j1.0$ pu, the phase a voltages at the various buses will be less and must be zero at the fault since the example was for a solid fault to ground. If there were no currents flowing in unfaulted phases, phase b voltage would be $1.0\angle -30°$, and phase c would be $1.0\angle -150°$ throughout the system. In this example it will be noted that the phase b and phase c voltage at the fault point are greater than 1.0. This will occur when the zero-sequence impedance to the fault is greater than the positive-sequence impedance. Using reactance values only with $X_1 = X_2$ and letting $X_0 = KX_1$, it can be shown that

$$V_{b\text{Fault}} = \left[\frac{\sqrt{3}\, K\angle -150° - j\sqrt{3}}{K + 2} \right] \quad V \text{ volts or pu depending on the unit of } V \qquad (6.51)$$

If $K = 1.0$, $V_{b\text{Fault}} = 1.0$ pu; similarly for the $V_{c\text{Fault}}$ voltage. In the system of Fig. 6.1 and for the two ground faults shown in Figs. 6.8 and 6.9, the unfaulted phase voltages at the fault points are:

Fault at station E:

$$X_1 = j0.1323 \qquad X_0 = j0.3246 \qquad V_{bE} \text{ and } V_{cE} = 1.20 \text{ pu}$$

Fault at station R:

$$X_1 = j0.1367 \qquad X_0 = j0.0621 \qquad V_{bR} \text{ and } V_{cR} = 0.91 \text{ pu}$$

A general review of the fault quantities appears in Chapter 8.

6.8 FAULT CALCULATIONS WITH AND WITHOUT LOAD

Consider a power system feeding a 30 MVA load over a 10.5 mile, 34.5-kV line, as shown in Fig. 6.10a. The positive-sequence

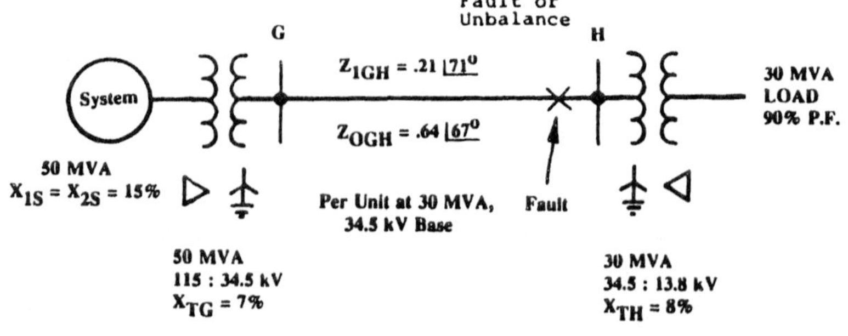

(a) Single line diagram

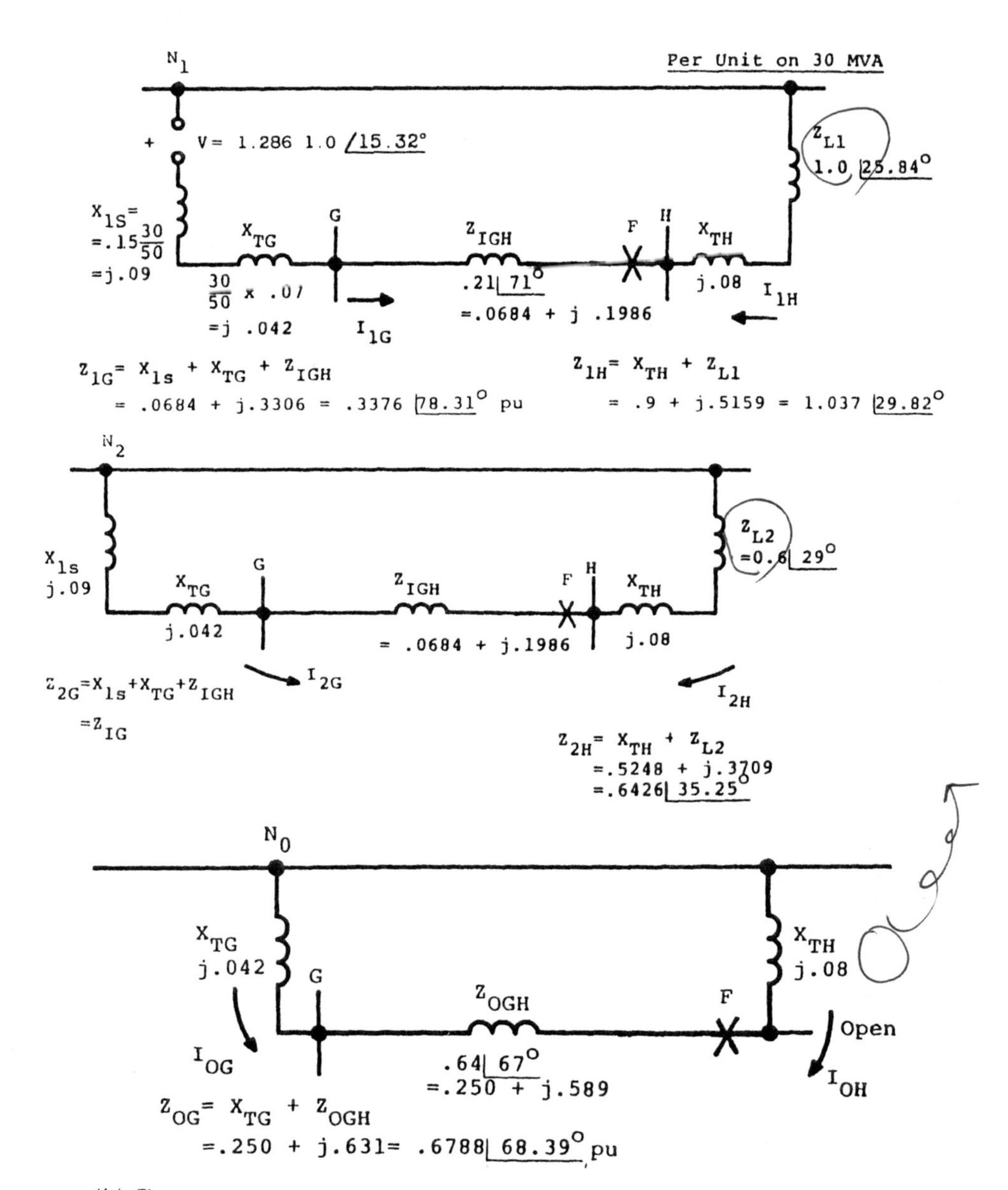

Figure 6.10 A 34.5-kV system supplying a 30-MVA load over a 10.5 mile line. (a) Single line diagram. (b) The positive-, negative-, and zero-sequence networks for shunt faults (F) or series unbalances ($X - Y$) on the line at bus H. With load considered, the values are as shown. With load neglected, $V = 1.0\angle 0°$ and the loads Z_{L1} and Z_{L2} are infinite (all values in pu at 30 MVA, 34.5 kV).

load impedance is $1.0\angle 25.84°$ pu corresponding to 30 MVA 90% PF. The negative-sequence load impedance is $0.60\angle 29°$ pu, reflecting the presence of some induction motor load (see 10.9).

The positive-, negative-, and zero-sequence networks for this system are shown in Figure 6.10b, with the impedance values reduced to a 30-MVA base and totaled for either a shunt fault (F) or a series unbalance ($X - Y$) both on the line at bus H.

Summarizing, the per unit values at 30 MVA, 34.5 kV to either the shunt fault or series unbalance are:

$$Z_{1G} = Z_{2G} = X_{1S} + X_{TG} + Z_{1GH} = 0.0684 + j0.3306$$
$$= 0.3376\angle 78.31°$$
$$Z_{1H}(\text{with load}) = Z_{L1} + X_{TH} = 0.90 + j0.5159 = 1.037\angle 29.82°$$
$$Z_{1H}(\text{neglecting load}) = \text{infinity}$$
$$Z_{2H}(\text{with load}) = Z_{L2} + X_{TH} = 0.5248 + j0.3709 = 0.6426\angle 35.25°$$
$$Z_{2H}(\text{neglecting load}) = \text{infinity}$$
$$Z_{0G} = X_{TG} + Z_{0GH} = 0.250 + j0.631 = 0.6788\angle 68.39°$$
$$Z_{0H} = X_{TH} = 0 + j0.080 = 0.080\angle 90° \tag{6.52}$$

Note that Z_{1G}, Z_{2G}, Z_{0G} and Z_{0H} are not affected by loads.

When loads are neglected, $V = V_S = 1.0\angle 0°$ pu

When loads are included, $V = V_S = 1.286\angle 15.315°$ pu, which provides approximately $1\angle 0°$ pu voltage at the 30-MVA load.

6.8.1 Phase-*a*-to-Ground Fault on Line at Bus *H*

The circuit of Fig. 6.11 shows the sequence interconnections for this fault. Considering load, the networks can be reduced by (1) Thévenin's theorem (Section 4.16) or (2) by combining and par-

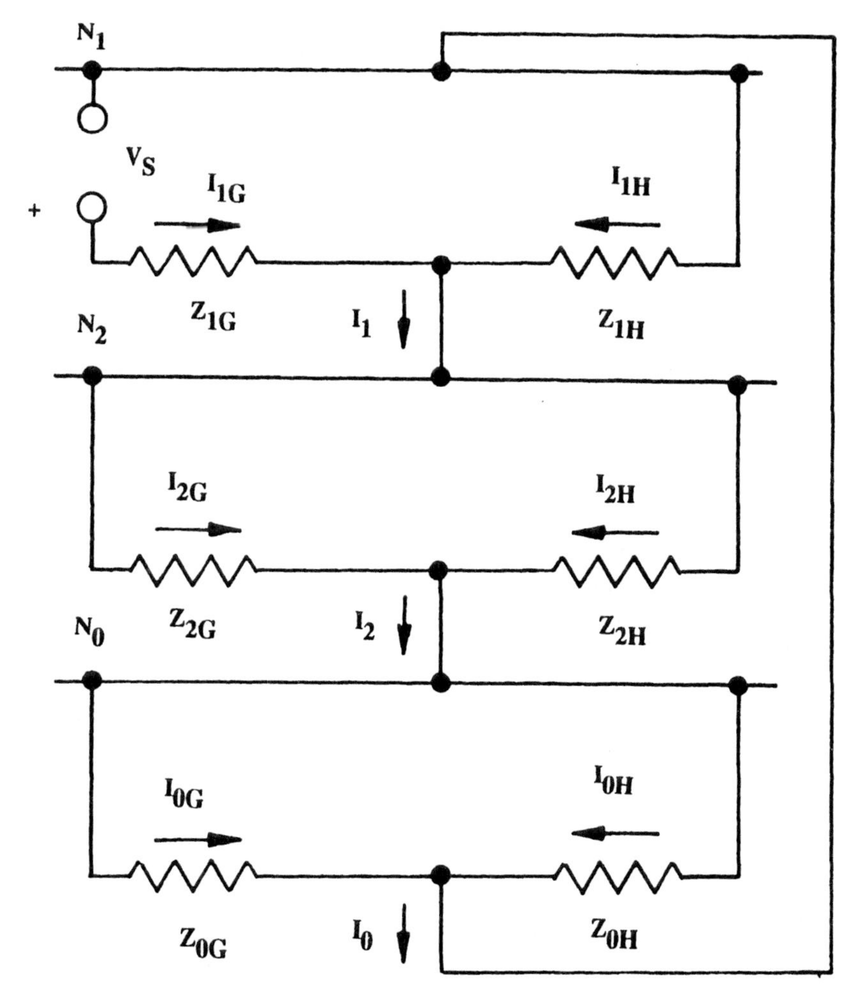

Figure 6.11 Sequence networks and interconnections for a phase-*a*-to-ground fault at bus H in the system of Fig. 6.10.

alleling, as shown in Fig. 6.13. Both methods will be outlined for information and comparison along with the no-load case.

6.9 SOLUTION BY THÉVENIN'S THEOREM

For a fault at bus *H* the voltage before the fault is the Thévenin voltage (V_H). The 30-MVA load on the base is 30 MVA is

$1.0\underline{/25.84^\circ}$ and the voltage at the load is $1.0\underline{/0^\circ}$. Thus the load current Z_{LZ} is $1.0\underline{/-25.84^\circ}$. $V_H = 1.0 + I_{\text{load}}(X_{TH}) = 1.037\underline{/3.98^\circ}$. This value could also be obtained by V_S minus the load drop to bus H, but in general the source or generator voltage for the particular load involved will not be known.

The Thévenin equivalent impedance is Z_{TH}, which is the parallel combination of Z_{1G} and Z_{1H}.

$$Z_{\text{TH}} = \frac{(Z_{1G})(Z_{1H})}{Z_{1G} + Z_{1H}} = 0.272\underline{/67.47^\circ} = 0.104 + j0.251 \qquad (6.53)$$

Similarly,

$$Z_2 = \frac{(Z_{1G})(Z_{2H})}{Z_{1G} + Z_{2H}} = 0.236\underline{/63.78^\circ} = 0.104 + j0.212 \qquad (6.54)$$

$$Z_0 = \frac{(Z_{0G})(Z_{0H})}{Z_{0G} + Z_{0H}} = 0.072\underline{/87.76^\circ} = 0.003 + j0.072 \qquad (6.55)$$

Adding these,

$$Z_{\text{TH}} + Z_2 + Z_0 = 0.211 + j0.535 = 0.575\underline{/68.48^\circ} \qquad (6.56)$$

$$\begin{aligned} I_1 = I_2 = I_0 &= \frac{V_H}{Z_{\text{TH}} + Z_2 + Z_0} = \frac{1.037\underline{/3.98^\circ}}{0.575\underline{/68.48^\circ}} \\ &= 1.803\underline{/-64.50^\circ} = 0.776 - j1.627 \end{aligned} \qquad (6.57)$$

These are fault currents only and in the positive-sequence network the load components must be added to obtain I_{1G} and I_{1H}. The fault component in the G branch is

$$\frac{Z_{1H}}{Z_{1G} + Z_{1H}}(I_1) = 0.362 - j1.405$$

$$\begin{aligned} &\text{Add:} \quad I_{\text{load}} &&= \underline{0.900 - j0.438} \\ &\qquad\quad I_{1G} &&= 1.262 - j1.843 = 2.234\underline{/-55.60^\circ} \end{aligned} \qquad (6.58)$$

$$\frac{Z_{1G}}{Z_{1G} + Z_{1H}}(I_1) = 0.422 - j0.215$$

$$\begin{aligned} \text{Add:} \quad -I_{\text{load}} &= \underline{-0.90 + j0.438} \\ I_{1H} &= -0.478 + j0.223 = 0.527\angle 154.99^\circ \end{aligned} \tag{6.59}$$

In a similar manner, I_{2G} and I_{2H} are determined from I_2 of Eq. (6.57), dividing per the negative-sequence network of Fig. 6.11: similarly, for I_{0G} and I_{0H} from I_0 of Eq. (6.57) and the zero-sequence network of Fig. 6.11. These values are

$$\begin{aligned} I_{2G} &= 1.268\angle -78.95^\circ = 0.243 - j1.244 \\ I_{2H} &= 0.656\angle -35.70^\circ = 0.533 - j0.383 \\ I_{0G} &= 0.190\angle -45.43^\circ = 0.133 - j0.135 \\ I_{0H} &= 1.625\angle -66.69^\circ = 0.643 - j1.492 \end{aligned} \tag{6.60}$$

With the values of Eqs. (6.58) through (6.60), the values for the phase currents can be calculated from Eqs. (4.5) through (4.7). The current up the station G transformer neutral is $3I_{0G}$ and up the station H transformer neutral, $3I_{0H}$. The current in the fault is $3I_0$ [Eq. (6.57)], equal to $I_{aG} + I_{aH}$. These per unit currents are shown in Fig. 6.12.

6.10 SOLUTION BY NETWORK REDUCTION

This method requires that the source or generator voltage behind X_{1S} be known for the particular load involved. In the example this is $V_S = 1.286\angle 15.315^\circ$, which for the 30-MVA load provides 1 pu voltage at the load.

By paralleling Z_{1G} and Z_{2H} to Z_2 [Eq. (6.54)] and also paralleling Z_{0G} and Z_{0H} to Z_0 [Eq. (6.55)], the networks of Fig. 6.11 reduce as shown in Fig. 6.13.

Paralleling $Z_2 + Z_0$ with Z_{1H} yields Z_{20}.

$$Z_{20} = \frac{(Z_2 + Z_0)(Z_{1H})}{Z_2 + Z_0 + Z_{1H}} = 0.120 + j0.213 \tag{6.61}$$

$$= \frac{0.065 + j0.331}{0.185 + j0.544} = 0.575\angle 71.22^\circ$$

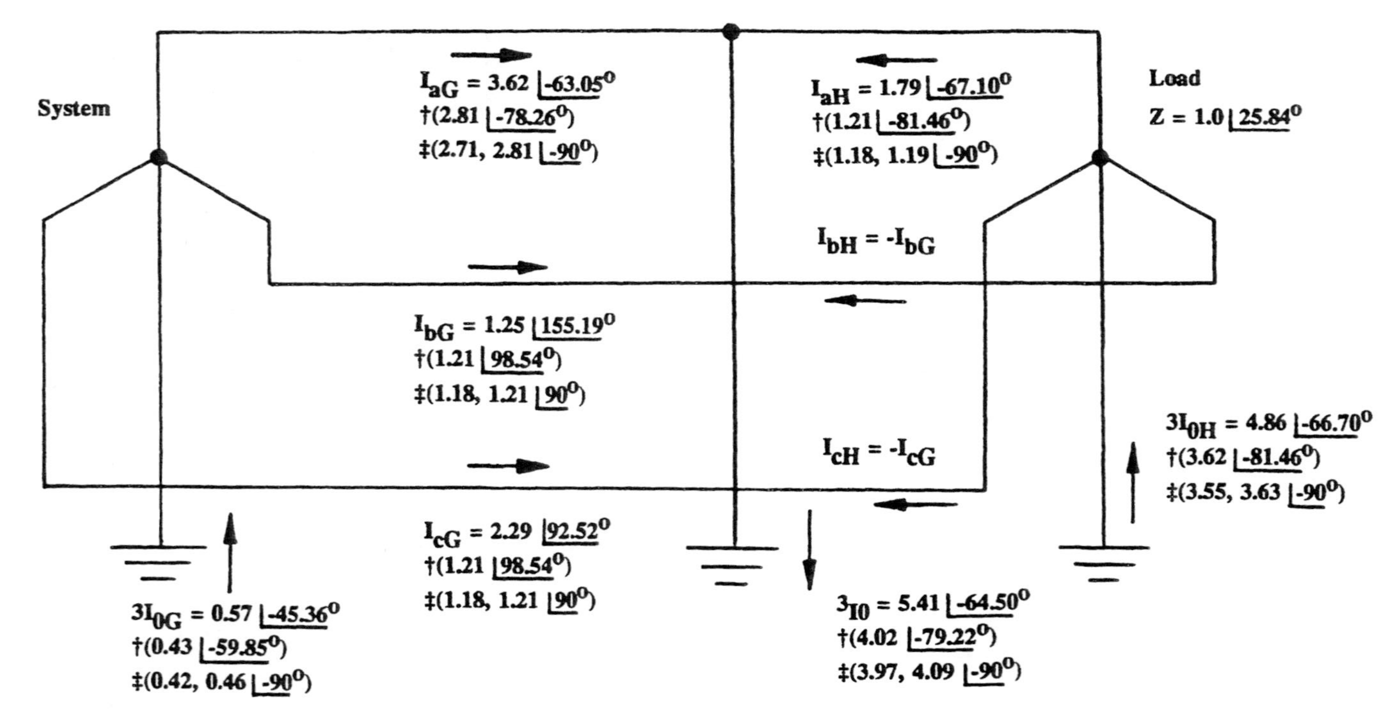

Figure 6.12 Per unit current distribution for a phase-*a*-to-ground fault at bus H in the system of Fig. 6.10. Top value is with 30-MVA load. Next value (†) in parentheses is for no-load condition. Lower two values (‡) in parentheses are for no load with first, $X + Z$ added directly, and second, the resistance neglected. Currents are on a 30-MVA base, so 1 pu = 502 A at 34.5 kV.

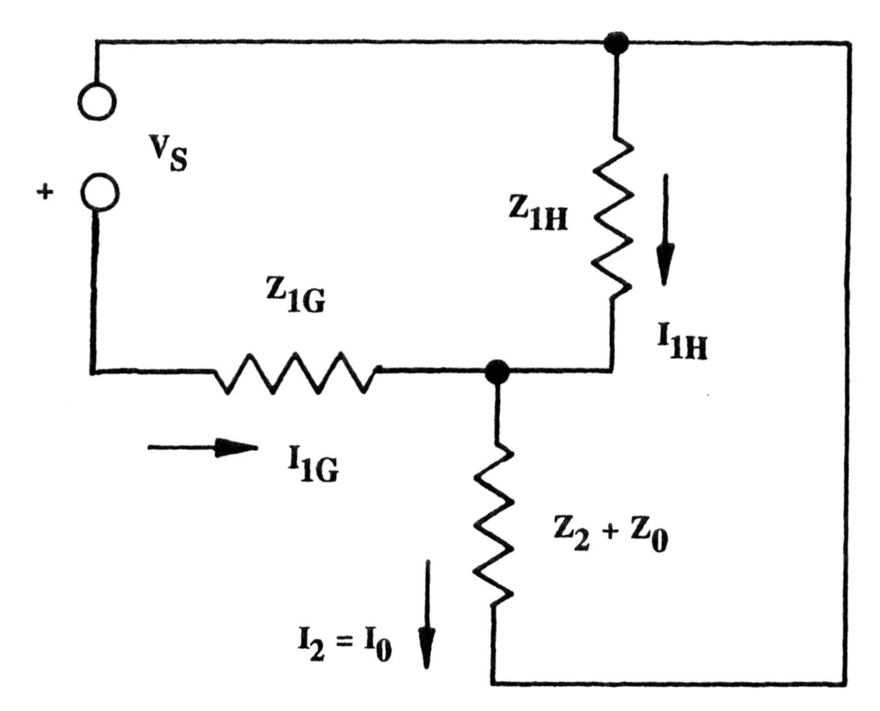

Figure 6.13 Reduction of Fig. 6.11 networks for fault calculation.

$$I_{1G} = \frac{1.286\angle 15.315^\circ}{0.575\angle 71.22^\circ} = 2.238\angle -55.90^\circ \tag{6.62}$$

This is the same as I_{1G} of Eq. (6.58). Load superposition is not required, as it was with the Thévenin theorem method.

$$I_0 = I_2 = \frac{Z_{1H}}{Z_2 + Z_0 + Z_{1H}}(I_{1G}) = 1.803\angle -64.43^\circ \tag{6.63}$$

and this is the same as the values derived in Eq. (6.57). From Fig. 6.13,

$$I_{1H} = I_2 - I_0 = -0.478 + j0.221 = 0.527\angle 155.18^\circ \tag{6.64}$$

This is the same as in Eq. (6.59).

With the values as in Eqs. (6.60), (6.62), and (6.64), the same zero-sequence and phase currents as shown in Fig. 6.12 can be determined. Thus the methods of Sections 6.9 and 6.10 produce the same results.

6.11 SOLUTION WITHOUT LOAD

For comparison, the calculation of a phase-a-to-ground fault at bus H for the example of Fig. 6.10 with the load ignored or not connected is as follows. Figure 6.11 and the values of Eqs. (6.52) apply with Z_{1H} and Z_{2H} omitted and $V_S = 1.0\angle 0°$. At no load the voltage of the source or generators would be reduced by their regulators. Thus for the ground fault at bus H,

$$I_{1F} = I_{2F} = I_{0F} = \frac{V_S}{Z_{1G} + Z_{2G} + Z_0} = 1.340\angle -79.22° \tag{6.65}$$

where Z_0 is per Eq. (6.55). With I_{1G} and $I_{2G} = I_{1F}$ since I_{1H} and $I_{2H} = 0$.

$$I_{0G} = \frac{Z_{0H}}{Z_{0G} + Z_{0H}}(I_{0F}) \quad \text{and} \quad I_{0H} = \frac{Z_{0G}}{Z_{0G} + Z_{0H}}(I_{0F}) \tag{6.66}$$

From these the phase and ground currents are calculated and are shown in Fig. 6.12.

6.12 SUMMARY

Figure 6.12 presents a summary of fault calculations for a line-to-ground fault at bus H of the 34.5-kV system of Fig. 6.10 under several conditions;

1. With a 30-MVA load
2. With no load
3. With load and all impedances considered as reactances ($X + Z$ added directly and at $+90°$)
4. As condition 3, but resistance neglected (only X of Z impedance considered)

Fault calculations in the last two cases are considerable simplified, yet as will be noted, with very little difference in fault currents.

In this particular example, the load resulted in an increase of the fault currents of about 35%—much of that because with load the source voltage must be about 29% higher in order to have a 1.0 per unit voltage at the load. The no-load study with the lower fault currents is desirable from a relay protection standpoint as well as being simpler to calculate. Protection is important at light or no-load conditions.

The phase currents are documented in Fig. 6.12, but as indicated earlier, the $3I_0$ values are the currents operating ground overcurrent protective relays. Note again that they are significantly different from the phase fault currents. The great value of the computer in calculating and documenting faults is apparent.

6.13 NEUTRAL INVERSION

In a power system with balanced voltages but unbalanced impedances, it is possible for the neutral to fall outside the voltage triangle. This is known as *neutral inversion*, and along with this ferroresonance can occur.

Practically, this situation can occur when a single voltage transformer is utilized to detect ground faults on an ungrounded system without sufficient resistor loading. An example is shown in Fig. 6.14. The exciting impedance of the unloaded voltage transformer is in parallel with the system shunt capacitance, as shown in Fig. 6.14b. The impedance X_a of phase a becomes

$$X_a = \frac{(-jX_c)(jX_e)}{jX_e - jX_c}$$

$$= \frac{jX_cX_e}{X_c - X_e} = jX_a$$

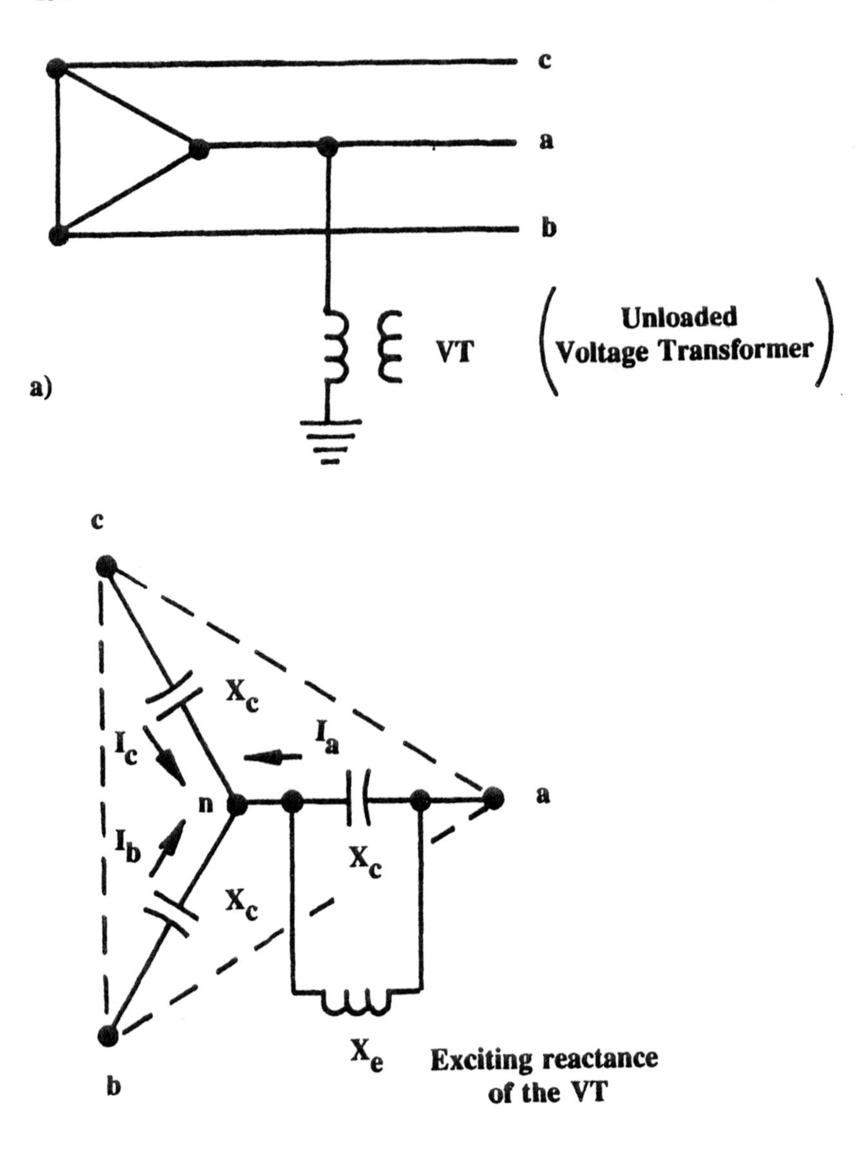

Figure 6.14 A single-voltage transformer provides an unbalance with possibility of neutral inversion.

From Fig. 6.14b voltage drops can be expressed as follows:

$$I_a(jX_a) - I_b(-jX_c) = V_{ab} = V_{LL}\underline{/30^\circ}$$

$$jI_aX_a + jI_bX_c = V_{LL}\underline{/30^\circ} \tag{6.67}$$

$$I_b(-jX_c) - I_c(-jX_c) = V_{bc} = V_{LL}\underline{/-90^\circ}$$

$$-jI_bX_c + jI_cX_c = V_{LL}\underline{/-90^\circ} \tag{6.68}$$

$$I_c(-jX_c) - I_a(jX_a) = V_{ca} = V_{LL}\underline{/150^\circ}$$

$$-jI_cX_c - jI_aX_a = V_{LL}\underline{/150^\circ} \tag{6.69}$$

and

$$I_a + I_b + I_c = 0 \qquad \text{or} \qquad I_c = -I_a - I_b \tag{6.70}$$

Substituting Eq. (6.70) in Eq. (6.68),

$$\begin{aligned} -jI_bX_c - jI_aX_c - jI_bX_c &= V_{LL}\underline{/-90^\circ} \\ -j2I_bX_c - jI_aX_c &= V_{LL}\underline{/-90^\circ} \end{aligned} \tag{6.71}$$

Multiply Eq. (6.67) by 2,

$$j2I_aX_a + j2I_bX_c = 2V_{LL}\underline{/30^\circ} \tag{6.72}$$

Add Eqs. (6.71) and (6.72),

$$\begin{aligned} jI_a(2X_a - X_c) &= \sqrt{3}\, V_{LL} \\ I_a &= \frac{\sqrt{3}\, V_{LL}}{2X_a - X_c} \end{aligned} \tag{6.73}$$

$$V_{an'} = I_aX_a \tag{6.74}$$

now substituting Eqs. (6.73) and $X_a = X_cX_e/(X_c - X_e)$ in Eq. (6.74) and reducing,

$$V_{an'} = \frac{\sqrt{3}\, V_{LL}}{3 - (X_c/X_e)} \tag{6.75}$$

If $X_e = 0.787X_c$, then, from Eq. (6.75), $V_{an'} = V_{LL}$. If X_e is less than $0.787X_c$, then $V_{an'}$ will be greater than V_{LL} and fall outside the delta voltage triangle–neutral inversion.

It should be recognized that Eq. (6.75) neglects resistance, which is always present although it can be relatively small compared to X_c and X_e. Also, the high voltage on the transformer results in saturation, with variable X_e and potential for ferro-resonance. Loading down the *VT* with resistance load is very important to negate these problems.

This neutral shift may also occur with unequal values of inductance and capacitance. This is discussed further in Westinghouse's *Electrical Transmission and Distribution Reference Book*, pp. 235–237. Figure 33, p. 236, in this reference illustrates neutral inversion with a wye-grounded transformer without a delta winding connected to an ungrounded system.

For a discussion of ferro-resonance, see Blackburn, *Protective Relaying*, Sec. 7.9.

6.14 EXAMPLE: GROUND FAULT ON AN UNGROUNDED SYSTEM

A 13.8-kV industrial plant is shown in Fig. 6.15. The following capacitances to ground were obtained from specific tests or estimating data (values in microfarads per phase):

Source transformer	0.004
Local generator	0.11
Motor	0.06
Power center transformers	0.008
Total connecting cables	0.13
Surge capacitor	0.25
Total capacitance to ground	0.562

$$X_C = -j\frac{10^6}{2\pi fC} = -j\frac{10^6}{2(3.1416)(60)(0.0562)}$$

$$= 4719.9\ \Omega/\text{phase} \qquad (6.76)$$

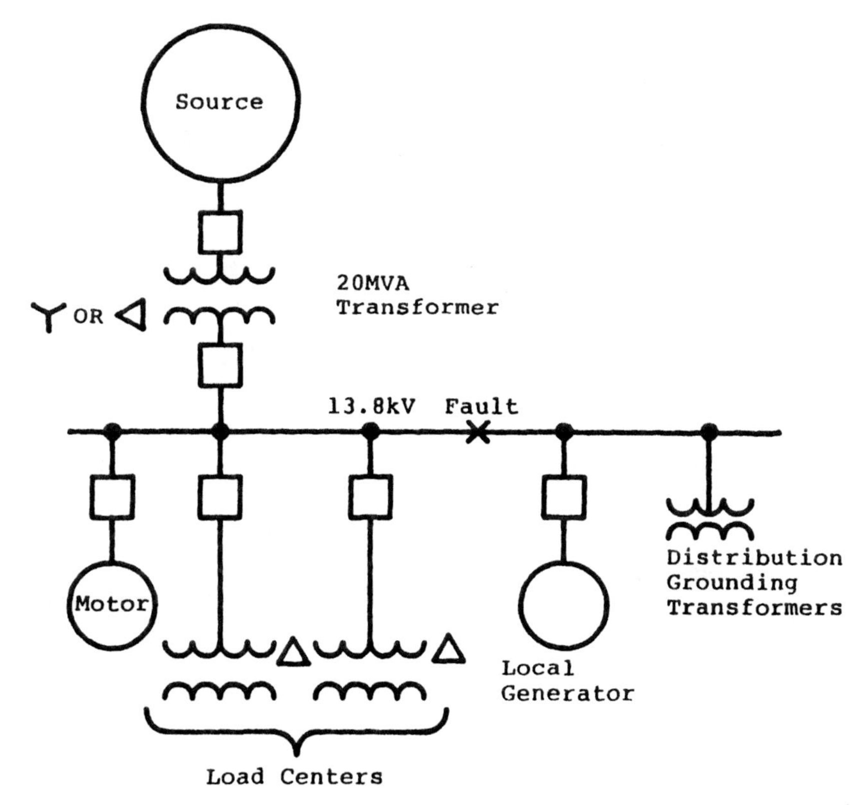

Figure 6.15 An ungrounded industrial system. Distribution grounding transformers used for high-resistance grounding.

Thus the charging current of this 13.8-kV system is

$$I_C = \frac{13{,}800}{\sqrt{3} \times 4719.9} = 1.69 \text{ A/phase at } 13.8 \text{ kV} \tag{6.77}$$

For a phase-a-to-ground fault and with X_C very large to X_1 and X_2,

$$\begin{aligned} I_1 &= I_2 = I_0 = \frac{13{,}800}{\sqrt{3}\, 4719.9} = 1.69 \text{ A} \\ I_a &= 3I_0 = 5.06 \text{ A at } 13.8 \text{ kV} \end{aligned} \tag{6.78}$$

6.15 EXAMPLE: GROUND FAULT WITH HIGH RESISTANCE ACROSS THREE DISTRIBUTION TRANSFORMERS

The total capacitance in per unit for the industrial system of Fig. 6.15 from Example 6.14 is

$$X_C = \frac{20(4719.9)}{13.8^2} = 495.68 \text{ pu} \tag{6.79}$$

As discussed in Section 5.10.2.2 select $R = X_C$ magnitude (Fig. 5.11); The zero sequence network is as in Fig. 6.16. The total zero sequence impedance is

$$Z_0 = \frac{(495.7)(-j495.7)}{495.7 - j495.7} = 350.5\angle -45^\circ \text{ pu} \tag{6.80}$$

For a line-to-ground fault the positive- and negative-sequence values of the system are very small and can be ignored. Thus

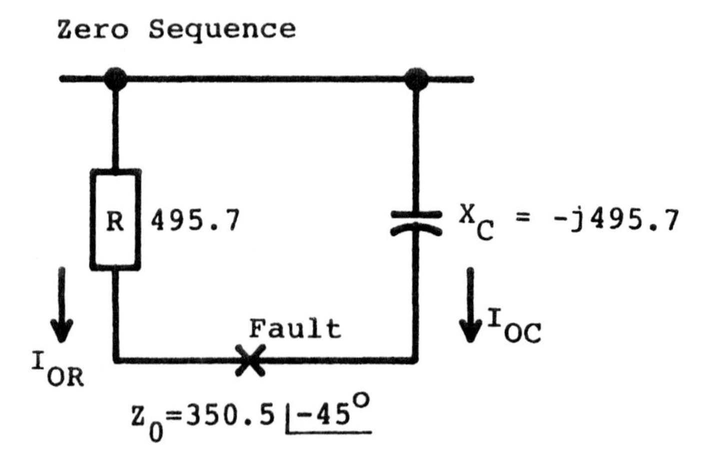

Figure 6.16 The zero sequence network for the example of Section 6.15.

for a line-to-ground fault on this 13.8-kV system,

$$I_1 = I_2 = I_0 = \frac{1.0}{350.5\angle -45^\circ} = 0.00285\angle 45^\circ \text{ pu} \tag{6.81}$$

The base per unit current is

$$I_{\text{base}} = \frac{20{,}000}{\sqrt{3} \times 13.8} = 836.74 \text{ A at } 13.8 \text{ kV} \tag{6.82}$$

Thus

$$I_1 = I_2 = I_0 = 0.00285(836.74) = 2.39 \text{ A at } 13.8 \text{ kV} \tag{6.83}$$

$$I_a = 3I_0 = 0.00856 \text{ pu} = 7.16 \text{ A at } 13.8 \text{ kV} \tag{6.84}$$

In Fig. 6.17:

$$I_{0R} = 0.00285 \cos 45^\circ = 0.00202 \text{ pu} = 1.69 \text{ A at } 13.8 \text{ kV} \tag{6.85}$$

The three distribution transformers have the ratio 13.8 kV : 120 = 115. Thus the secondary current for the ground fault is

$$I_{0R(\text{sec})} = 1.69(115) = 194.13 \text{ A} \tag{6.86}$$

The resistor was sized at 495.68 pu or 4719.9 Ω at 13.8 kV. Reflected to the secondary, the resistor value becomes

$$3R = 3 \times \frac{4719.0}{115^2} = 1.071\ \Omega \text{ secondary} \tag{6.87}$$

This is the resistance value for installation. Alternatively, this value can be calculated directly from the system values:

$$3R = \frac{(\sqrt{3}\ V_{\text{sec}})^2}{2\pi f C (V_{\text{pri LL}})^2} = \frac{(\sqrt{3} \times 120)^2}{377(0.562)(13.8)^2} \tag{6.88}$$

$$= 1.071\ \Omega \text{ secondary}$$

$$3V_0 = 194.13(1.071) = 207.85 \text{ V/secondary} \tag{6.89}$$

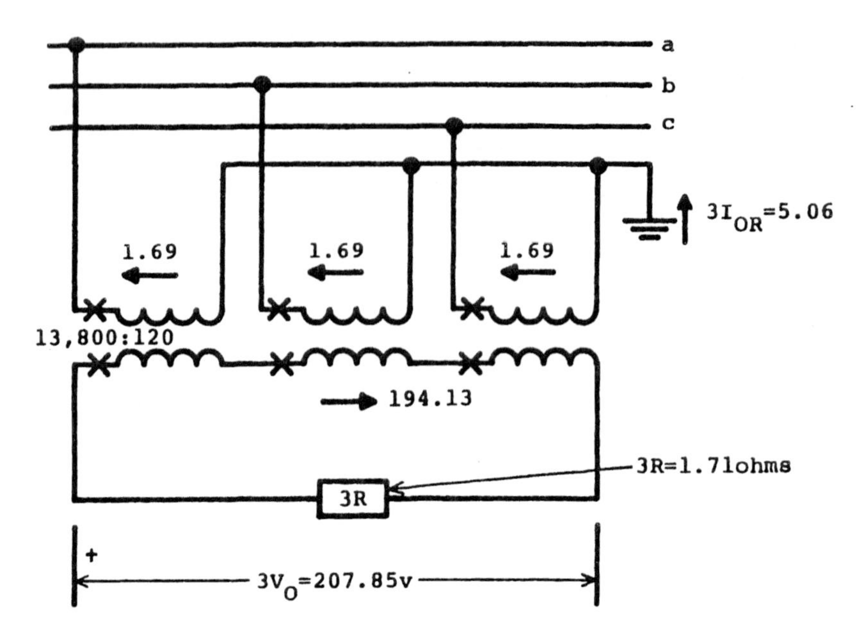

Figure 6.17 The currents in the grounding resistor for the example of Section 6.15.

The continuous resistor and transformer ratings are

$$\text{Resistor:} \quad I^2(3R) = \frac{(194.13)^2(1.071)}{1000} = 40.36 \text{ kW} \tag{6.90}$$

$$\text{Transformer:} \quad VI = \frac{1.69(13{,}800)}{1000} = 23.3 \text{ kVA} \tag{6.91}$$

The line-to-line voltage was used as during a fault; this voltage appears essentially across the primary winding. If relays are used to trip, short-time ratings may be used for the resistor and transformer.

6.16 EXAMPLE: GROUND FAULT WITH HIGH RESISTANCE IN NEUTRAL

A common application for this type of grounding is for large utility unit type generators. Consider a ground fault on a 160-MVA, 18-kV unit, shown in Fig. 6.18a.

The area of ground protection is from the generator to the low-voltage winding of the power transformer and to the high-voltage winding of the unit auxiliary transformer. In this area the following capacitances to ground (microfarads per phase) must be considered:

Generator windings	0.24
Generator surge capacitor	0.25
Generator to transformer leads	0.004
Power transformer low-voltage winding	0.03
Station service transformer high-voltage winding	0.004
Voltage transformer windings	0.0005
Total capacitance to ground	0.5285

$$X_C = -j\frac{10^6}{2\pi fC} = -j\frac{10^6}{2(3.1416)(60)(0.5285)} = 5019.08\ \Omega/\text{phase} \qquad (6.92)$$

This capacitive reactance, in per unit on a 100-MVA 18-kV base, is

$$\frac{100(5019)}{18^2} = 1549.1\text{ pu} \qquad (6.93)$$

or on the generator MVA base, $160(5019)/18^2 = 2478.56$ pu. Selecting the grounding resistor to be equal to the capacitive reactance and using the convenient 100-MVA base, $3R$ in the zero-sequence network would be 1549.1 pu. For a solid fault in this area,

$$Z_0 = \frac{1549.1(1549.1\underline{/-90^\circ})}{1549.1 - j1549.1} = 1095.38\underline{/-45^\circ}\text{ pu} \qquad (6.94)$$

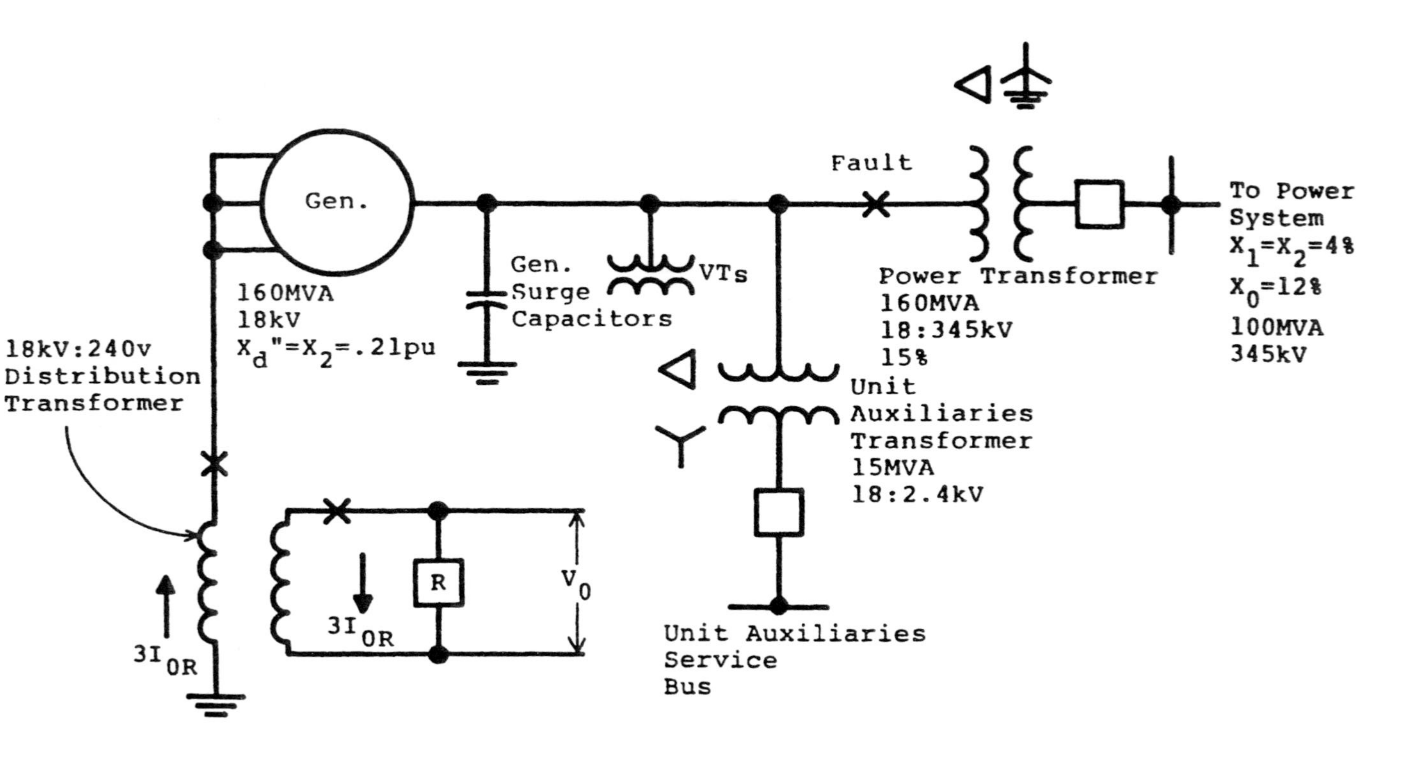

(a) The Unit Generator System

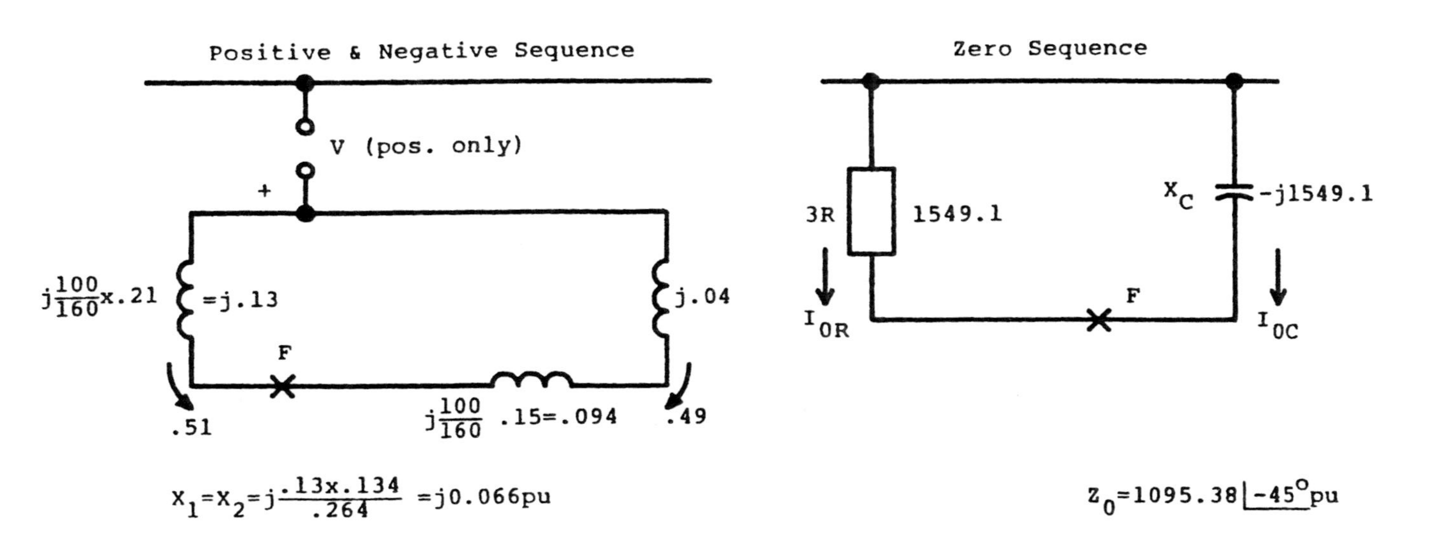

(b) the Sequence Networks. Values in per unit at 100 MVA, 18kV

Figure 6.18 Typical example of high-resistance neutral grounding.

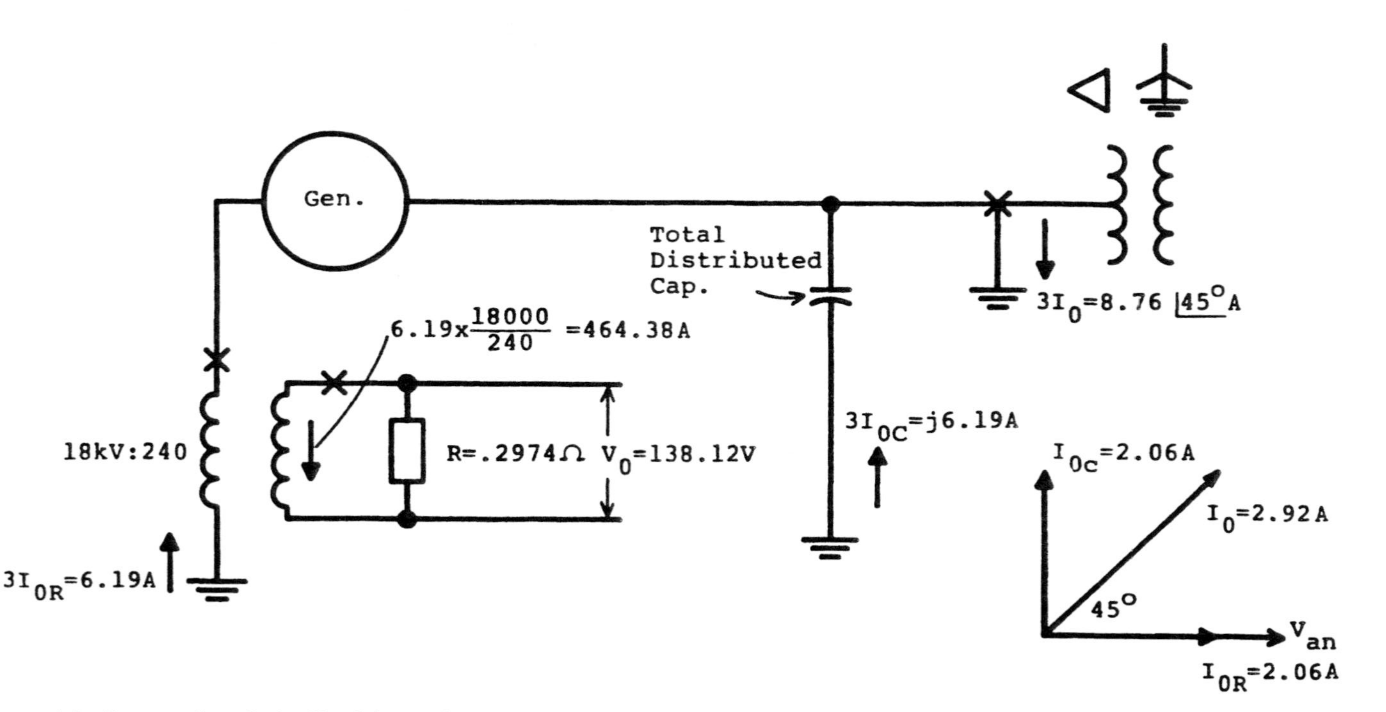

(c) Current Distribution for a Phase-a-to-ground Fault

Figure 6.18 (*Continued*)

In contrast, the positive- and negative-sequence reactance for this system is $j0.066$ pu and so is quite negligible. From Eq. 5.15 and 5.16:

$$I_1 = I_2 = I_0 = \frac{1.0}{1095.38\angle -45^\circ} = 0.00091\angle 45^\circ \text{ pu} \tag{6.95}$$

$$1.0 \text{ pu } I = \frac{100{,}000}{\sqrt{3} \times 18} = 3207.5 \text{ A at 18 kV} \tag{6.96}$$

so the fault currents are

$$I_1 = I_2 = I_0 = 0.00091(3207.5) = 2.92 \text{ A at 18 kV} \tag{6.97}$$

$$I_a = 3I_0 = 3(2.92) = 8.76 \text{ A at 18 kV} \tag{6.98}$$

The distribution of these fault currents is shown in Fig. 6.18c. The resistor selected with its primary resistance $3R$ equal to X_C provides a value of 5019.08/3 = 1673.03 Ω at 18 kV. The actual resistor value connected to the secondary of the distribution transformer will be

$$R = 1673.03 \left(\frac{240}{18{,}000}\right)^2 = 0.2974\ \Omega \tag{6.99}$$

With a secondary current of 6.19(18,000/240) = 464.38 A in the distribution transformer secondary, the V_0 available for a primary line-to-ground fault will be

$$V_0 = (464.38)(0.2974) = 138.12 \text{ V} \tag{6.100}$$

The wattage in the resistor during the fault is

$$\frac{(464.38)^2(0.2974)}{1000} = 64.14 \text{ kW} \tag{6.101}$$

Similarly, the distribution transformer kVA is

$$6.19\left(\frac{18}{\sqrt{3}}\right) = 64.33 \text{ kVA} \tag{6.102}$$

equal within decimal-point accuracy. When this grounding is used for generator units, tripping the unit is recommended, so these ratings can be short-time rather than continuous ratings.

The normal charging current for this system would be

$$I_C = \frac{18{,}000}{\sqrt{3} \times 5019} = 2.07 \text{ A/phase at 18 kV} \tag{6.103}$$

The use of a distribution transformer and a secondary resistor rather than a resistor directly connected in the neutral is an economic consideration. With high-resistance grounding it is generally less expensive to use the resistor in the secondary, as shown.

In contrast, a through-phase fault on the generator of Fig. 6.18 where $X_1 = j0.066$ pu is

$$\begin{aligned} I_1 = I_{a3\phi} &= \frac{1}{0.066} = 15.15 \text{ pu} \\ &= 15.15 \times \frac{100{,}000}{\sqrt{3}\, 18} \\ &= 48{,}610 \text{ amperes at 18 kV} \end{aligned} \tag{6.104}$$

compared to $I_{a\phi G} = 6.19$ A at 18 kV.

The flow of current through the system for a ground fault is sometimes hard to visualize from the zero-sequence quantities, such as shown in Fig. 6.18c. Although positive- and negative-sequence impedances are quite negligible in high-resistance grounded system, the three sequence currents are equal at the fault [Eq. (6.95)] and flow through the system. Since there is a positive-sequence source at either end of Fig. 6.18a, the positive- and negative-sequence currents divide as shown by the 0.51 and 0.49 distribution factors in the positive- and negative-sequence networks. The approximate currents flowing through this system are documented in Fig. 6.19. They are approximate since the capacitance normally distributed is shown lumped. Before the fault a charging current of 2.07 A [Eq. (6.103)] flows symmet-

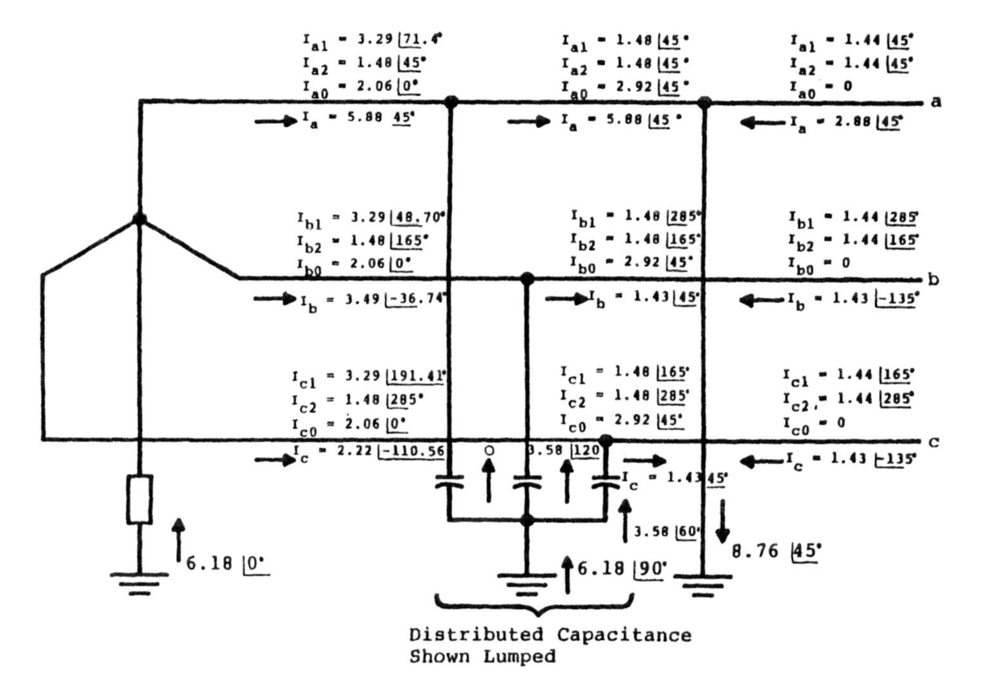

Figure 6.19 Three-phase and sequence current distribution for the system of Fig. 6.18 during a solid phase-a-to-ground fault.

rically in the three phases. Since this is of the same order of magnitude as the fault currents, Thévenin's theorem and superposition must be used to determine the currents flowing during the ground fault. Thus, in phase a from the generator to the lumped capacitance, I_{a1} is the sum of the prefault charging current plus the fault component, or $2.07\underline{/90^\circ} + 0.51 \times 2.92\underline{/45^\circ}$ which is $I_{a1} = 3.29\underline{/71.4^\circ}$. Similarly, $I_{b1} = 2.07\underline{/-30^\circ} + 0.51 \times 2.92\underline{/-75^\circ} = 3.29\underline{/-48.7^\circ}$ and $I_{c1} = 2.07\underline{/210^\circ} + 0.51 \times 2.92\underline{/165^\circ} = 3.29\underline{/191.41^\circ}$. The negative- and zero-sequence components are as normally determined by the fault. In the lumped shunt capacitance, the charging current of 2.07 A cancels the zero-sequence phase a component of 2.07 A to give zero current, since this branch is essentially shorted out by the solid phase-

a-to-ground fault. In the unfaulted phases the charging currents add to the zero-sequence component, providing currents as shown in Fig. 6.19. Figure 6.19 is similar and consistent with Fig. 5.7. The source voltage for Fig. 5.7 is $1\angle 90°$, while for Fig. 6.19 it is $1\angle 0°$.

Again this assumes that none of the distributed capacitance is in the area to the right of the fault location. As has been indicated, the total fault value would not change for different fault locations; similarly, the distribution will not change essentially, as no series impedance is considered between the generator and the power transformer.

6.17 EXAMPLE: PHASE-a-TO-GROUND FAULT CURRENTS AND VOLTAGES ON BOTH SIDES OF A WYE–DELTA TRANSFORMER

A 25-MVA, 69:13.8-kV transformer is connected wye–delta as shown in Fig. 6.20. For a fault at F on the 13.8-kV side, the three sequence reactances to the fault on a 25-MVA base are:

$$X_1 = X_2 = j16 + J7 = j23\% = j0.23 \text{ pu} \tag{6.105}$$

$$X_0 = j7\% = j0.07 \text{ pu} \tag{6.106}$$

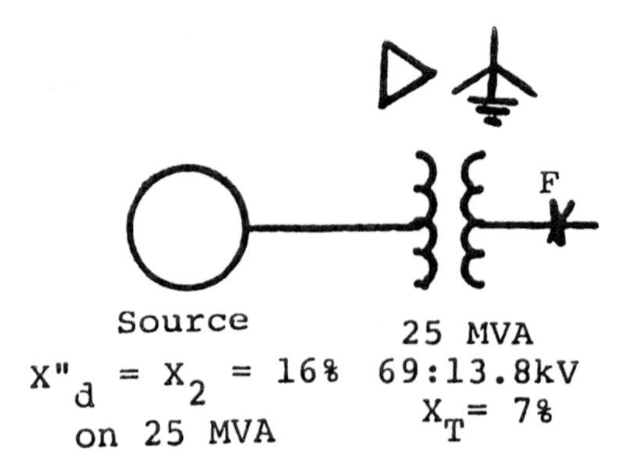

Figure 6.20 Transformer and associated source.

and for a phase-*a*-to-ground fault:

$$I_1 = I_2 = I_0 = j1.0/j(0.23 + 0.23 + 0.07) = 1.0/0.53 \tag{6.107}$$
$$= 1.89\,\text{pu}$$

$$I_a = 3I_0 = 5.66\,\text{pu} = 5.66(25{,}000)/\sqrt{3}(13.8) \tag{6.108}$$
$$= 5920.33\,\text{A at } 13.8\,\text{kV}$$

The sequence voltages at the fault are;

$$V_1 = V - I_1X_1 = j1.0 - 1.89(j0.23) = j0.566\ \text{pu} \tag{6.109}$$

$$V_2 = 0 - I_2X_2 = 0 - 1.89(j0.23) = -j0.434\ \text{pu} \tag{6.110}$$

$$V_0 = 0 - I_0X_0 = 0 - 1.89(j0.07) = -j0.132\ \text{pu} \tag{6.111}$$

The phase-to-neutral voltages at the fault are:

$$V_a = V_1 + V_2 + V_0 = 0$$
$$V_b = a^2V_1 + aV_2 + V_0 \tag{6.112}$$

$$= 0.566\underline{/90° + 240°} + 0.434\underline{/-90° + 120°} + 0.132\underline{/-90°}$$

$$= 0.866 - j0.198 = 0.888\underline{/-12.88°}\,\text{pu} \tag{6.113}$$

$$= 0.888(13{,}800)/\sqrt{3} = 7{,}077.84\ \text{V}$$

$$V_c = aV_1 + a^2V_2 + V_0$$

$$= 0.566\underline{/90° + 120°} + 0.434\underline{/-90° + 240°} + 0.132\underline{/-90°}$$

$$= -0.866 - j0.198 = 0.888\underline{/-167.12°}\,\text{pu} \tag{6.114}$$

$$= 0.888(13{,}800)/\sqrt{3} = 7{,}077.84\ \text{V}$$

Now for the faults currents on the 69-kV delta side. As developed in Section 4.13, the sequence currents on the delta side are as follows:

$$I_1' = I_1\underline{/+30°} = 1.89\underline{/+30°} \tag{6.115}$$

$$I_2' = I_2\underline{/-30°} = 1.89\underline{/-30°} \tag{6.116}$$

$$I_0' = 0 \tag{6.117}$$

and the phase currents are:

$$I_A = I_1' + I_2'$$
$$= 1.89\underline{/+30^\circ} + 1.89\underline{/-30^\circ} = 3.27\,\text{pu} \tag{6.118}$$
$$= 3.27(25{,}000)/\sqrt{3}(69) = 683.62\,\text{A at } 69\,\text{kV}$$

$$I_B = a^2I_1' + aI_2'$$
$$= 1.89\underline{/30^\circ + 240^\circ} + 1.89\underline{/-30^\circ + 120^\circ} = 0 \tag{6.119}$$

$$I_C = aI_1' + a^2I_2'$$
$$= 1.89\underline{/30^\circ + 120^\circ} + 1.89\underline{/-30^\circ + 240^\circ} = -3.27\,\text{pu} \tag{6.120}$$
$$= -3.27(25{,}000)/\sqrt{3}(69) = -683.62\,\text{A at } 69\,\text{kV}$$

Since the source voltage for the fault calculation was $1.0\underline{/90^\circ}$, this voltage for the 69-kV side of the transformer will be shifted $+30^\circ$ or $1.0\underline{/120^\circ}$. The sequence voltages are:

$$V_1' = V\underline{/120^\circ} - jI_1'X_d''$$
$$= 1.0\underline{/120^\circ} - 1.89(0.16)\underline{/30^\circ + 90^\circ} \tag{6.121}$$
$$= 0.698\underline{/120^\circ}\ \text{pu}$$

$$V_2' = 0 - I_2'X_d'' \tag{6.122}$$
$$= -1.89(0.16)\underline{/-30^\circ + 90^\circ} = 0.302\underline{/-120^\circ}$$

Correspondingly, the phase voltages are:

$$V_A = 0.698\underline{/120^\circ} + 0.302\underline{/-120^\circ}$$
$$= -0.50 + j0.343 = 0.606\underline{/145.55^\circ}\ \text{pu} \tag{6.123}$$
$$= 0.606(69{,}000)/\sqrt{3} = 24{,}154.9\underline{/145.55^\circ}\ \text{V}$$

$$V_B = 0.698\underline{/120^\circ + 240^\circ} + 0.302\underline{/-120^\circ + 120^\circ} \tag{6.124}$$
$$= 1.0\ \text{pu} = 39{,}837\ \text{V}$$

$$V_C = 0.698\underline{/120^\circ + 120^\circ} + 0.302\underline{/-120^\circ + 240^\circ}$$
$$= -0.50 - j0.343 = 0.606\underline{/-145.55^\circ}\ \text{pu} \tag{6.125}$$
$$= 24{,}154.9\underline{/-145.55^\circ}\ \text{V}$$

The currents and voltages on both sides of the transformer are shown in Fig. 6.21.

An alternative method to determine the voltages on the 69-kV side is to add the voltage drop across the transformer to the 13.8-kV sequence voltages. This technique is illustrated in Fig. 6.22. An ideal or perfect transformer is used to provide the phase

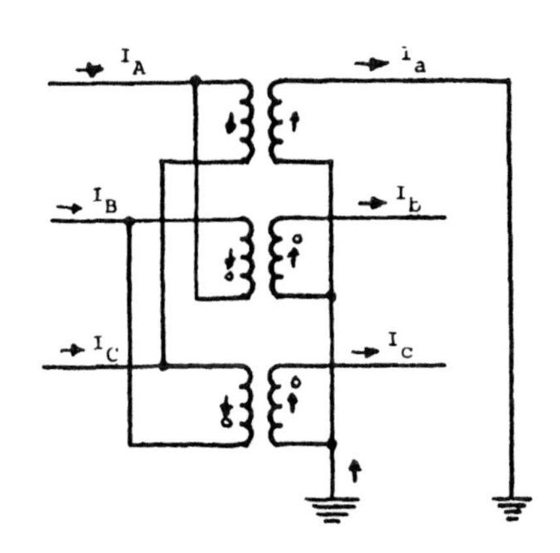

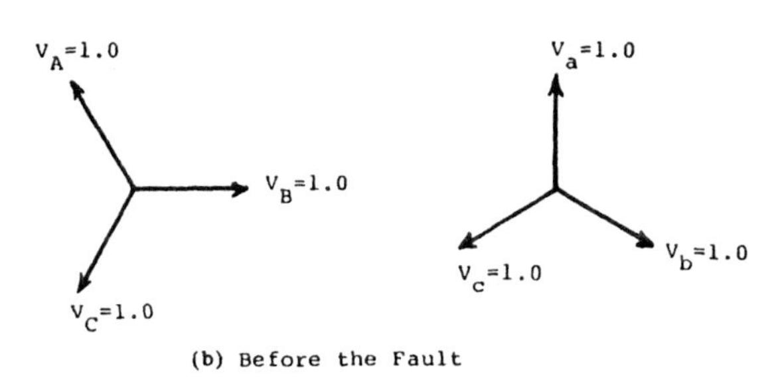

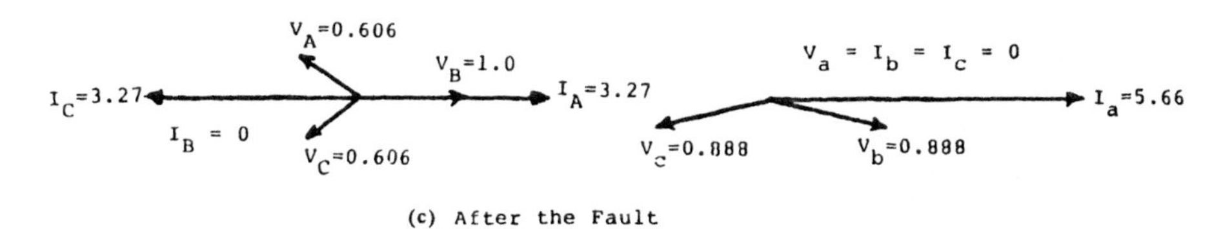

Figure 6.21 Current and voltage before and after a phase-*a*-to-ground fault.

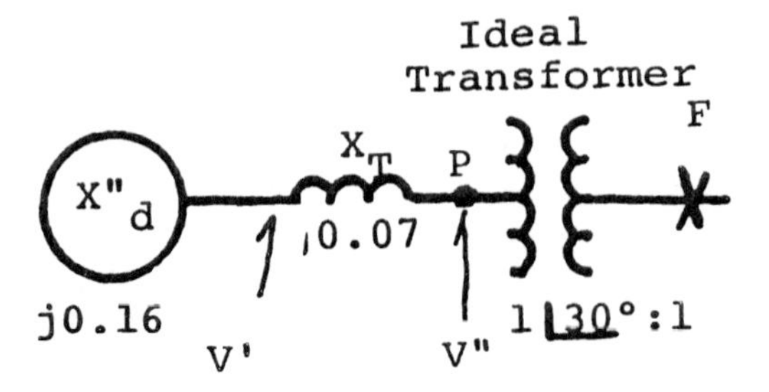

Figure 6.22 Alternative method to determine voltages on the delta side of the bank of Fig. 6.20.

shifts through the bank so the voltages at point P are:

$$V_1'' = 0.566\angle 90° + 30° = 0.566\angle 120° \text{ pu} \tag{6.126}$$

$$V_2'' = 0.434\angle -90° - 30° = 0.434\angle -120° \text{ pu} \tag{6.127}$$

Then the sequence voltages at the 69-kV transformer are:

$$V_1' = V_1'' + I_1'(X_T) = 0.698\angle 120° \text{ pu} \tag{6.121}$$

$$V_2' = V_2'' + I_2'(X_T) = 0.302\angle -120° \text{ pu} \tag{6.122}$$

These are the same as the earlier (6.121) and (6.122).

7

Series and Simultaneous Unbalance Sequence Network Interconnections

7.1 INTRODUCTION

A general representation of power systems and their sequence networks was introduced in Section 5.2. The right-hand representations of Fig. 5.1 represent, at the top, the power system, which is assumed symmetrical up to the area or point of unbalance, which can be documented between the phases *a*, *b*, *c* and *n* (neutral) terminals shown. Below these are the three sequence networks with their terminal points: the neutral or zero potential bus (*n*), (*x*) the left side, and (*y*) the right side of the unbalance point or area.

7.2 SERIES UNBALANCE SEQUENCE INTERCONNECTIONS

A variety of potential series unbalances that can occur in a power system is shown in Fig. 7.1. A common one is a blown fuse or broken conductor, represented by Fig. 7.1j. Other problems are:

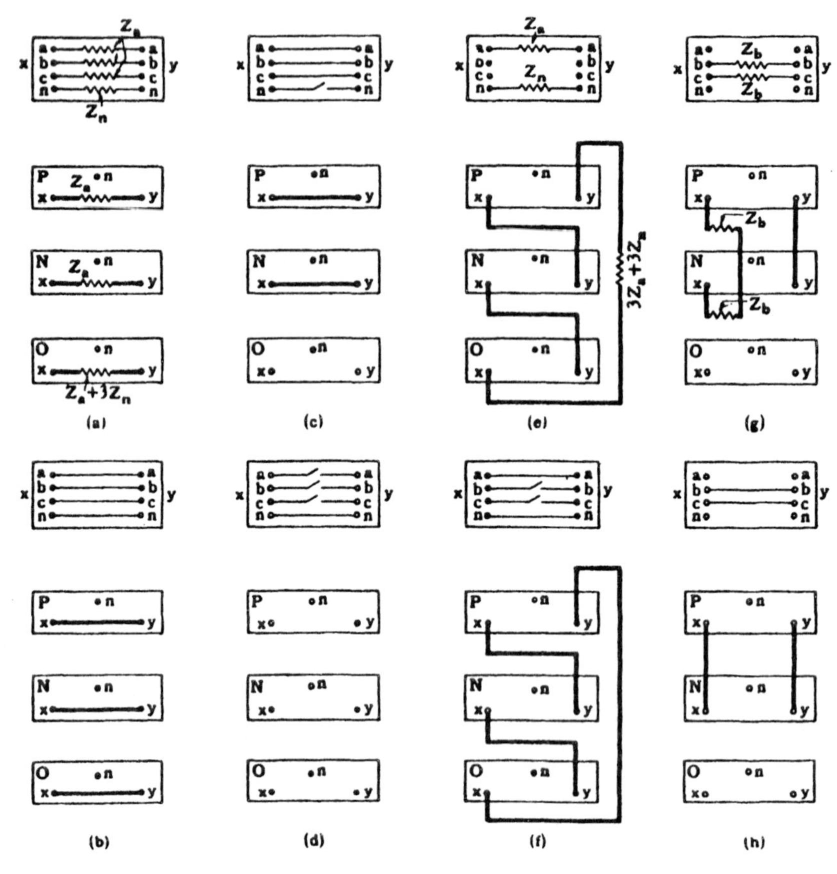

Figure 7.1 Box sequence connections for series unbalanced conditions: (a) equal impedances in three phases; (b) normal balanced conditions; (c) neutral circuit open; (d) any three or four phases open; (e) phases b and c open and impedances in phase a and neutral; (f) phases b and c open; (g) phases a and neutral open and impedances in phases b and c; (h) phase a and neutral open; (i) phase a open and impedances

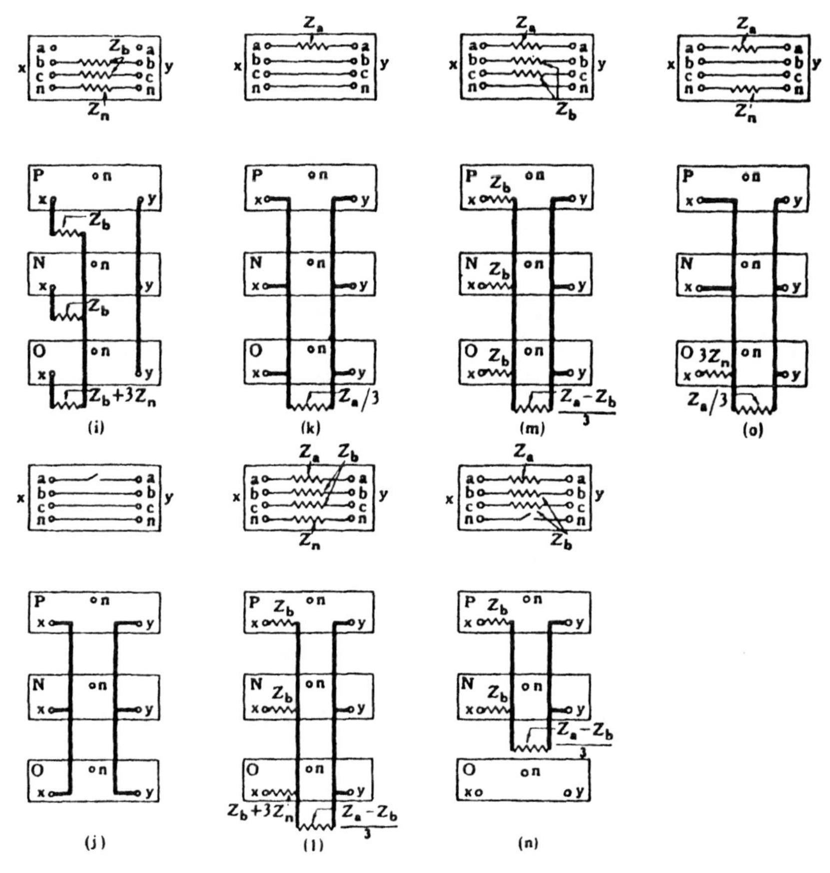

in phases b, c, and neutral; (j) phase a open; (k) impedance in phase a; (l) equal impedances in phases b and c, impedance in neutral; (m) equal impedances in phases b and c; (n) equal impedances in phases b and c, neutral open; (o) impedances in phase a and neutral. (From E. L. Harder, Sequence Network Connections for Unbalanced Load and Fault Conditions, *The Electrical Journal*, December 1937.)

two blown fuses or open conductors (Figure 7.1f); unequal impedances in the phases in several combinations (Figure 7.1k through o), and so on. Applications of these are best discussed with examples.

7.3 ONE PHASE OPEN: BROKEN CONDUCTOR OR BLOWN FUSE

This is Fig. 7.1j. The application to a power system is shown in Fig. 7.2. Load or the equivalent represented by a difference between the source voltages must be included for these series unbalances, as the current is dependent on the load or difference between the system voltages on either side of the open. If load (Z_L) was omitted as is common for the shunt faults, no current can flow in the networks in Fig. 7.2. This makes sense, as opening a unloaded circuit does not change the current. Such opening will cause a transient arc as the system capacitance current is interrupted. This also indicates that the current flow during the open will be on the order of the load current.

Including load now indicates that impedance values should be used rather than reactance only as for the shunt faults. The system impedances will be mostly reactance except for low voltage lines, while the load will be mostly resistance. Thus ($r + jx$) calculations are required.

The system shown in Fig. 7.2 is grounded on both sides of the open. If one side is ungrounded, it will be seen that the zero-sequence network is open. Thus no zero-sequence current can flow. This is typical of a motor load. This type of unbalance is shown in Fig. 7.1g or h.

7.4 EXAMPLE: OPEN PHASE CALCULATION

7.4.1 Power System Grounded on Both Sides of the Open Conductor

Using the power system of Fig. 6.10 with the fault now being the phase *a* conductor open in the line at bus *H*, the sequence

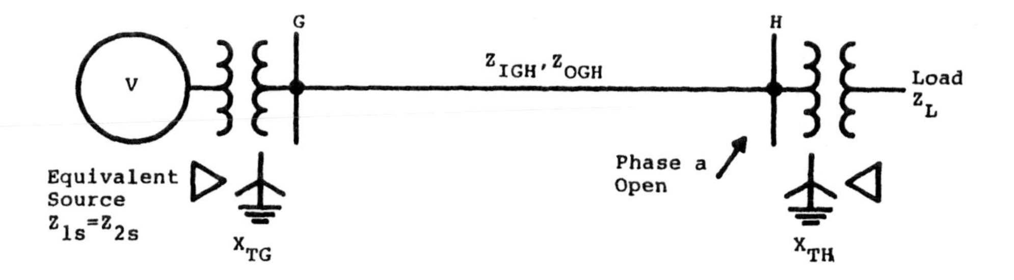

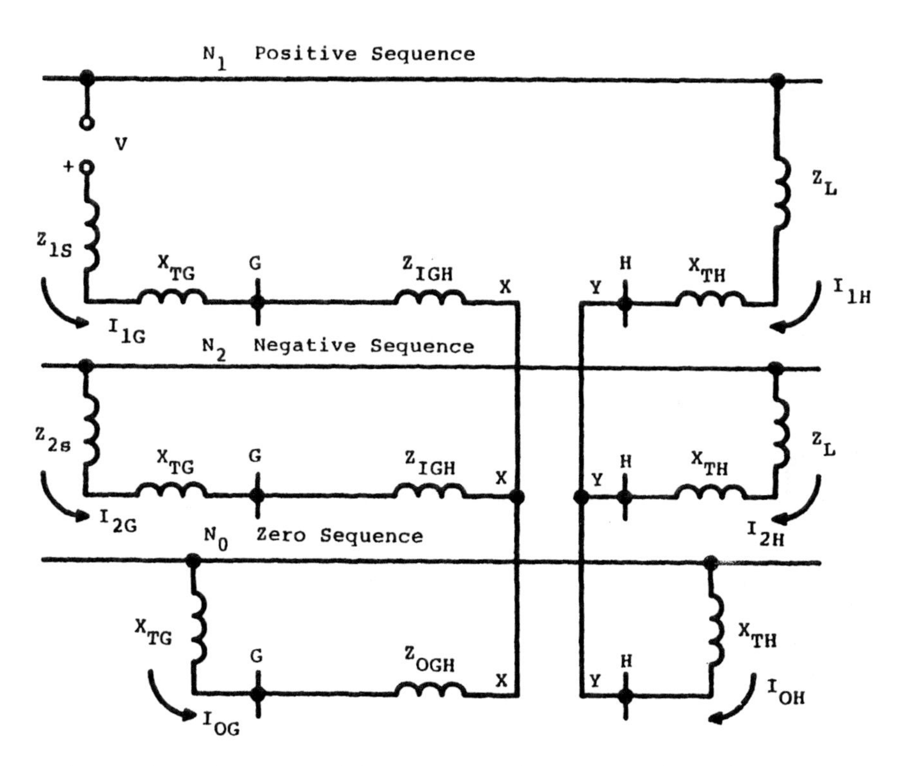

Figure 7.2 Example of the sequence interconnections for phase *a* open at bus *H* for a typical power system.

networks and interconnections of Fig. 7.1j are shown in Fig. 7.3. The values are those of Eqs. (6.52). With induction motors their negative-sequence impedance is less than the positive-sequence impedance, so $Z_{1\text{Load}}$ and $Z_{2\text{Load}}$ are not equal. This is discussed in Chapter 10. From the interconnections, $Z_{1G} + Z_{2H} = Z_2$ is in parallel with $Z_{0G} + Z_{0H} = Z_0$. Thus

$$Z_2 = 0.5932 + j0.7015 = 0.9187\angle 49.78^\circ$$

$$Z_0 = 0.250 + j0.711 = 0.7537\angle 70.63^\circ$$

$$\frac{Z_2Z_0}{Z_2 + Z_0} = \frac{0.9187 \times 0.7537\angle 49.78^\circ + 70.63^\circ}{0.8432 + j1.4125}$$

$$= 0.4209\angle 61.245^\circ = 0.2025 + j0.3690$$

The total impedance is

$$Z_{1G} + \frac{Z_2Z_0}{Z_2 + Z_0} + Z_{1H} = 1.171 + j1.216 = 1.688\angle 46.08^\circ \tag{7.1}$$

The equivalent system voltage to provide a voltage at the load of $1.0\angle 0^\circ$ with I in the load of $1.0\angle -25.84^\circ$ would be

$$-V + I(Z_{1G} + X_{TH}) + 1.0 = 0$$

$$V = 1.0 + 1.0\angle -25.84^\circ\,(0.0684 + j0.4106)$$

$$= 1.0 + 1 \times 0.4163\angle -25.84^\circ + 80.54^\circ \tag{7.2}$$

$$= 1.0 + 0.4163\angle 54.70^\circ = 1.286\angle 15.315^\circ \text{ pu}$$

From Eqs. (7.1) and (7.2),

$$I_{1G} = \frac{1.286\angle 15.315^\circ}{1.688\angle 46.08^\circ} = 0.762\angle -30.765^\circ \text{ pu} = 0.655 - j0.390 \text{ pu} \tag{7.3}$$

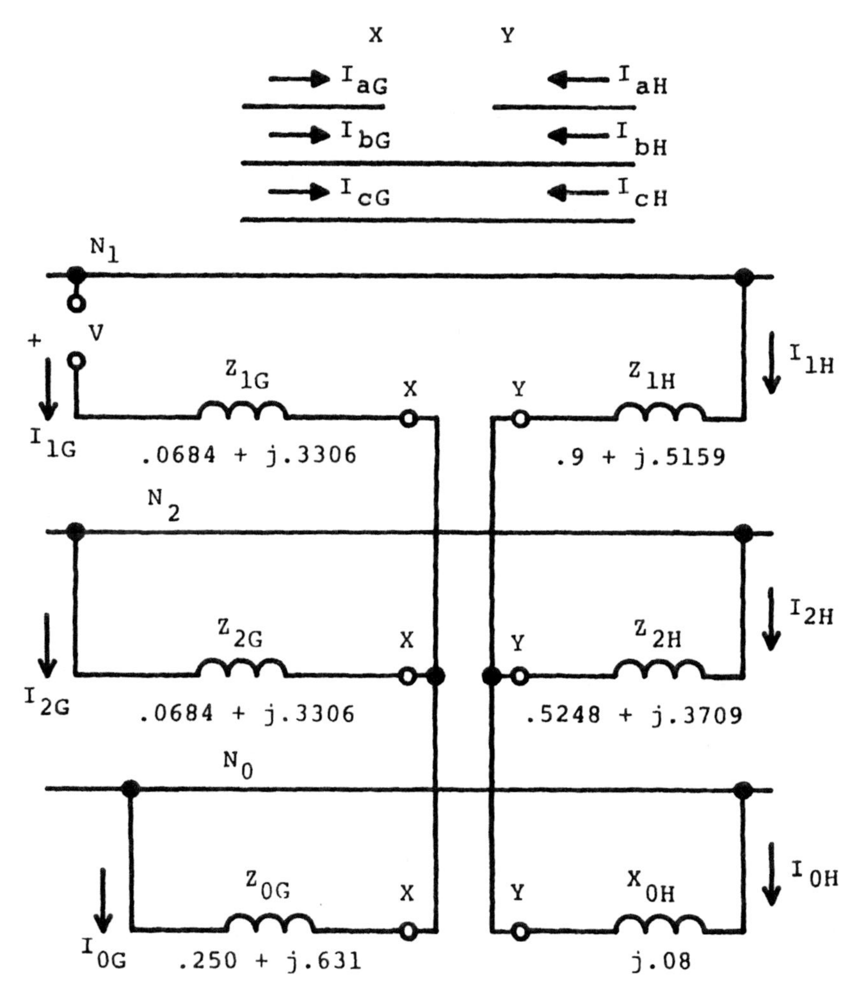

Figure 7.3 Sequence interconnections for a phase *a* open on the line at bus *H* of the power system of Fig. 6.10. Values are in per unit at 30 MVA, 34.5 kV.

$$I_{2G} = -I_{1G}\frac{Z_0}{Z_2 + Z_0} = -0.762\angle{-30.76°}\frac{0.7537\angle{70.63°}}{1.645\angle{59.165°}}$$

$$= 0.349\angle{160.70°} = -0.329 + j0.115\,\text{pu} \tag{7.4}$$

$$I_{0G} = -I_{1G}\frac{Z_2}{Z_2 + Z_0} = -0.762\angle{-30.76°}\frac{0.9187\angle{49.78°}}{1.645\angle{59.165°}}$$

$$= 0.426\angle{139.85°} = -0.325 + j0.274\,\text{pu} \tag{7.5}$$

From these,

$$I_{aG} = I_{1G} + I_{2G} + I_{0G} = 0 \tag{7.6}$$

$$I_{bG} = a^2 I_{1G} + aI_{2G} + I_{0G}$$

$$= 0.762\angle{209.24°} + 0.349\angle{280.7°} + 0.426\angle{139.85°} \tag{7.7}$$

$$= -0.925 - j0.441 = 1.025\angle{-154.53°}\,\text{pu}$$

$$I_{cG} = aI_{1G} + a^2 I_{2G} + I_{0G}$$

$$= 0.762\angle{89.24°} + 0.349\angle{40.7°} + 0.426\angle{139.85°} \tag{7.8}$$

$$= -0.050 + j1.264 = 1.265\angle{92.285°}\,\text{pu}$$

These per unit currents as they flow in the power system are shown in Fig. 7.4a. Their magnitudes are around load current with the open phase a current essentially flowing through the ground. Figure 6.12 shows the currents for ground faults on this same system for comparison.

7.4.2 Power System Ungrounded on One Side of the Open Conductor

If the 30-MVA transformer at station H (Fig. 6.10) is not grounded, the zero-sequence X_{TH} circuit is open in Fig. 7.2, and neither I_{0G} nor I_{0H} can flow. Then only the total negative-sequence impedance is across X and Y. From the previous cal-

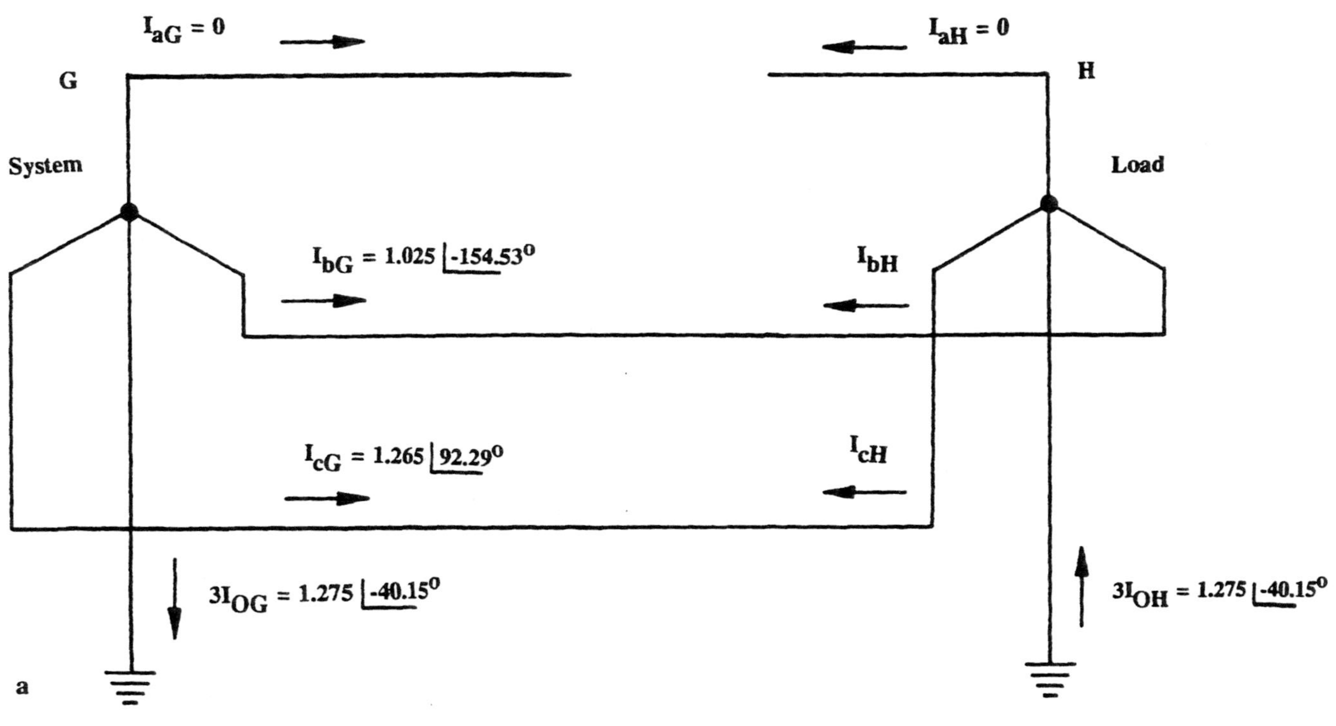

Figure 7.4 Per unit current distribution for an open phase *a* in the power system of Fig. 6.10: (a) power system grounded on either side of the open phase. *Figure continues*

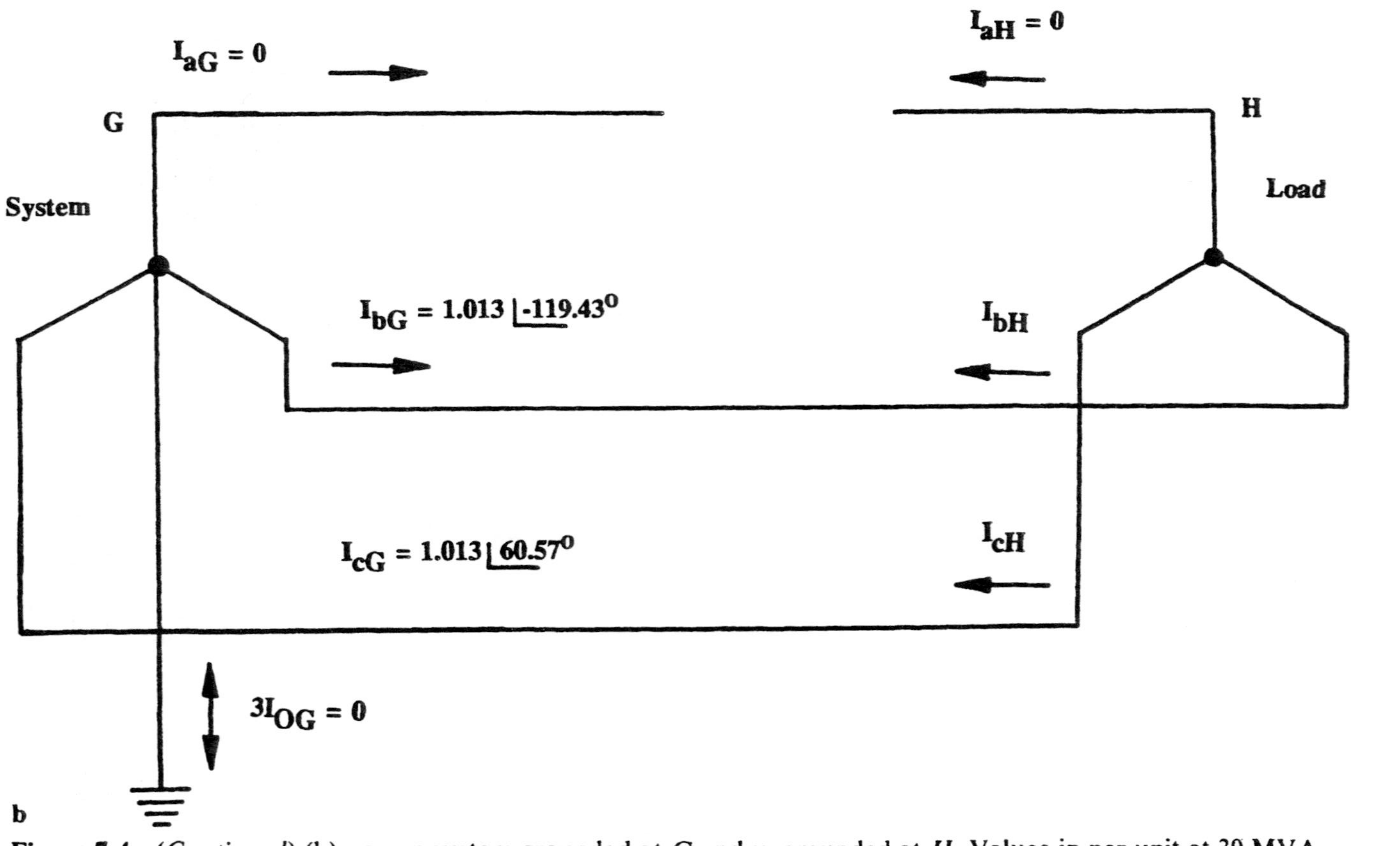

Figure 7.4 (*Continued*) (b) power system grounded at *G* and ungrounded at *H*. Values in per unit at 30 MVA where 1 pu = 502 A at 34.5 kV. Load current is 1 pu.

culations the total impedance around the positive and negative sequence is the sum of values in per unit at 30 MVA.

$$Z_1 + Z_2 = 1.562 + j1.548 = 2.199\underline{/44.75^\circ} \tag{7.9}$$

$$I_{1G} = -I_{2G} = -I_{1H} = I_{2H} = \frac{1.286\underline{/15.315^\circ}}{2.199\underline{/44.75^\circ}} \tag{7.10}$$

$$= 0.5849\underline{/-29.434^\circ}\,\text{pu}$$

$$I_{aG} = I_{aH} = 0$$

$$I_{bG} = a^2 I_{1G} - aI_{1G} = 1.013\underline{/-119.434^\circ} \tag{7.11}$$

$$I_{cG} = aI_{1G} - a^2 I_{1G} = 1.013\underline{/60.566^\circ}$$

These are shown in Fig. 7.4b.

7.5 SIMULTANEOUS UNBALANCE SEQUENCE INTERCONNECTIONS

Typical simultaneous unbalances that occur in a power system are a broken conductor falling to ground and faults occurring in different parts of the system at the same time. The sequence interconnections for a number of cases are shown in Fig. 7.5. Since the sequence connections for the unbalances are different either in type or location, it is necessary to isolate one from the other. This can be done with an ideal or perfect transformer connected as shown. These transformers have no impedances or losses and so are mathematical "fictions" used for the purpose of isolation.

In applying these ideal transformers, which have no impedance, the voltage drops across their windings cannot be expressed by the current in their windings. If the unbalances are related to the same phase, the ratios of the three ideal transformers are 1:1. If different phases are involved, the ratios involve the phasor operators a and a^2 as shown. The application of these sequence network interconnections is best illustrated by typical cases and examples.

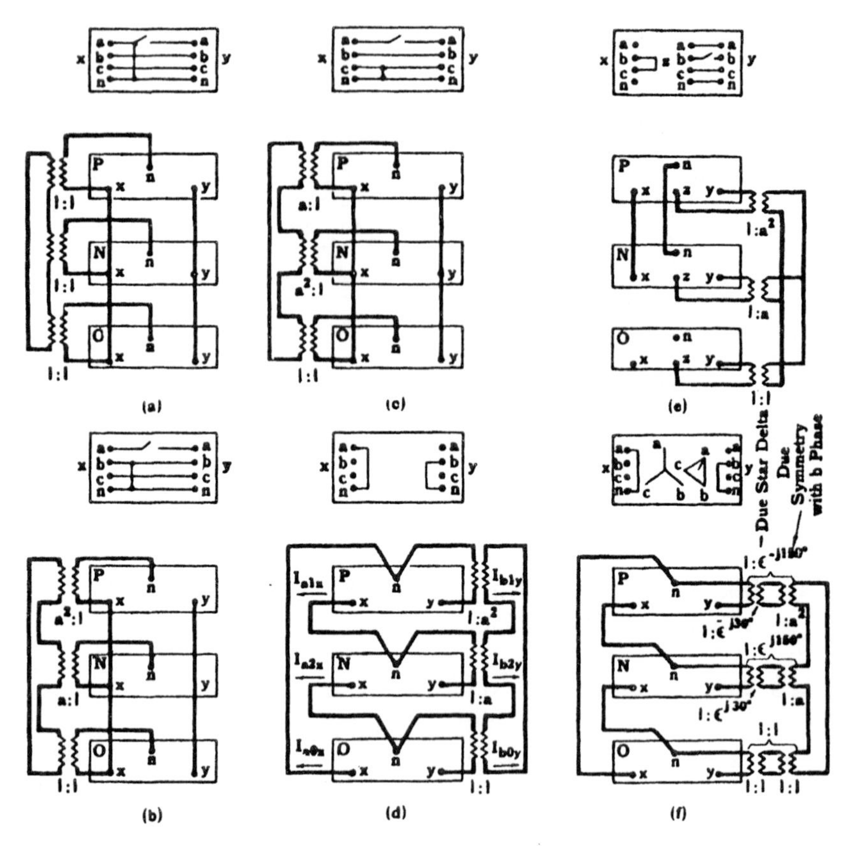

Figure 7.5 Box sequence connections for simultaneous unbalanced conditions: (a) phase *a* open and phase-*a*-to-ground fault on the *x* side; (b) phase *a* open and phase-*b*-to-ground fault; (c) phase *a* open and phase-*c*-to-ground fault; (d) phase-*a*-to-ground fault at *x* and phase-*b*-to-ground fault at *y*; (e) phase-*b*-to-*c* fault at *x* and phase *b* open *z* to *y*; (f) phase-*a*-to-neutral fault at *x*, phase-*b*-to-neutral fault on the other side of a wye–delta transformer bank at *y*, with *x* taken as the reference point. (From E. H. Harder, Sequence Network Connections for Unbalanced Load and Fault Conditions, *The Electrical Journal*, December 1937.)

7.6 EXAMPLE: BROKEN CONDUCTOR FALLING TO GROUND ON BUS SIDE

The sequence interconnection for a conductor falling to ground on the load or bus side are shown in Fig. 7.5a. For the system of Figs. 7.2 and 6.10 the network interconnections are as in Fig. 7.6 if the broken conductor falls to ground on the bus side of the open. Unless a computer program is available, the solutions of these types of interconnections generally involve the solution of three equations with three unknowns, as discussed in Section 4.18.

7.6.1 Power System Grounded on Both Sides of the Unbalances; Calculations with 30-MVA Load

The drops around the positive- and negative-sequence networks from Fig. 7.6 after combining impedances as in Eq. (6.52) are

$$-V + I_{1G}Z_{1G} - I_{2G}Z_{1G} + I_{2H}Z_{2H} - I_{1H}Z_{1H} = 0 \qquad (7.12)$$

$$\text{Positive-sequence network: } I_{1G} + I_{1H} + I_0 = 0 \qquad (7.13)$$

$$\text{Negative-sequence network: } I_{2G} + I_{2H} + I_0 = 0 \qquad (7.14)$$

$$\text{Zero-sequence network: } I_{0G} + I_{0H} + I_0 = 0 \qquad (7.15)$$

Also from the interconnections,

$$-I_0 = \tfrac{1}{3}(I_{1H} + I_{2H} + I_{0H}) \qquad (7.16)$$

Substituting Eq. (7.16) in Eqs. (7.13), (7.14), and (7.15):

$$I_{1G} = -\tfrac{2}{3}I_{1H} + \tfrac{1}{3}I_{2H} + \tfrac{1}{3}I_{0H} \qquad (7.17)$$

$$I_{2G} = \tfrac{1}{3}I_{1H} - \tfrac{2}{3}I_{2H} + \tfrac{1}{3}I_{0H} \qquad (7.18)$$

$$I_{0G} = \tfrac{1}{3}I_{1H} + \tfrac{1}{3}I_{2H} - \tfrac{2}{3}I_{0H} \qquad (7.19)$$

Substituting Eqs. (7.17) and (7.18) in Eq. (7.12) yields

$$-I_{1H}(Z_{1G} + Z_{1H}) + I_{2H}(Z_{1G} + Z_{2H}) = V \qquad (7.20)$$

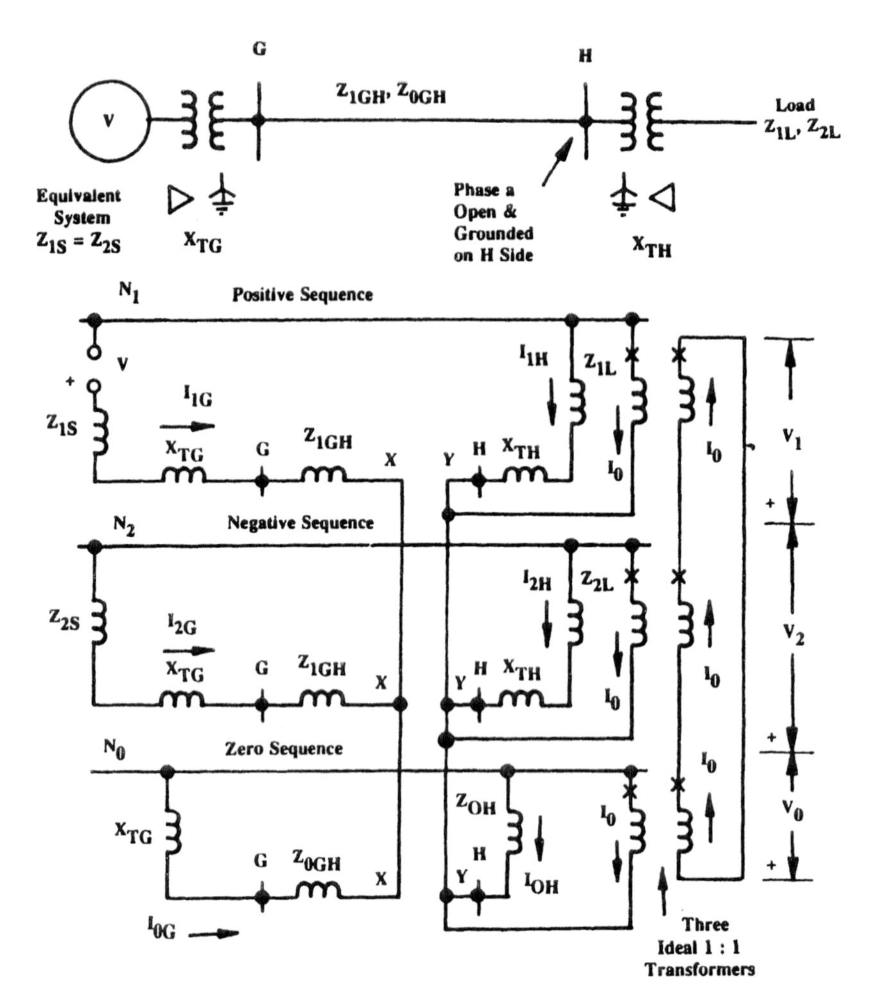

Figure 7.6 Sequence network interconnections for phase *a* open and grounded at bus *H* (broken conductor fallen to ground on the bus side.

Also from the network (Fig. 7.6),

$$V_1 + I_{1H}Z_{1H} = 0 \quad V_2 + I_{2H}Z_{2H} = 0 \quad V_0 + I_{0H}Z_{0H} = 0 \tag{7.21}$$

and since $V_1 + V_2 + V_0 = 0$, adding the three parts of Eq. (7.21) yields

$$I_{1H}Z_{1H} + I_{2H}Z_{2H} + I_{0H}Z_{0H} = 0 \tag{7.22}$$

Now the drops around the negative- and zero-sequence networks:

$$I_{2G}Z_{1G} - I_{0G}Z_{0G} + I_{0H}Z_{0H} - I_{2H}Z_{2H} = 0 \tag{7.23}$$

Substituting Eqs. (7.18) and (7.19) in Eq. (7.23), reducing, and multiplying by 3, Eq. (7.23) becomes

$$I_{1H}(Z_{1G} - Z_{0G}) - I_{2H}(2Z_{1G} + Z_{0G} + 3Z_{2H}) + I_{0H}(Z_{1G} + 2Z_{0G} + 3Z_{0H}) = 0 \tag{7.24}$$

The three equations with three unknowns I_{1H}, I_{2H}, and I_{0H} are Eqs. (7.20), (7.22), and (7.24). If a computer program is not available, the three currents can be determined from Eqs. (4.67) using the system constants of Fig. 6.10 and Eqs. (6.52).

Alternatively, the three equations can be solved by multiplying Eq. (7.22) by $(Z_{1G} + 2Z_{0G} + 3Z_{0H})$ and Eq. (7.24) by Z_{0H}. Then if the new equation (7.24) is subtracted from the new equation (7.22), the I_{0H} terms are canceled to provide

$$I_{1H}Z_x + I_{2H}Z_y = 0 \tag{7.25}$$

where

$$Z_x = Z_{1H}(Z_{1G} + 2Z_{0G} + 3Z_{0H}) + Z_{0H}(Z_{0G} - Z_{1G}) \tag{7.26}$$

$$Z_y = Z_{2H}(Z_{1G} + 2Z_{0G}) + Z_{0H}(2Z_{1G} + Z_{0G} + 6Z_{2H}) \tag{7.27}$$

Rewriting Eq. (7.25), we have

$$I_{2H} = -I_{1H}\frac{Z_x}{Z_y} \tag{7.28}$$

Substituting Eq. (7.28) in Eq. (7.20) and solving for

$$I_{1H} = \frac{-VZ_y}{Z_x(Z_{1G} + Z_{2H}) + Z_y(Z_{1G} + Z_{1H})} \tag{7.29}$$

from Eq. (7.22),

$$I_{0H} = \frac{-I_{1H}Z_{1H} - I_{2H}Z_{2H}}{Z_{0H}} \tag{7.22}$$

The current I_0 can be determined from Eq. (7.16); then

$$I_{1G} = -I_{1H} - I_0 \tag{7.30}$$

$$I_{2G} = -I_{2H} - I_0 \tag{7.31}$$

$$I_{0G} = -I_{0H} - I_0 \tag{7.32}$$

From these sequence currents the various phase currents can be determined from Eqs. (4.5) through (4.7). Thus for this example [Fig. 6.10, power system; Fig. 7.6, sequence interconnections; and constants per Eqs. (6.52)], the sequence currents from the equations above are in per unit at 30 MVA:

$$\begin{aligned} I_{1H} &= 0.502\angle 154.93^\circ & I_{1G} &= 0.881\angle -40.53^\circ \\ I_{2H} &= 0.697\angle -35.25^\circ & I_{2G} &= 0.354\angle 173.11^\circ \\ I_{0H} &= 1.039\angle -58.94^\circ & I_{0G} &= 0.618\angle 120.98^\circ \end{aligned}$$

$$I_0 = 0.421\angle 121.19^\circ \tag{7.33}$$

From these sequence currents the phase and neutral currents are calculated from basic equations (4.5) through (4.7). These currents are shown in Fig. 7.7a.

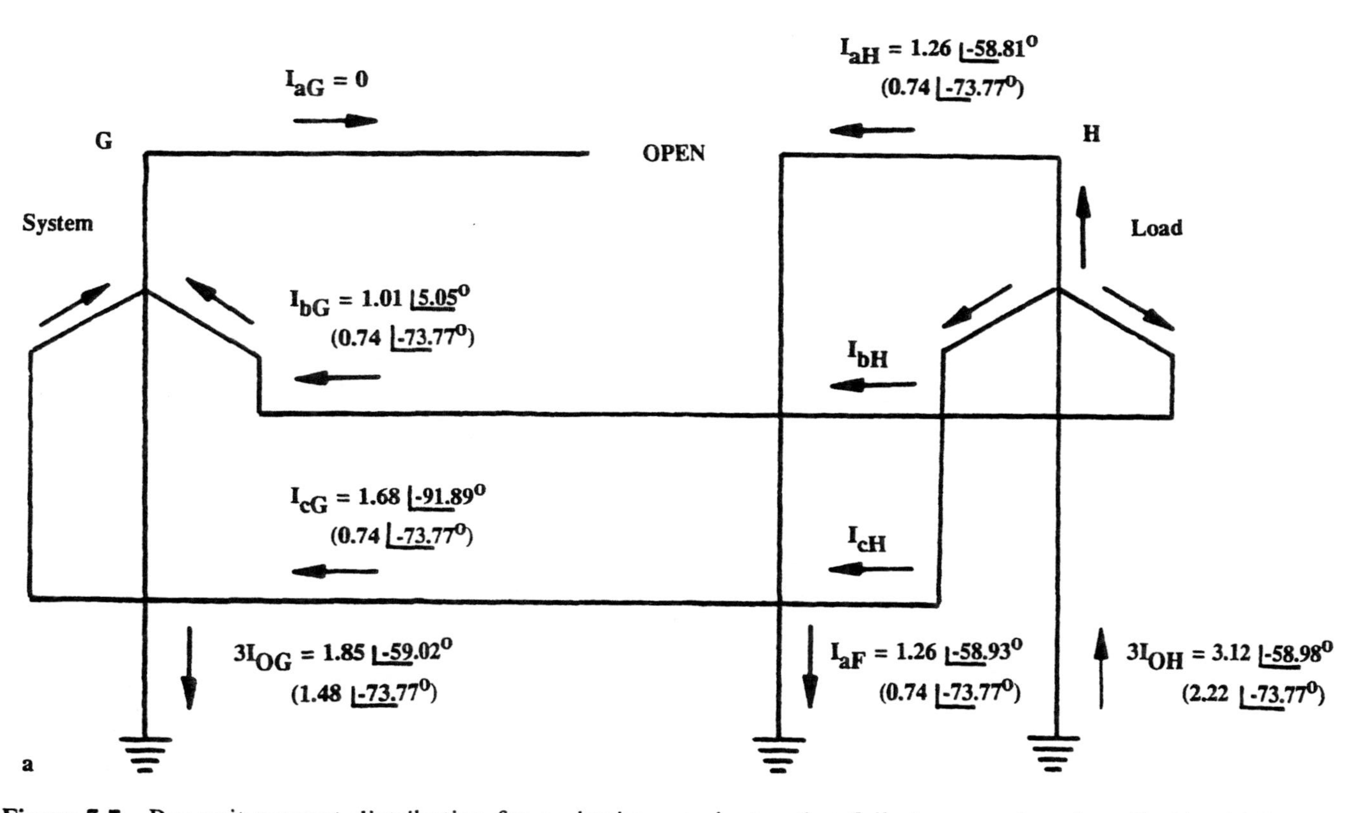

Figure 7.7 Per unit current distribution for an broken conductor that falls to ground on bus *H* side. (a) Power system grounded on either side of the unbalance area. Top values with 30 MVA load. Values in parenthesis with load neglected (no load). *Figure continues*

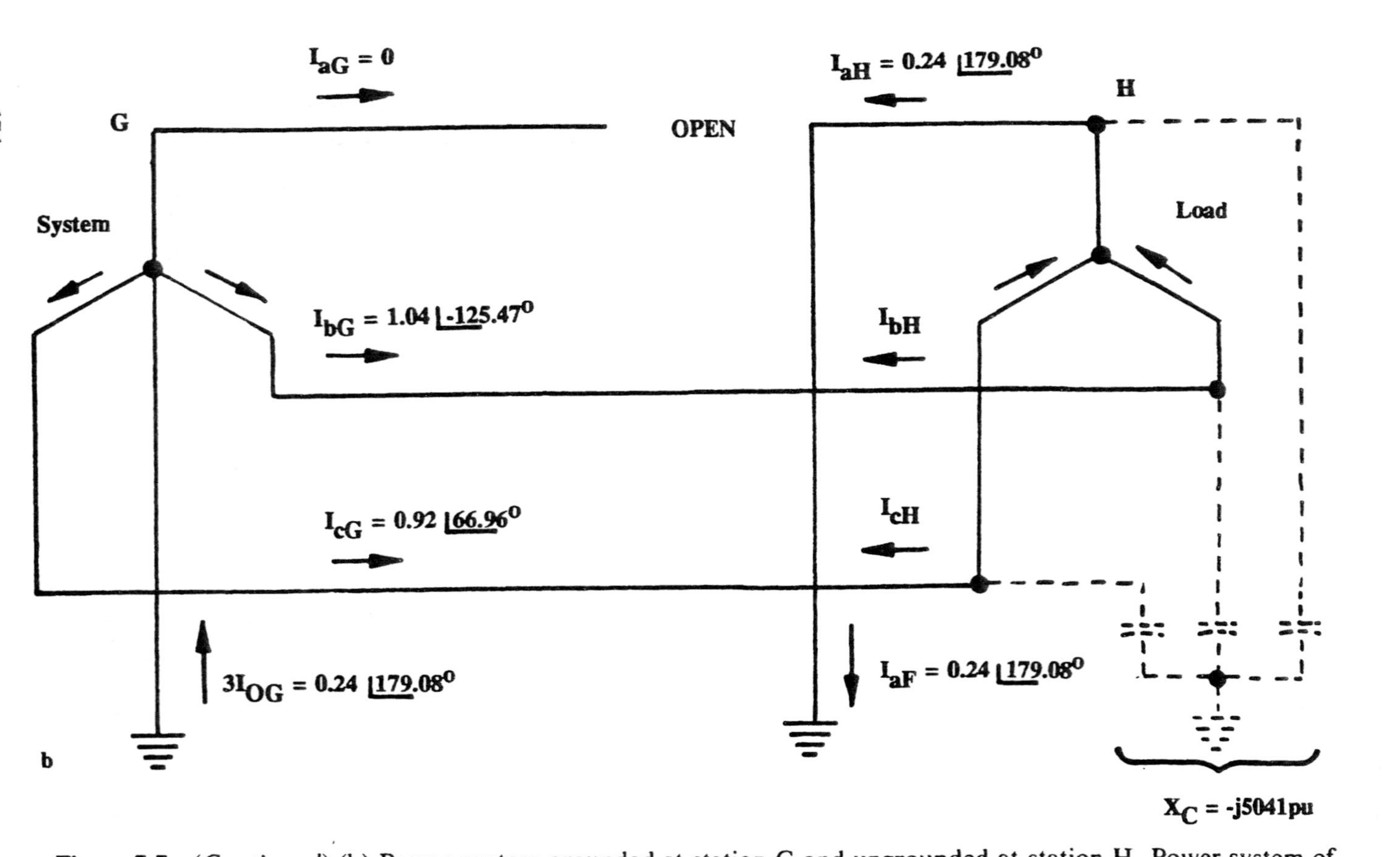

Figure 7.7 (*Continued*) (b) Power system grounded at station G and ungrounded at station H. Power system of Fig. 6.10 with sequence networks connected as in Fig. 7.8. Per unit at 30 MVA where 1 pu = 502 A at 34.5 kV.

7.6.2 Power System Grounded on Both Sides of the Unbalances; Calculations Without Load

From the interconnections of Fig. 7.6 it can be seen that current can flow without the load. If the load branches $(Z_{1L} + X_{TH})$ and $(Z_{2L} + X_{TH})$ are omitted, I_{1H} and I_{2H} do not exist. This is illustrated in Fig. 7.8. The network and calculations are much simplified and are as follows:

$$V_1 + V_2 + V_0 = 0 \tag{7.34}$$

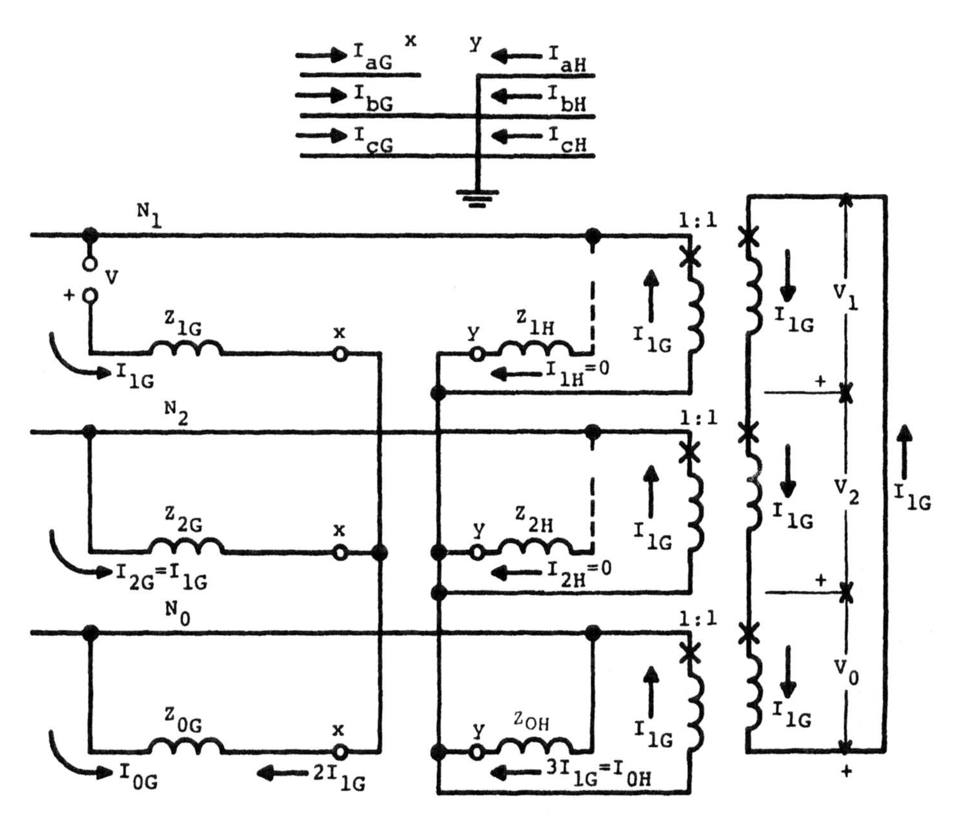

Figure 7.8 Sequence network interconnections for phase *a* open that falls to ground on the bus side at bus *H* of the power system of Fig. 6.10 with the load neglected.

Since the perfect or ideal transformers have no impedance and with the load omitted, voltages V_1 and V_2 cannot be expressed in terms of impedance drops. However, for the zero-sequence network,

$$3I_{1G}X_{TH} + V_0 = 0 \tag{7.35}$$

Subtracting Eq. (7.34) from Eq. (7.35) yields

$$3I_{1G}X_{TH} - V_1 - V_2 = 0 \tag{7.36}$$

The drops around the positive- and negative-sequence networks are

$$-V + I_{1G}Z_{1G} - I_{2G}Z_{1G} - V_2 + V_1 = 0$$

Since $I_{1G} = I_{2G}$, the above becomes

$$-V - V_2 + V_1 = 0 \tag{7.37}$$

Subtracting Eq. (7.37) from Eq. (7.36) yields

$$3I_{1G}X_{TH} + V - 2V_1 = 0 \tag{7.38}$$

Around the positive- and zero-sequence networks,

$$-V + I_{1G}Z_{1G} + 2I_{1G}Z_{0G} + 3I_{1G}X_{TH} + V_1 = 0$$

Multiplying by 2 gives

$$-2V + I_{1G}(2Z_{1G} + 4Z_{0G} + 6X_{TH}) + 2V_1 = 0 \tag{7.39}$$

Adding Eq. (7.38) and Eq. (7.39) yields

$$-V + I_{1G}(2Z_{1G} + 4Z_{0G} + 9X_{TH}) = 0$$

$$I_{1G} = I_{2G} = \frac{V}{2Z_{1G} + 4Z_{0G} + 9X_{TH}} \tag{7.40}$$

$$I_{0G} = -2I_{1G} \qquad I_{1H} = I_{2H} = 0 \qquad I_{0H} = 3I_{1G}$$

Now substituting the per unit values of Eqs. (6.52) with $V = 1.0\angle 0^\circ$ for this no-load case, we have

$$I_{1G} = I_{2G} = \frac{1.0\angle 0^\circ}{1.137 + j3.91} = 0.246\angle -73.77^\circ \tag{7.41}$$

$$I_{0G} = 0.492\angle 106.23^\circ \tag{7.42}$$

$$I_{0H} = 0.738\angle -73.72^\circ \tag{7.43}$$

From the sequence values the phase and neutral currents are calculated using the basis equations (4.5) through (4.7). These currents are shown in parentheses in Fig. 7.7 for comparison with the previous calculations that included load. As seen, load has more impact on the fault currents even though the load impedance is much larger than the grounded transformer impedances. Still, no or light load is an operating condition that must be covered by the protection.

Note that the open phase has caused the current to flow down the transformer neutral at bus G, a reversal from the normal direction of up-the-bank neutrals. Also, the current in the fault is the smallest value rather the largest value as would be the case without the open phase.

7.6.3 Power System Not Grounded on the Load Side of the Unbalances

If the 30-MVA transformer at station H was not grounded or was connected delta on the high side, then with reference to Fig. 7.6, X_{TH} would not be connected in the zero-sequence network. As a result there is no zero-sequence impedance connected to the Y point, so no zero-sequence current can flow in all the zero-sequence network. Thus the power system essentially is ungrounded even though the station G transformer is grounded and the phase is connected to ground. However, no system is ungrounded completely since there is capacitance between the phases and ground. Assume that the total capacitance of station

H is 200,000 Ω at 34.5 kV. The charging current would be

$$I_{\text{charging}} = \frac{34{,}500}{\sqrt{3}\ 200{,}000} = 0.10 \text{ A at } 34.5 \text{ kV} \tag{7.44}$$

With the system of Fig. 6.10 and the constants of Eqs. (6.52) but with $Z_{0H} = -j5041$ pu in Fig. 7.6, the networks can be solved with Eqs. (7.22) and (7.28) through (7.32). These yield

$$I_{1G} = 0.598\underline{/-27.77^\circ} \qquad I_{1H} = 0.671\underline{/155.33^\circ}$$

$$I_{2G} = 0.528\underline{/148.27^\circ} \qquad I_{2H} = 0.461\underline{/-36.85^\circ}$$

$$I_{0G} = 0.080\underline{/179.11^\circ} \qquad I_{0H} = 0.00008\underline{/100^\circ}$$

$$I_0 = 0.080\underline{/-0.94^\circ} \tag{7.45}$$

The phase currents are determined from Eqs. (4.5) through (4.7). These are shown in Fig. 7.7b.

7.7 EXAMPLE: BROKEN CONDUCTOR FALLING TO GROUND ON LINE SIDE

For contrast, if the conductor falls to ground on the line side, the ideal transformers would be connected on that side as shown in Fig. 7.9. As can be observed, load Z_L cannot be neglected, as there would be no path for current flow except through the system capacitance to ground. As this capacitance is large and normally neglected in grounded systems, fault currents will be very small.

Using the system of Fig. 6.10 and constants of Eqs. (6.52), a set of three simultaneous equations with three unknowns can be written and solved by computer, by Eqs. (4.54) or by reduction of the three equations.

From Fig. 7.9 the drops involving the perfect transformers are

$$-V + I_{1G}Z_{1G} + V_1 = 0 \tag{7.46}$$

$$I_{2G}Z_{1G} + V_2 = 0 \tag{7.47}$$

$$I_{0G}Z_{0G} + V_0 = 0 \tag{7.48}$$

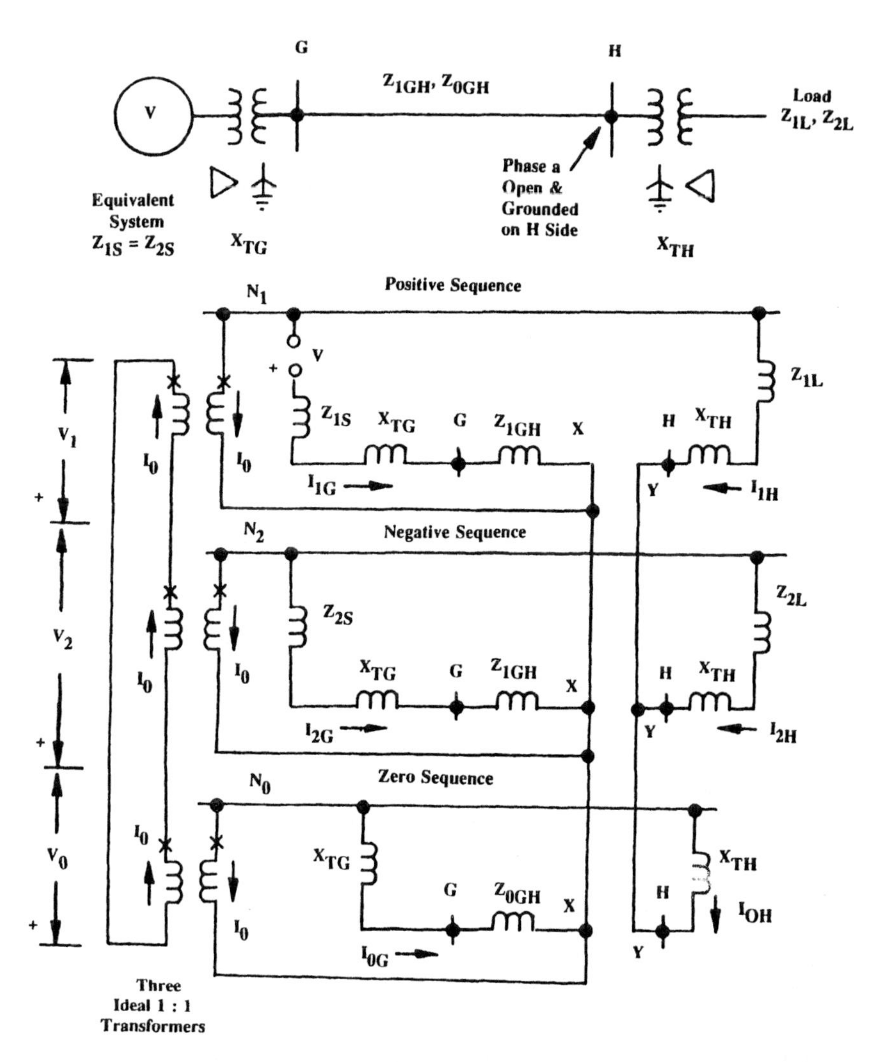

Figure 7.9 Sequence network interconnections for phase *a* open and grounded at bus *H* (broken conductor fallen to ground on the line side).

Adding the three equations above with $V_1 + V_2 + V_0 = 0$ gives

$$I_{1G}Z_{1G} + I_{2G}Z_{1G} + I_{0G}Z_{0G} = V \tag{7.49}$$

From the drops around the negative- and zero-sequence network, the following can be developed:

$$I_{1G}\frac{X_{TH} - Z_{2H}}{3} + I_{2G}\left(Z_{1G} + Z_{2H} + \frac{X_{TH} - Z_{2H}}{3}\right) - I_{0G}\left(Z_{0G} + X_{TH} - \frac{X_{TH} - Z_{2H}}{3}\right) = 0 \tag{7.50}$$

and from the drops around the positive- and negative-sequence networks,

$$I_{1G}\left(Z_{1G} + \frac{Z_{2H} + 2Z_{1H}}{3}\right) - I_{2G}\left(Z_{1G} + \frac{2Z_{2H} + Z_{1H}}{3}\right) + I_{0G}\frac{Z_{2H} - Z_{1H}}{3} = V \tag{7.51}$$

From Eqs. (7.49), through (7.51), I_{1G}, I_{2G}, and I_{0G} are determined from which the phase and neutral currents are calculated from Eqs. (4.5) through (4.7). These are shown in Fig. 7.10.

7.8 EXAMPLE: OPEN CONDUCTOR ON HIGH SIDE AND GROUND FAULT ON LOW SIDE OF A DELTA–WYE TRANSFORMER

Distribution substations transformers commonly are delta on the high-voltage side and wye-grounded on the low side. If a low-side ground fault occurs and the high-side fuse opens, the calculation of this simultaneous unbalance involves the wye–delta phase shifts as discussed in Section 4.13. An example is shown in Fig. 7.11 with the sequence networks and interconnections. For the delta–wye phase shifts, ideal (perfect) transformers are required. The transformer in the positive-sequence network

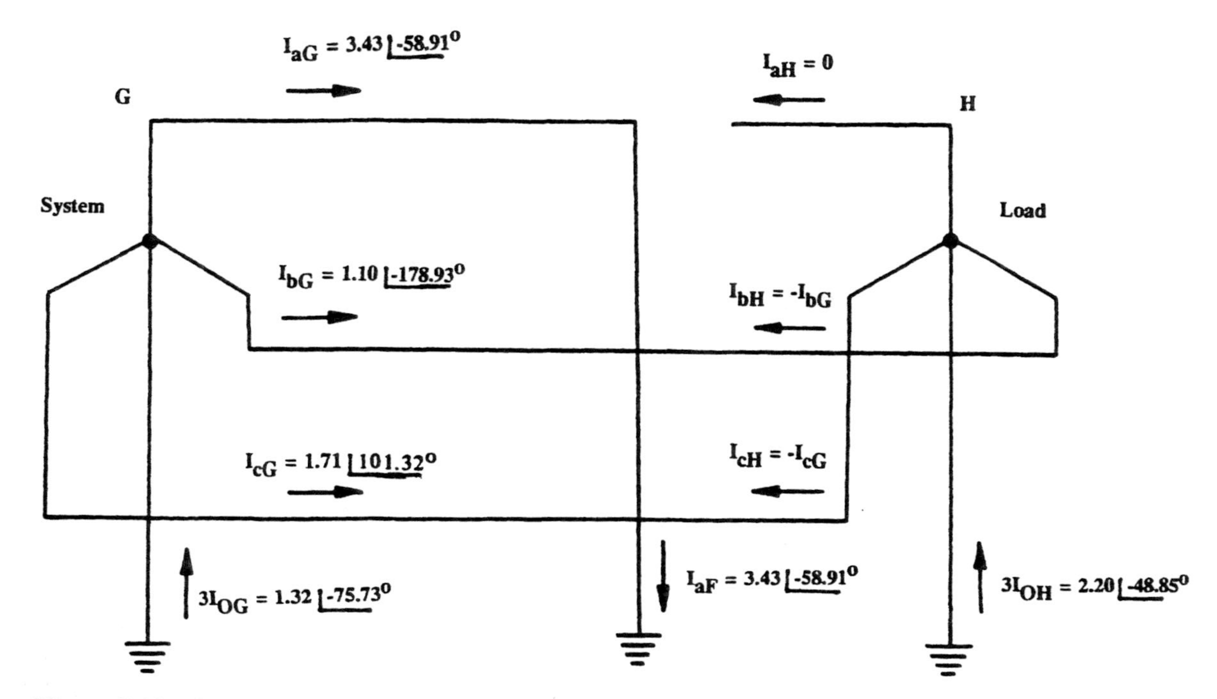

Figure 7.10 Per unit current distribution for an open phase *a* that falls to ground on the line side at bus *H* in the power system of Fig. 6.10. Network interconnection as in Fig. 7.9. Per unit at 30 MVA where 1 pu = 502 A at 34.5 kV.

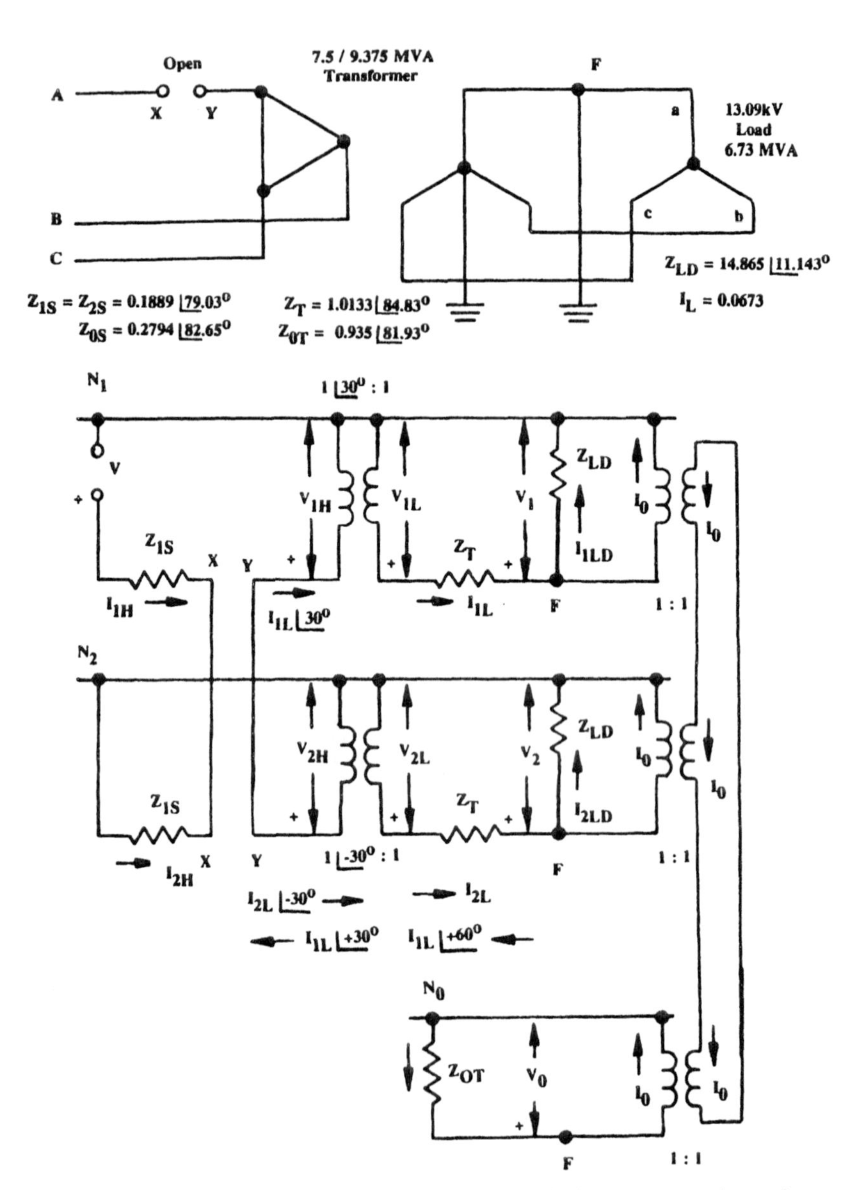

Figure 7.11 System and sequence network interconnections for a phase-*a*-to-ground fault on the 13.09-kV secondary and phase *a* open on the 115-kV primary of a 7.5-MVA delta–wye grounded transformer with 6.73 MVA load. Values are in per unit at 100 MVA, 115 or 13.09 kV.

shifts the secondary $+30°$, while the transformer in the negative-sequence network shifts the secondary $-30°$ according to ANSI standards. Thus

$$I_{1H} = I_{1L}\angle 30° \quad \text{and} \quad V_{1H} = V_{1L}\angle 30° \tag{7.52}$$

$$I_{2H} = I_{2L}\angle -30° \quad \text{and} \quad V_{2H} = V_{2L}\angle -30° \tag{7.53}$$

Also from Fig. 7.11,

$$-I_{2H} = I_{1H} \quad \text{and} \quad I_{2L}\angle -30° = -I_{1L}\angle 30° \tag{7.54}$$

The drops around the left positive-sequence network are

$$-V + I_{1H}Z_{1S} - I_{2H}Z_{1S} - V_{2H} + V_{1H} = 0$$

or

$$2Z_{1S}I_{1H} - V_{2H} + V_{1H} = V$$

$$2Z_{1S}I_{1L}\angle 30° - V_{2L}\angle -30° + V_{1L}\angle 30° = V \tag{7.55}$$

The drops around the right part of the positive-sequence network:

$$-V_{1L} + I_{1L}Z_T + I_{1LD}Z_{LD} = 0$$

$$V_{1L} = I_{1L}Z_T + I_{1L}Z_{LD} - I_0Z_{LD}$$

$$V_{1L} = I_{1L}(Z_T + Z_{LD}) - I_0Z_{LD} \tag{7.56}$$

The drops around the right part of the negative-sequence network:

$$-V_{2L} - I_{1L}Z_T\angle 60° + I_{2LD}Z_{LD} = 0$$

$$-V_{2L} - I_{1L}Z_T\angle 60° - I_{1L}Z_{LD}\angle 60° - I_0Z_{LD} = 0$$

$$-V_{2L} = I_{1L}(Z_T + Z_{LD})\angle 60° + I_0Z_{LD} \tag{7.57}$$

Substituting Eqs. (7.56) and (7.57) in Eq. (7.55) yields

$$2Z_{1S}I_{1L}\underline{/30^\circ} + I_{1L}(Z_T + Z_{LD})\underline{/60^\circ - 30^\circ} + I_0 Z_{LD}\underline{/-30}$$
$$+ I_{1L}(Z_T + Z_{LD})\underline{/30^\circ} - I_0 Z_{LD}\underline{/30^\circ} = V$$
$$2(Z_{1S} + Z_T + Z_{LD})I_{1L}\underline{/30^\circ} - I_0 Z_{LD}\underline{/90^\circ} = V \qquad (7.58)$$

Equation (7.58) is one with two unknowns. A second exists by the voltage drops of $V_1 + V_2 + V_0 = 0$. Thus

$$V_1 = I_{1LD}Z_{LD} = I_{1L}Z_{LD} - I_0 Z_{LD} \qquad (7.59)$$
$$V_2 = I_{2LD}Z_{LD} = -I_{1L}Z_{LD}\underline{/60^\circ} - I_0 Z_{LD} \qquad (7.60)$$
$$V_0 = -I_0 Z_{0T} \qquad (7.61)$$

Adding Eqs. (7.59) through (7.61) gives us

$$I_{1L}(Z_{LD} - Z_{LD}\underline{/60^\circ}) - I_0(2Z_{LD} + Z_{0T}) = 0 \qquad (7.62)$$

Solving for I_0 with $(1\underline{/0^\circ} - 1\underline{/60^\circ}) = 1\underline{/-60^\circ}$ yields

$$I_0 = I_{1L}\frac{Z_{LD}\underline{/-60^\circ}}{2Z_{LD} + Z_{0T}} \qquad (7.63)$$

Substituting Eq. (7.63) in Eq. (7.58) and solving for I_{1L}, we have

$$I_{1L} = \frac{V\underline{/-30^\circ}}{2(Z_{1S} + Z_T + Z_{LD}) - (Z_{LD})^2/(2Z_{LD} + Z_{0T})} \qquad (7.64)$$

Substituting numerical values from Fig. 7.11 in Eqs. (7.63) and (7.64) yields

$$I_{1L} = 0.04420\underline{/-43.04^\circ} \qquad (7.65)$$
$$I_0 = 0.02186\underline{/-104.72^\circ} \qquad (7.66)$$

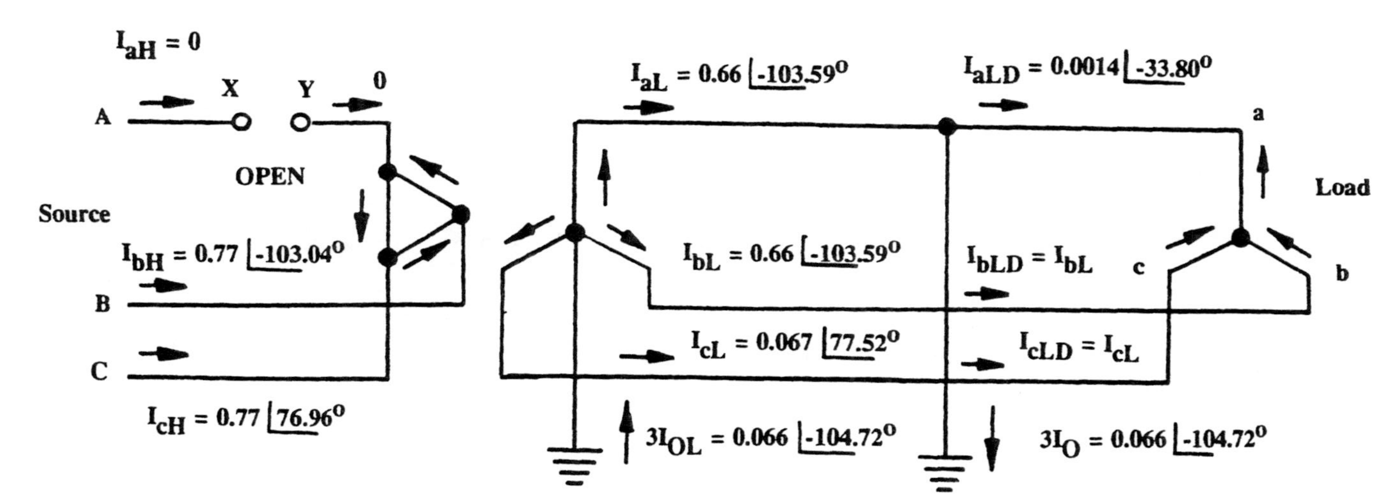

Figure 7.12 Per unit current distribution for the open 115-kV phase *a* and a phase-*a*-to-ground 13.09-kV fault for the system of Fig. 7.11. Source voltage is $1.027\angle 4.31^\circ$ pu to provide $1.0\angle 0^\circ$ pu voltage at the load. Values in per unit at 100 MVA, 115 kV, or 13.09 kV where 1 pu = 502 A at 115 kV or 4410.62 A at 13.09 kV.

From these two values the other quantities can be determined either from the equations above or from inspection of Fig. 7.11. The phase and neutral currents are documented in Fig. 7.12.

I_{aL}, I_{bL}, and I_{cL} of Fig. 7.12 are the sums of I_{1L}, I_{2L} and I_0 from Eqs. (4.5), (4.6), and (4.7). $3I_{0L} = 3I_0$; and I_{aLD}, I_{bLD}, and I_{cLD} are from I_{1LD} and $I_{2LD}(I_{0LD} = 0)$ using Eqs. (4.5), (4.6), and (4.7).

I_{AH}, I_{BH}, and I_{CH} are the sum of I_{1H} and I_{2H} ($I_{0H} = 0$). Alternatively,

$$I_{AH} = \frac{1}{\sqrt{3}}(I_{aL} - I_{bL}) \tag{7.67}$$

$$I_{BH} = \frac{1}{\sqrt{3}}(I_{bL} - I_{cL}) \tag{7.68}$$

$$I_{CH} = \frac{1}{\sqrt{3}}(I_{cL} - I_{aL}) \tag{7.69}$$

7.9 GROUND FAULT ON LOW SIDE OF A DELTA–WYE TRANSFORMER

For comparison with the simultaneous unbalances of Section 7.8, the currents for a phase-*a*-to-ground fault (no open phase) for the system of Fig. 7.11 are shown in Fig. 7.13. The 30° phase shifting transformers are not required, so the determination of the fault currents is the same, as illustrated in Section 6.8 and Fig. 6.11. The phase shift through the delta–wye transformer is not involved in the calculation of low-side faults unless the delta-side phase currents are desired.

The delta-side currents can be determined using the phase shifts of the positive- and negative-sequence currents as outlined in Section 4.13 or Eqs. (7.67) through (7.69). Similarly, the voltages on the high side can be determined by properly shifting the positive- and negative-sequence voltages.

The 30° ideal transformers could be used for the ground fault only, but generally this adds complications. They must be uti-

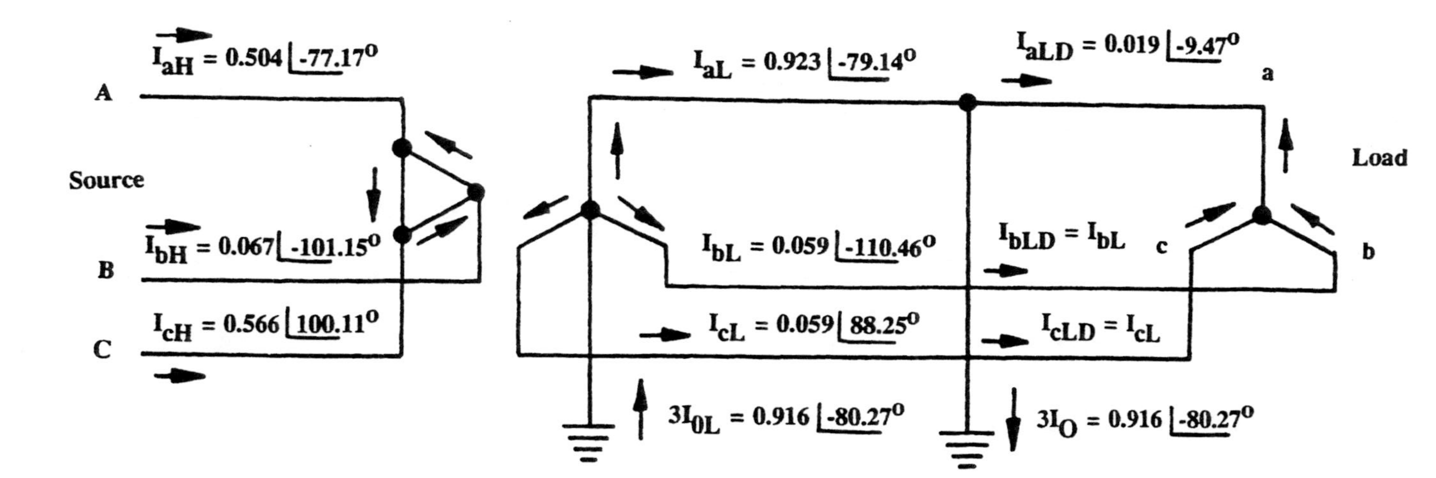

Figure 7.13 Per unit current distribution for phase-*a*-to-ground 13.09-kV fault for the system of Fig. 7.11. Source voltage is $1.027\angle 4.31^\circ$ pu to provide $1.0\angle 0^\circ$ pu voltage at the load. Values in per unit at 100 MVA, 115 kV, or 13.09 kV where 1 pu = 502 A at 115 kV or 4410.62 A at 13.09 kV.

lized for any unbalances occurring simultaneously on both sides of a transformer bank phase shift.

7.10 EXAMPLE: OPEN CONDUCTOR ON HIGH SIDE AND GROUND FAULT ON LOW SIDE OF A WYE-GROUNDED/DELTA-WYE-GROUNDED TRANSFORMER

Another application of the simultaneous unbalance technology of Fig. 7.5 with impedance between the two unbalances is shown in Fig. 7.14. This was an actual occurrence on a large utility system, and the impedances are from that system. A phase-*a*-

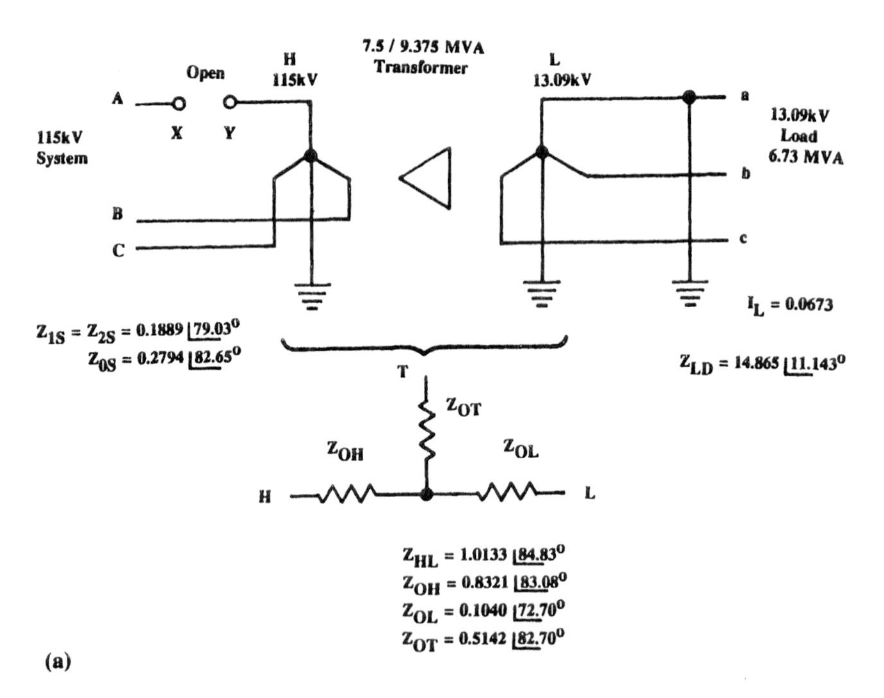

Figure 7.14 System and sequence network interconnections for a phase-*a*-to-ground fault on the 13.09-kV secondary and phase *a* open on the 115-kV primary of a 7.5-MVA wye-grounded/delta-wye-grounded transformer with 6.73 MVA load: (a) Power system, location

to-ground fault in the 13.09-kV system was accompanied by a blown phase *a* fuse in the 115-kV transformer primary. The 13.09-kV load at the time was 6.6 MW + 1.3 MVAR.

Figure 7.14b shows the sequence networks and their interconnections. Writing the drops around the positive- and negative-sequence networks, and around the positive- and zero-sequence networks, expressions for V_1, V_2, and V_0 (remembering

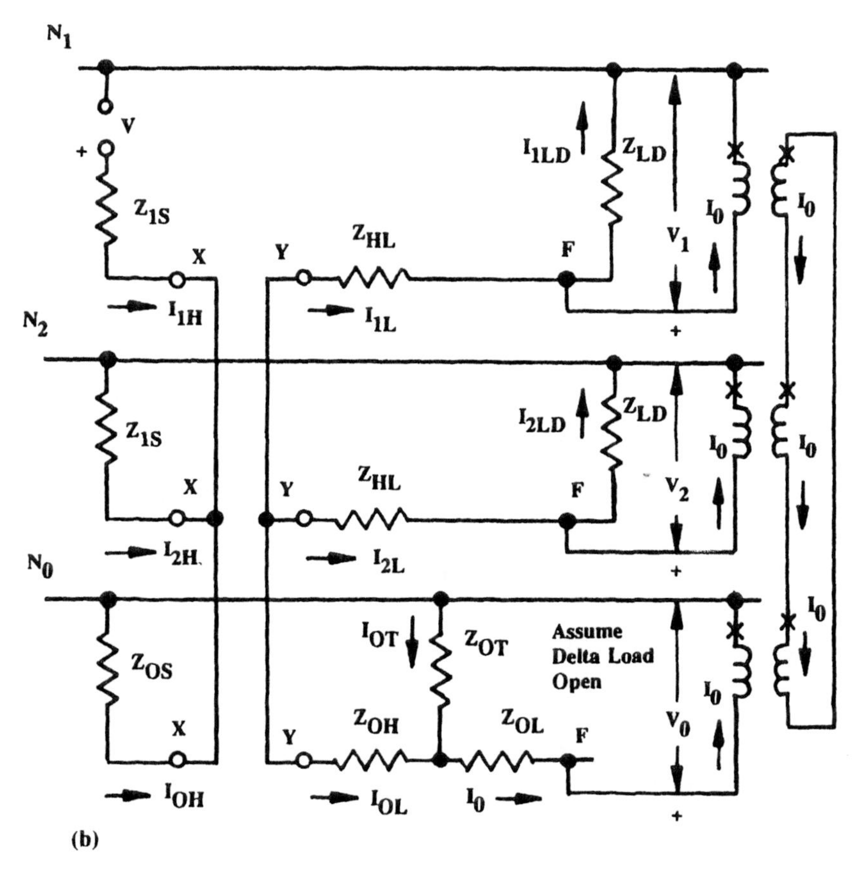

of the unbalances, and the system characteristic; (b) sequence networks and interconnections. Values are in per unit at 100 MVA, 115 or 13.09 kV.

that $V_1 + V_2 + V_3 = 0$), three equations with three unknowns, can be developed as follows:

$$I_{1LD}Z_X + I_0Z_Y + I_{0T}Z_Z = V \tag{7.70}$$

$$2I_{1LD}Z_{LD} - I_0Z_{0L} - I_{0T}Z_{0T} = V\frac{Z_{LD}}{Z_X} \tag{7.71}$$

$$2I_{1LD} + 3I_0 - I_{0T} = \frac{V}{Z_X} \tag{7.72}$$

where

$$Z_X = Z_{1S} + Z_{HL} + Z_{LD} \tag{7.73}$$

$$Z_Y = Z_{1S} - Z_{0S} - Z_{0H} + Z_{HL} \tag{7.74}$$

$$Z_Z = Z_{0S} + Z_{0T} + Z_{0H} \tag{7.75}$$

These can be solved by computer, by Eqs. (4.67), or by substitutions. The source voltage V to provide 1 pu load voltage is $V = 1.027\angle 4.31^\circ$ per unit. The results obtained are

$$I_{1LD} = 0.0385\angle -10.79^\circ \tag{7.76}$$

$$I_0 = 0.0874\angle -81.25^\circ \tag{7.77}$$

$$I_{0T} = 0.2652\angle -79.23^\circ \tag{7.78}$$

The other sequence quantities, and the various phase and ground values, can be obtained from the relations shown in Fig. 7.14 and basic equations (4.5) through (4.7). The currents are documented in Fig. 7.15.

The load impedance Z_{LD} can be omitted (no load), as can be observed from Fig. 7.14. This simplifies the calculations. The no-load current values are shown in parentheses in Fig. 7.15. For comparison purposes Fig. 7.16 shows the currents in the system of Fig. 7.14 for a solid 13.09-kV phase-*a*-to-ground fault without the open phase. For further comparisons the impedance Z_{HL} of the Fig. 7.14 transformer is the same as the Z_T of the Fig. 7.11 transformer.

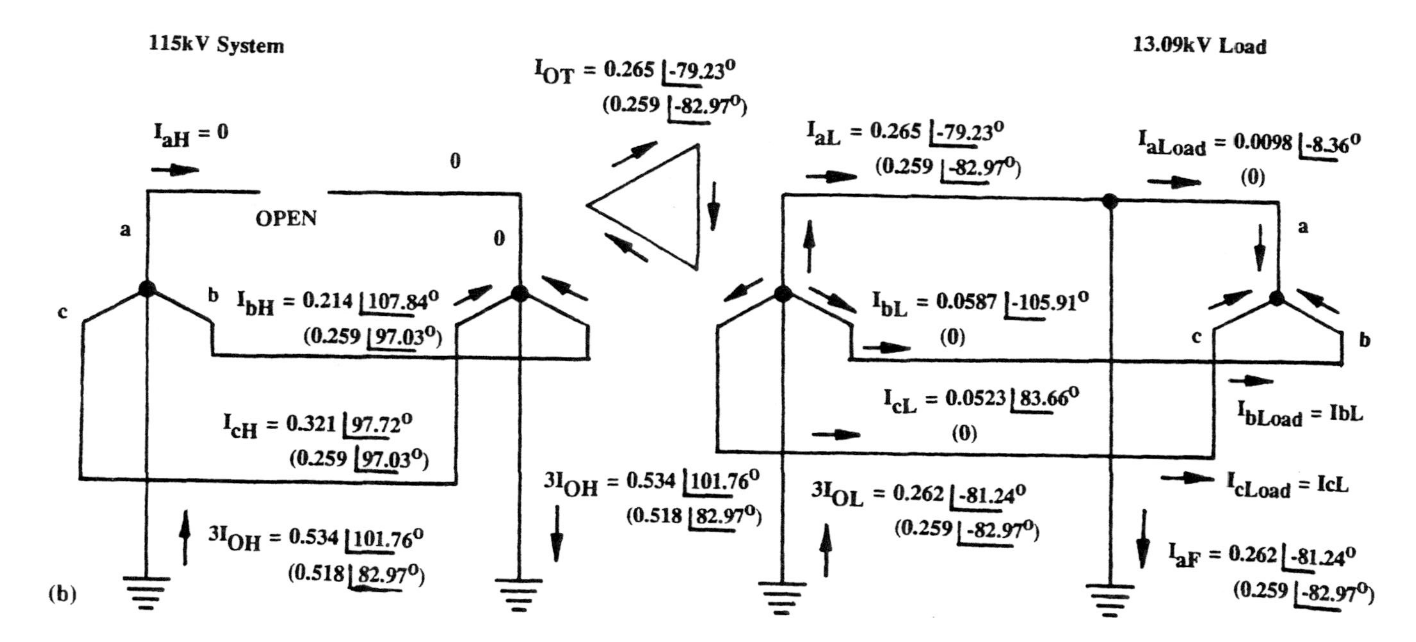

Figure 7.15 Per unit current distribution for the open 115-kV phase *a* and a phase-*a*-to-ground 13.09-kV fault for the system shown in Fig. 7.14. Upper values are per unit currents with 6.73 MVA load on the 13.09-kV side. Source voltage is $1.027\angle 4.31°$ pu to provide $1.0\angle 0°$ pu at the load. Lower values are with load neglected (no load) and source voltage $1.0\angle 0°$. Values at 100 MVA, 115 or 13.09 kV, where 1 pu = 502 A at 115 kV, or 4410.62 A at 13.09 kV.

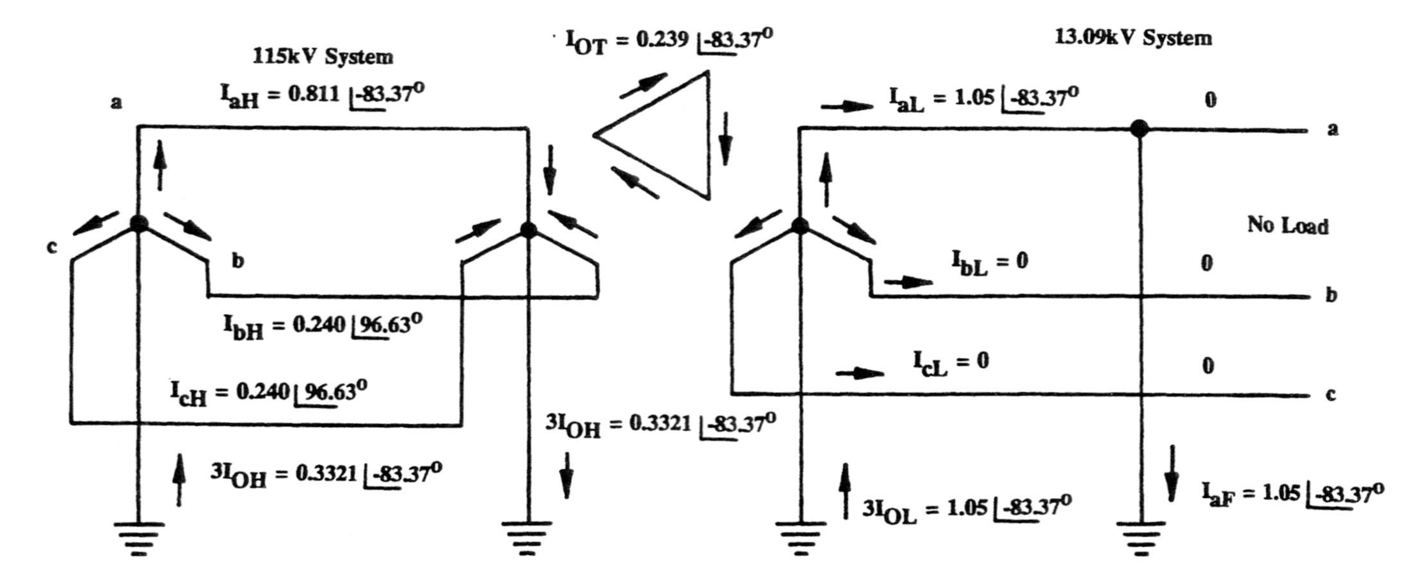

Figure 7.16 Per unit current distribution for the system of Fig. 7.14 without the 115-kV phase open. Source voltage is $1.0\angle 0^\circ$ pu. Values at 100 MVA, 115 or 13.09 kV, where 1 pu = 502 A at 115 kV or 4410.62 A at 13.09 kV.

7.11 GROUND FAULT CALCULATION FOR A MID-TAPPED GROUNDED DELTA SECONDARY TRANSFORMER

Transformers with mid-tapped delta secondaries are used in distribution and industrial applications to supply power at two voltages, such as 240 and 120 V. Usually, one of the mid-taps is grounded. The calculation for a phase-to-ground fault on one of the phases of the delta is complicated by two sets of unbalances: the unbalance of the ground and the unbalance of the ground fault. Thus at least two sets of ideal (perfect) transformers are required to provide isolation and necessary voltage/phase angle shifts.

Several cases will be covered by examples. The *Westinghouse Distribution Reference Book*, chapter 6, pp. 232–234 presented one example which is somewhat difficult to follow. The following examples may be clearer. As a simplification and convenience, the transformer impedances in the Westinghouse book, Fig. 30, will be used. These are $j20$ ohms primary (the $\frac{1}{3}$ converts the delta value to equivalent line to neutral) and $j\frac{1}{4}$ secondary with $N = 480/240 = 2$. Only the reactance was used, load was neglected, and the source was considered infinite ($Z_{1S} = 0$). The determination of impedances for this type of transformer is given in Sec. 9.14. At this voltage level the resistance usually is significant and normally should be considered.

Consider several cases;

Case 1. Delta–Delta Bank, a' Mid-tap Grounded, Secondary Phase a-to-Ground Fault

The circuit is shown in Fig. 7.17, where the transformers in the positive, negative, and zero sequence networks provide isolation with ratios and phase shifts as required. Since the fault is on phase a, the upper transformer provides only a 2:1 ratio for the 480/240 voltages. The lower transformer in the positive sequence network has a ratio of 4:1 (480/120) with a shift of $-60°$ since the a' ground to neutral voltage lags 60°. Thus V_x' is 60° lag $= -V_x'$ at $+120$ on the a' secondary.

In the negative sequence network similar conditions exist except that the shifts are opposite: $+60°$ instead of $-60°$, etc., as shown.

With the fault at the transformer terminals, $Z_{1\ \text{line}}$ and $Z_{0\ \text{line}}$ are zero, so $V'_x = V_x$, $V'_y = V_y$, and $V'_0 = V_0$. The drops around the positive sequence primary circuit are

$$4V_x\underline{/0°} + I_x(\tfrac{1}{2}\underline{/0°} - \tfrac{1}{4}\underline{/+60°})(j20/3) = 480/\sqrt{3}\underline{/0°}$$

$$V_x = 69.282 + 0.722I_x\underline{/240°} \tag{7.79}$$

The drops around the negative sequence primary circuit are

$$4V_y\underline{/0°} + I_x(\tfrac{1}{2}\underline{/0°} - \tfrac{1}{4}\underline{/-60°})(j20/3) = 0$$

$$V_y = 0.722I_x\underline{/-60°} \tag{7.80}$$

The drops around the a positive, negative, and zero sequence circuits are $V_0 = -NV_x - NV_y = -2V_x - 2V_y$ with $Z_{1\ \text{line}}$ and Z_0 line $= 0$. Then the drops around the a' positive, negative,

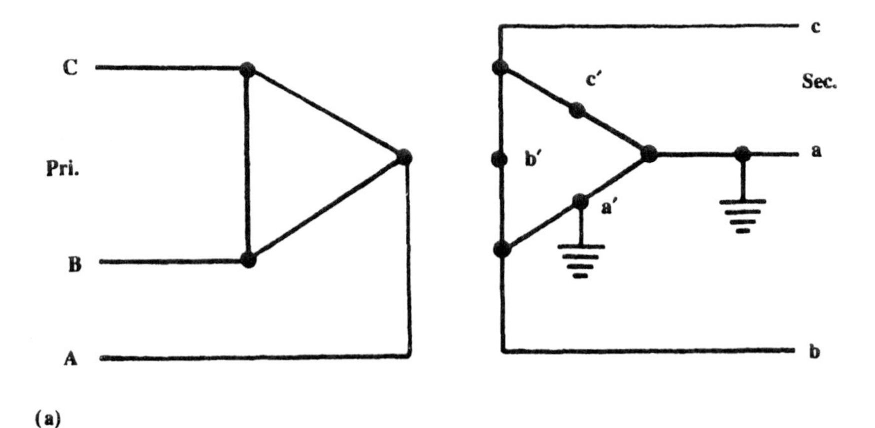

Figure 7.17 Equivalent circuit for a phase a-to-ground fault on a delta–delta distribution transformer with a' secondary mid-tap grounded. (a) Three-line transformer diagram; (b) the equivalent circuit, case 1.

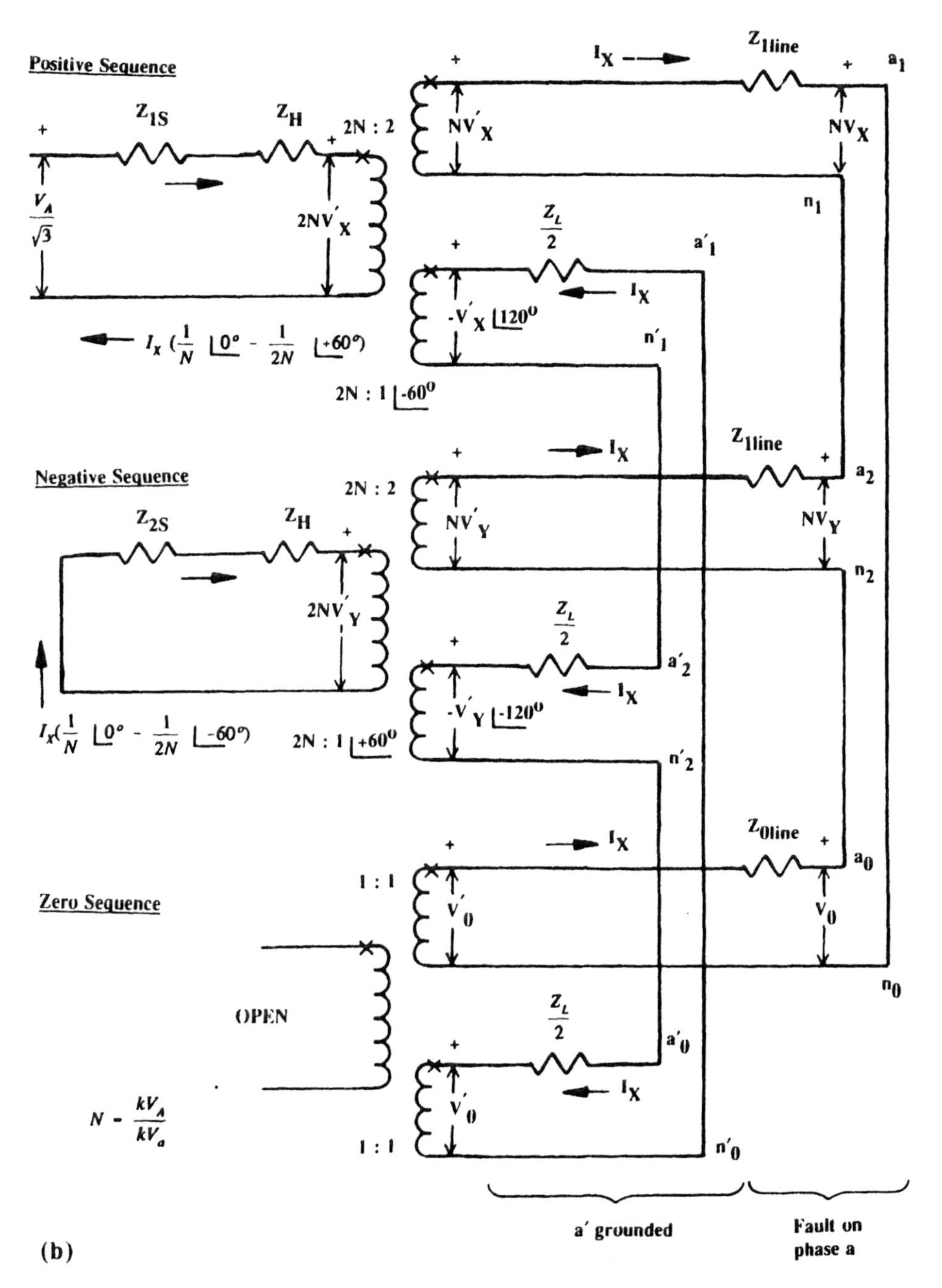

(b)

Figure 7.17 (*Continued*)

and zero sequence circuits are

$$j\tfrac{1}{4}I_x - V_x\angle{+120°} + j\tfrac{1}{4}I_x - V_y\angle{-120°} + \tfrac{1}{4}I_x + V_0 = 0$$

$$j0.75I_x - V_x\angle{+120°} - V_y\angle{-120°} - 2V_x - 2V_y = 0$$

$$j0.75I_x - 1.732\angle{+30°}\,V_x - 1.732\angle{-30°}\,V_y = 0 \qquad (7.81)$$

Substitute Eqs. (7.79) and (7.80) in Eq. (7.81) and solve for I_x:

$$\begin{aligned} I_x &= 36.92\angle{-60°} \text{ A secondary} \\ 3I_x &= 110.77\angle{-60°} \text{ A in the fault} \end{aligned} \qquad (7.82)$$

In the primary

$$I_{A1} = I_x(\tfrac{1}{2} - \tfrac{1}{4}\angle{+60°}) = 0.433I_x\angle{-30°} = 16.0\angle{-90°} \text{ A}$$

$$I_{A2} = I_x(\tfrac{1}{2} - \tfrac{1}{2}\angle{-60°} = 0.433I_x\angle{+30°} = 16.0\angle{+30°} \text{ A}$$

The primary phase currents are

$$I_A = I_{A1} + I_{A2} = 27.69\angle{-60°} \text{ A} \qquad (7.83)$$

$$I_B = a^2I_{A1} + aI_{A2} = 27.69\angle{120°} \text{ A} \qquad (7.84)$$

$$I_C = aI_{A1} + a^2I_{A2} = 0 \qquad (7.85)$$

The distribution of these fault currents is shown in Fig. 7.18.

Case 2. Wye–Delta Bank, *a′* Mid-tap Grounded, Secondary Phase *a*-to-Ground Fault

With the primary in wye rather than delta and assuming that the transformer impedances are the same, the primary phase voltages will be 30°, leading from those in Fig. 7.17, as shown in Fig. 7.19. Thus, the positive sequence on the wye side leads +30° and the negative sequence lags −30° as discussed in Sec.

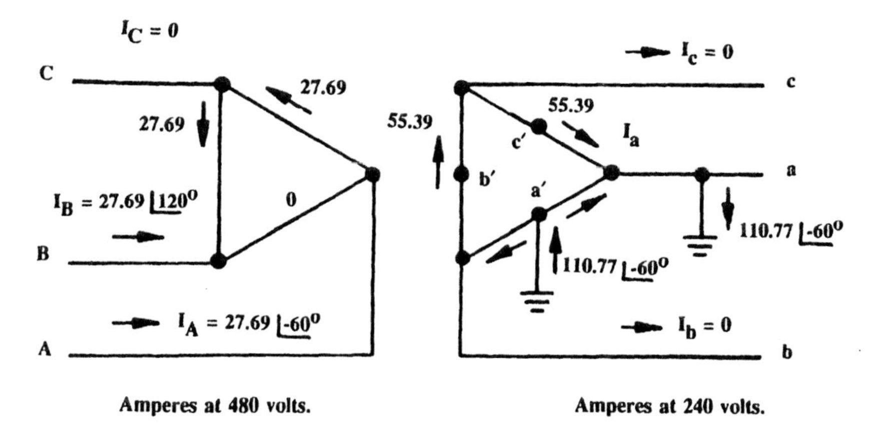

Figure 7.18 Fault current distribution for the example of case 1.

4.13. In the positive sequence primary with the values shifted +30° from those in Fig. 7.17,

$$4V_x\angle 30^\circ + I_x(\tfrac{1}{2}\angle 30^\circ - \tfrac{1}{4}\angle +90^\circ)(j20/3) = 480/\sqrt{3}\angle 30^\circ$$

$$V_x = 69.282 + 0.722I_x\angle 240^\circ \qquad (7.86)$$

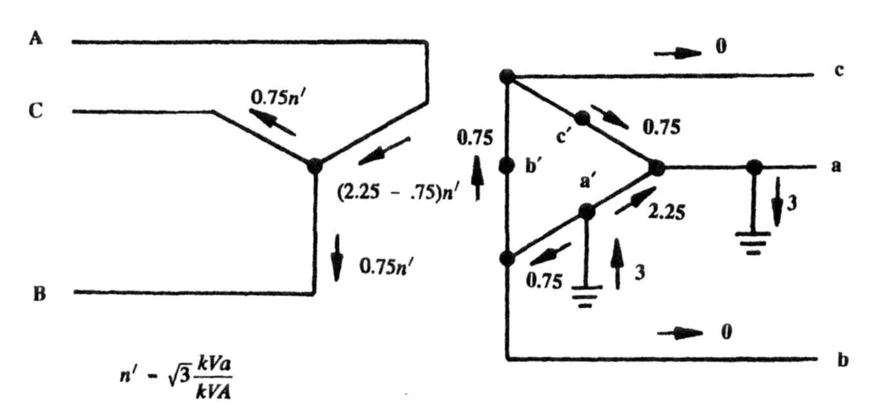

Figure 7.19 The wye primary mid-tapped delta secondary transformer with per unit current distribution for a-to-ground fault with a' grounded, case 2.

The drops around the negative sequence primary with the values shifted $-30°$ from those in Fig. 7.17,

$$4V_y + I_x(\tfrac{1}{2}\underline{/-30°} - \tfrac{1}{4}\underline{/-90°})(j20/3) = 0$$

$$V_y = 0.722I_x\underline{/-60°} \tag{7.87}$$

Substituting Eqs. (7.86) and (7.87) in Eq. (7.81) and solving for I_x gives

$$I_x = 36.92\underline{/-60°} \text{ A secondary}$$

as in Eq. (7.82), and the fault current

$$3I_x = 110.77\underline{/-60°} \text{ A in the fault}$$

With the same impedances for the bank, the fault current is the same as for case 1.

The wye-connected primary results in a different pattern of fault current flow, which is shown in per unit of I_x in Fig. 7.19. In the *ab* delta secondary with *a'* grounded, the net per unit current of $2.25 - 0.75 = 1.50$ must be one-half of the 0.75 per unit currents in the *bc* and *ca* windings. These are reflected in the primary wye as 1.3 per unit in phase *A* and 0.65 in phases *B* and *C*. n' is the ratio of the secondary and primary winding normal voltages; $n' = 240/480/\sqrt{3}$ or $240/277.13 = 0.866$.

Alternatively, in the primary,

$$I_{A1} = 1.5I_x(\tfrac{1}{2}\underline{/+30°} - \tfrac{1}{4}\underline{/+90°}) = 0.65I_x = 23.98\underline{/-60°}$$

$$I_{A2} = 1.5I_x(\tfrac{1}{2}\underline{/-30°} - \tfrac{1}{4}\underline{/-90°}) = 0.65I_x = 23.98\underline{/-60°}$$

and the primary phase currents are

$$I_A = I_{A1} + I_{A2} = 47.98\underline{/-60°} \text{ A} \tag{7.88}$$

$$I_B = a^2I_{A1} + aI_{A2} = 23.98\underline{/-60°} \text{ A} \tag{7.89}$$

$$I_C = aI_{A1} + a^2I_{A2} = 23.98\underline{/-60°} \text{ A} \tag{7.90}$$

The distributions of the currents are shown in Fig. 7.20.

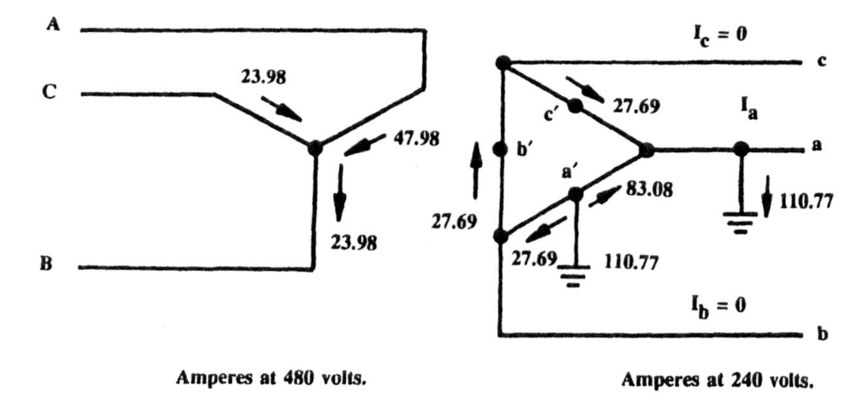

Figure 7.20 Fault current distribution for the example of case 2.

Case 3. Delta–Delta Bank, a' Mid-tap Grounded, Secondary Phase c-to-Ground Fault

Moving the ground fault to phase c requires another set of ideal transformers with phase shifting ratios as shown at the far right of Fig. 7.21.

The drops around the primary positive sequence circuit are

$$4V_x\underline{/-120^\circ} + I_x(\tfrac{1}{2}\underline{/-120^\circ} - \tfrac{1}{4}\underline{/+60^\circ})(j20/3) = 480/\sqrt{3}\underline{/+0^\circ}$$

$$V_x = 69.28\underline{/+120^\circ} + 1.25I_x\underline{/-90^\circ} \tag{7.91}$$

The drops around the primary negative sequence circuit are

$$4V_y\underline{/+120^\circ} + I_x(\tfrac{1}{2}\underline{/+120^\circ} - \tfrac{1}{4}\underline{/-60^\circ})(j20/3) = 0$$

$$V_y = 1.25I_x\underline{/-90^\circ} \tag{7.92}$$

In the a positive, negative, and zero sequence circuits,

$$V_0 = -NV_x - NV_y = -2V_x - 2V_y$$

Since $Z_{1\text{ line}}$ and $Z_{0\text{ line}}$ are zero with the fault at the transformer terminal, $V_x' = V_x$, $V_y' = V_y$ and $V_0' = V_0$. The drops around

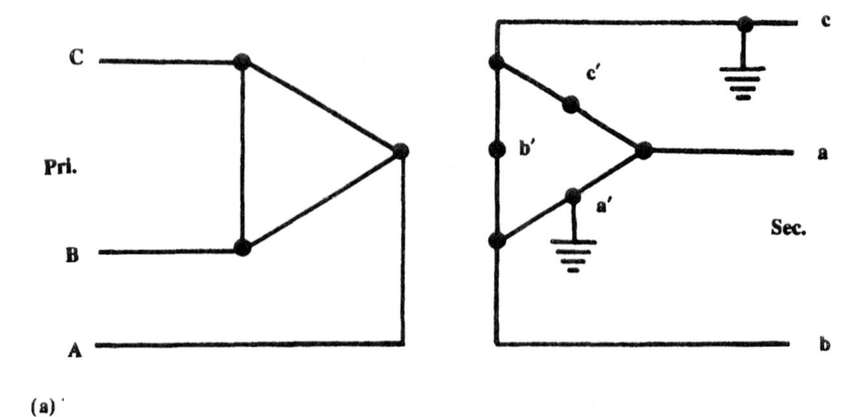

Figure 7.21 Equivalent circuit for a phase c-to-ground fault on a delta–delta distribution transformer with a' secondary mid-tap grounded, case 3.

the a' positive, negative, and zero sequence circuits are

$$\begin{aligned} +j\tfrac{1}{4}I_x - V_x + j\tfrac{1}{4}I_x - V_y + j\tfrac{1}{4}I_x - 2V_x - 2V_y &= 0 \\ +j0.75I_x - 3V_x - 3V_y &= 0 \end{aligned} \tag{7.93}$$

Substituting Eq. (7.91) and (7.92) in Eq. (7.93) and solving for I_x,

$$I_x = 25.19\angle{+30^\circ}\ \text{A} \tag{7.94}$$

The fault current is $3I_x = 75.57\angle{+30^\circ}$ A.

On the primary,

$$I_{A1} = I_x(\tfrac{1}{2}\angle{-120^\circ} - \tfrac{1}{4}\angle{+60^\circ}) = 18.9\angle{-90^\circ}\ \text{A}$$

$$I_{A2} = I_x(\tfrac{1}{2}\angle{+120^\circ} - \tfrac{1}{4}\angle{-60^\circ}) = 18.9\angle{+150^\circ}\ \text{A}$$

from which

$$I_A = I_{A1} + I_{A2} = 18.9\angle{-150^\circ}\ \text{A} \tag{7.95}$$

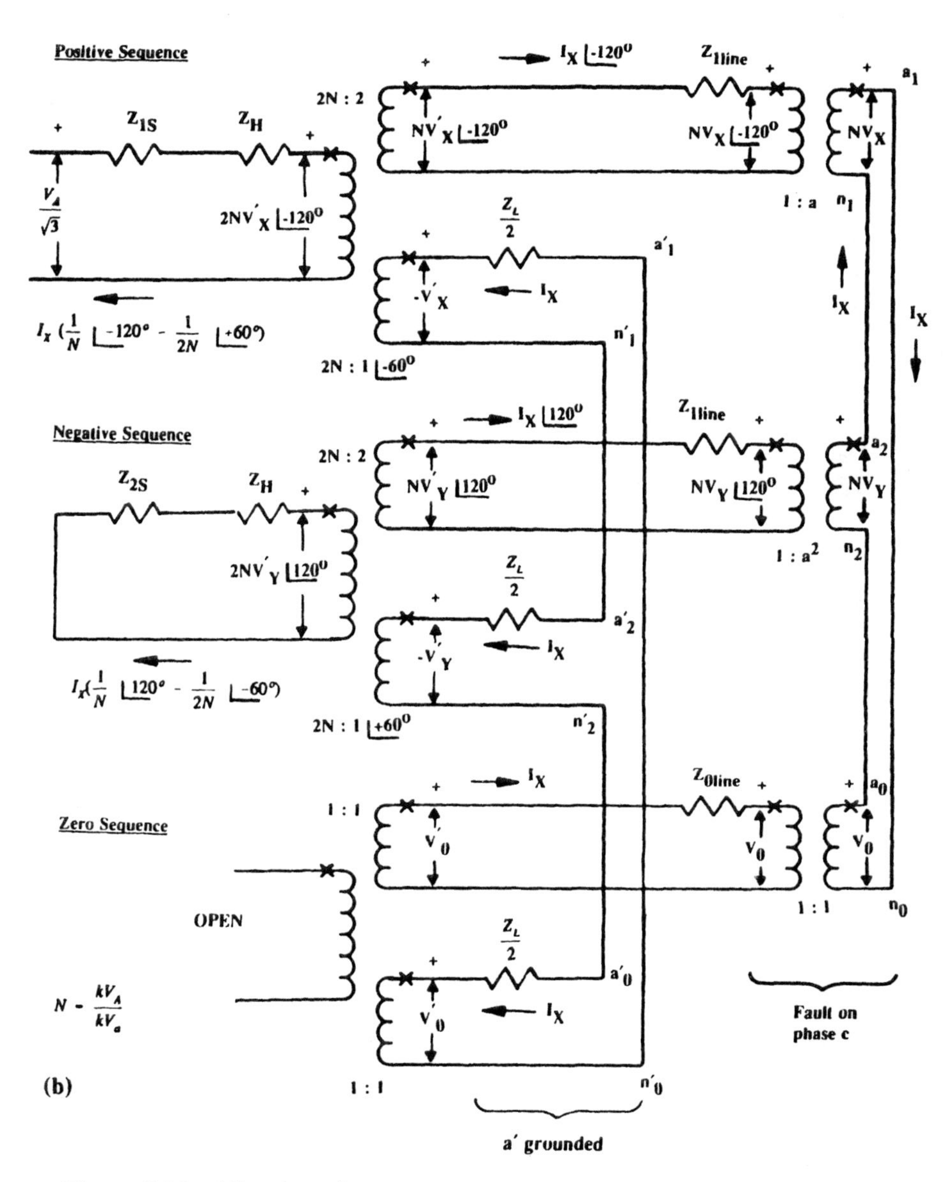

Figure 7.21 (*Continued*)

$$I_B = a^2 I_{A1} + a I_{A2} = 18.9\underline{/-150^\circ} \text{ A} \quad (7.96)$$

$$I_C = a I_{A1} + a^2 I_{A2} = 37.8\underline{/+30^\circ} \text{ A} \quad (7.97)$$

These currents are shown in Fig. 7.22. They are the same values as in the *Westinghouse Distribution Reference Book*, (Fig. 30, p. 234) except that the fault and grounding are in different locations.

Case 4. Wye-Delta Bank, *a'* Mid-tap Grounded, Secondary Phase *c*-to-Ground Fault

With the primary in wye rather than delta and assuming that the transformer impedances are the same, the primary phase voltages will be 30° leading from those in Fig. 7.21. Thus, the positive sequence on the wye primary leads +30° and the negative sequence lags −30° (Sec. 4.13). (See Fig. 7.23.)

In the positive sequence primary with the values shifted +30° Eq. (7.91) becomes

$$4V_x\underline{/-90^\circ} + I_x(\tfrac{1}{2}\underline{/-90^\circ} - \tfrac{1}{2}\underline{/+90^\circ})(j20/3) = 277.13\underline{/+30^\circ}$$

$$V_x = 1.25 I_x\underline{/-90^\circ} + 69.28\underline{/+120^\circ} \quad (7.98)$$

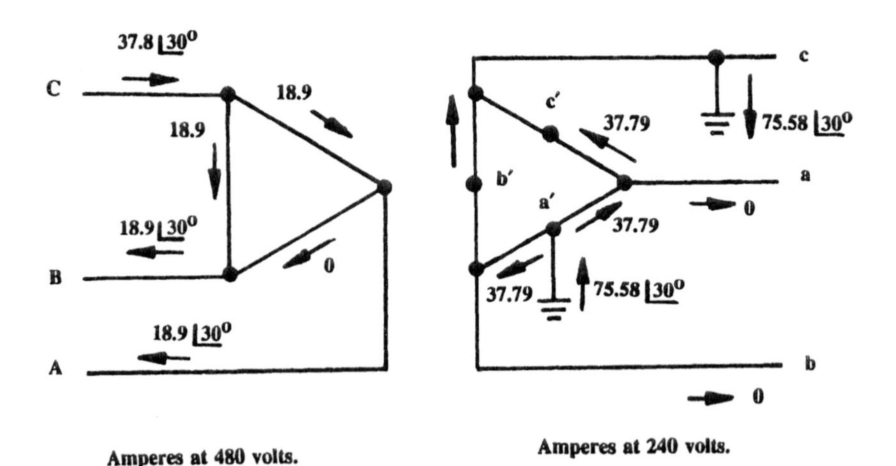

Figure 7.22 Fault current distribution for the example of case 3.

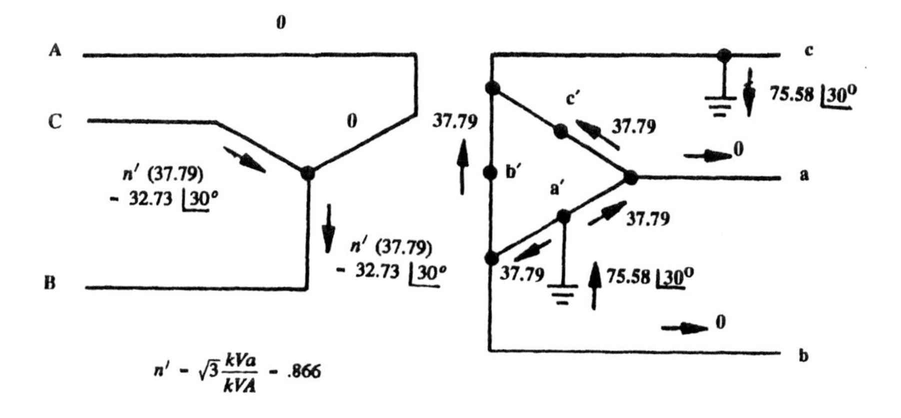

Figure 7.23 Fault current distribution for the example of case 4.

Similarly, the negative sequence primary is shifted $-30°$ from Eq. (7.92) and is

$$4V_y\underline{/+90°} + I_x(\tfrac{1}{2}\underline{/+90°} - \tfrac{1}{4}\underline{/-90°})(j20/3) = 0$$

$$V_y = 1.25I_x\underline{/-90°} \tag{7.99}$$

The drops around the a' positive, negative, and zero sequence network are as in Eq. (7.93). Substituting Eqs. (7.98) and (7.99) in (7.93) and solving for I_x gives

$$I_x = 25.19\underline{/+30°}\text{ A} \tag{7.100}$$

and a fault current of $3I_x = 75.58\underline{/+30°}$ A, as in case 3.

The primary currents are

$$I_{A1} = I_x(\tfrac{1}{2}\underline{/-90°} - \tfrac{1}{4}\underline{/+90°}) = 18.9\underline{/-60°}\text{ A}$$

$$I_{A2} = I_x(\tfrac{1}{2}\underline{/+90°} - \tfrac{1}{4}\underline{/-90°}) = 18.9\underline{/+120°}\text{ A}$$

and the primary currents are

$$I_A = I_{A1} + I_{A2} = 0 \tag{7.101}$$

$$I_B = a^2 I_{A1} + aI_{A2} = 32.73\underline{/-150^\circ} \text{ A} \tag{7.102}$$

$$I_C = aI_{A1} + a^2 I_{A2} = 32.73\underline{/+30^\circ} \text{ A} \tag{7.103}$$

The problem with the primary currents in case 2 does not occur here because the currents in transformer secondary *ab* add to zero.

7.12 SUMMARY

The techniques of calculating unbalances such as an open phase (broken conductor or blown fuse) with and without a simultaneous shunt fault have been presented with a number of examples. By utilizing the same power system with different combinations of unbalances, comparisons can be made.

In general, series and simultaneous unbalance calculations are more tedious unless they are programmed in the computer fault study. Load is a major factor for all open phases. As a result, the currents resulting from the open phase are small and in the order of load current. $r + jx$ calculations usually are required, as load is mostly resistive while the power system is mostly reactive.

Ground current ($3I_0$) may reverse from normal when the open phase in a feeder circuit is between a ground fault and a grounded power system (Figs. 7.4a and 7.7a). In these cases the current is *down* the power system neutral, and in Fig. 7.7a the current in the fault is smaller than either transformer neutral currents. Unusual!

In Fig. 7.15, $3I_{0H}$ is shown at angle 101.76°. Reversing the current gives an angle of $-78.24°$ more compatible with the $3I_{0L}$ angle of $-82.24°$. Load had more impact on the currents for the Fig. 7.7a case than for the Fig. 7.15 case. Hence load effect for simultaneous unbalances will vary with the system as well as the load level. A ground fault without an open phase (Figs. 7.13 and 7.16) provides significantly higher fault currents as expected.

An open phase on one side of the ground fault is a different case from the open on the other side. This is illustrated in Figs. 7.7a and 7.10. Methods and examples of simultaneous unbalances on different sides of a delta–wye transformer are presented. The results for a system of Fig. 7.11 are shown in Fig. 7.12. Figure 7.13 is the same system without the open phase for comparison.

Four cases are given for ground faults on the delta secondary of distribution transformers where one of the secondary midtaps is grounded. These are simultaneous unbalance examples. Figures 7.18, 7.20, 7.22, and 7.23 show the fault currents in the primary and secondary of a typical transformer for wye–delta and delta–delta connections for two fault locations.

8

Overview of Sequence Currents and Voltages During Faults

8.1 INTRODUCTION

Faults and sequence quantities can be visualized and perhaps better understood by several "helicopter" views in contrast to the specific representations and calculations. Accordingly, several overview "flights" are presented for the shunt faults, which are the most common occurring types.

The faults are for reactive systems (resistance neglected), no load, neglecting fault resistance and assuming that $X_1 = X_2 = X_0$. As has been indicated previously, X_0 is never equal to X_1 or X_2, but this simplification with the others make presentation easier and does not change the trends or basic fundamentals.

8.2 VOLTAGE AND CURRENT PHASORS FOR SHUNT FAULTS

The collapse of voltage and increase of current for the four shunt faults experienced in a power system are shown in Fig. 8.1. Dia-

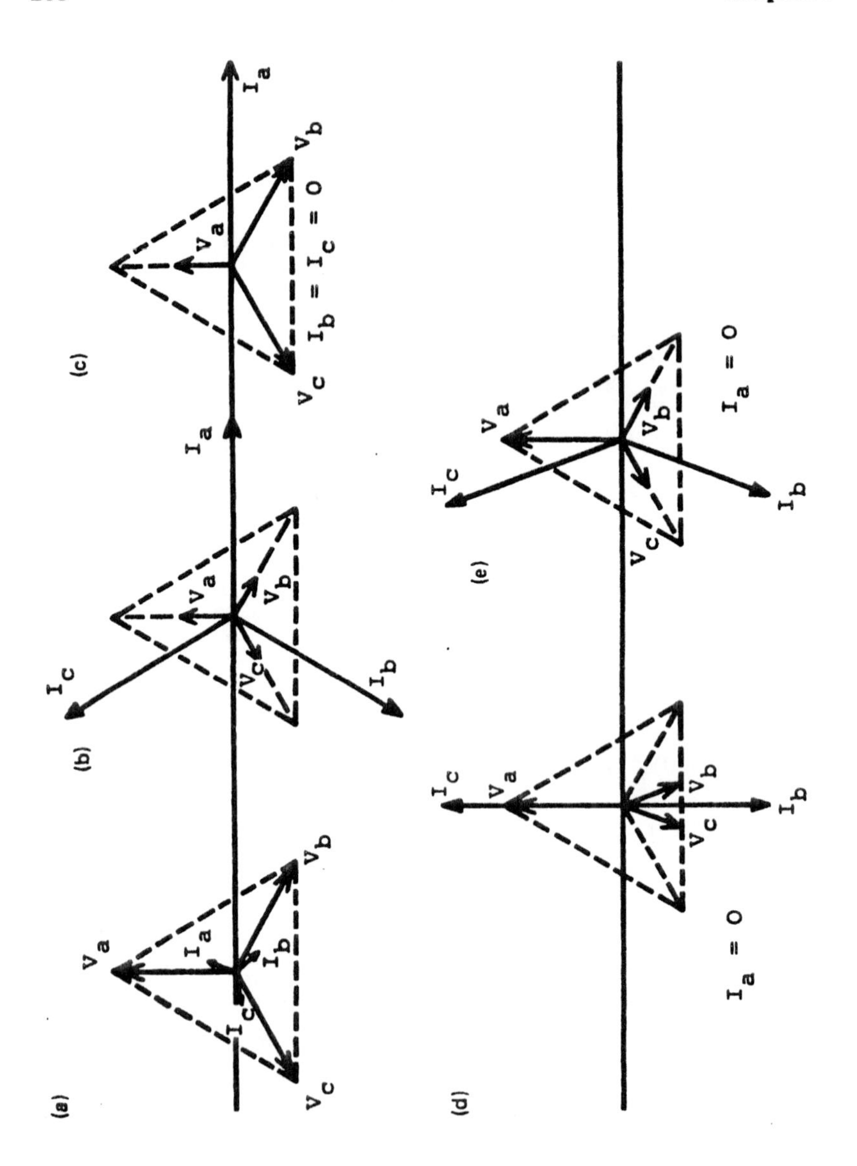

(a)
V_a
V_b
V_c
I_a
I_b
I_c
(b)
I_c
V_a
V_b
V_c
I_b
I_a
(c)
V_a
V_b
V_c
I_a
$I_b = I_c = 0$
(d)
I_c
V_a
V_b
V_c
I_b
$I_a = 0$
(e)
I_c
V_a
V_b
V_c
I_b
$I_a = 0$

gram (a) is for normal operation; all voltages are normal and symmetrical (equal in magnitude and 120° displaced); small load currents are symmetrical and normally from unity power factor to lagging up to around 30°. With capacitors at light load, the currents may lead slightly.

When faults occur the internal voltage of the generators or synchronous equipment does not change: that is, unless the fault is left on far too long and the voltage regulators act to increase the fault-reduced terminal voltage. Faults cause instantaneous changes in the voltages and current.

A three-phase fault (Fig. 8.1b) reduces all three phase voltages and causes a large symmetrical increase in the three phase currents usually at a highly lagging angle. In the figure the current is shown lagging 90° for the X-only system. Otherwise, the angle is determined by the angle of the total positive-sequence impedance to the fault (angle of Z_{1F} and Z_F if applicable in Fig. 5.2).

The most common single-phase-to-ground fault (Fig. 8.1c) is for the fault on phase a, so the phase a voltage collapses and the phase a current increases and lags to 90° for the assumed X system. The angle of I_a is determined by the positive-, negative-, and zero-sequence impedances to the fault point (angles of Z_{1F}, Z_{2F}, and Z_{0F} and Z_F if applicable in Fig. 5.4).

With load current neglected, $I_b = I_c = 0$; otherwise, small currents will flow in these unfaulted phases. As indicated previously, fault current will flow in the unfaulted phases on loop-type systems where the current distribution for the three sequence networks are different (see Figs. 6.8 and 6.9).

The phase-to-phase fault (phase-b-to-phase-c) is illustrated in Fig. 8.1d. Neglecting load V_a is normal and $I_a = 0$. V_b and V_c

←

Figure 8.1 Typical current and voltage phasors for common shunt faults. Currents shown for 90° lagging or for a power system with $Z = X$, neglecting resistance. Load current is assumed negligible. (a) Normal phasors for a balanced system; (b) three-phase fault; (c) single-phase-to-ground fault (a-gnd.); (d) phase-to-phase fault (bc fault); (e) two-phase-to-ground fault (bc-gnd.).

collapse from their normal position to small vertical phasors where $V_{bc} = 0$ for a solid fault. I_b and I_c are equal and opposite and at 90° for the X system. Otherwise, the angle of the currents is determined by the fault impedances as shown in Fig. 5.5.

The two-phase-to-ground fault (phase-*b*-to-phase-*c*-to-ground; Fig. 8.1e) results in the faulted phase voltages collapsing along their normal position until for a solid fault, $V_b = V_c = 0$. This is in contrast with the solid phase-to-phase fault, where $V_b = V_c = V_{\phi N}$, as shown in Fig. 8.1d. $I_a = 0$ for no load. I_b and I_c will be in the general area as shown. The angle of these currents is a function of the sequence impedances to the fault (Fig. 5.6). An increasing amount of zero-sequence current will cause I_b and I_c to swing closer to each other; contrarily a very low zero-sequence (high-Z_0) current component will result in the phasors approaching the phase-to-phase fault of Fig. 8.1d. This can be seen from the sequence network connections of Fig. 5.6. If Z_0 becomes infinite (essentially ungrounded system), the interconnections become as a phase-to-phase fault (Fig. 5.5). On the other hand, for a very solidly grounded system, Z_0 approaches zero relative to Z_1 and Z_2. If Z_0 were zero, the negative-sequence network would be shorted out as for a three-phase fault (Fig. 5.2).

In some parts of a loop system it is possible for the zero-sequence current to flow in the opposite direction from the positive- and negative-sequence currents. In this case I_c may lag V_a rather than lead it. Correspondingly, I_b will lead the position shown.

In summary:

1. Three-phase faults are characterized by all three phase voltages reduced, and all three phase currents increased and are generally more lagging. There is no $3I_0$ current.
2. Single-phase-to-ground faults have one phase voltage collapsed relative to the other two voltages, and one increased and generally more lagging phase current relative to the other two phase currents. $3I_0$ current exists.
3. Phase-to-phase faults have two phase voltages reduced

relative to the other phase voltage, and two phase currents essentially equal and opposite. These currents are increased and generally more lagging than the third phase current. There is no $3I_0$ current.

4. Double-phase-to-ground faults are similar to phase-to-phase faults in the voltage and current changes except that the two increased currents are not equal and opposite and $3I_0$ current exists.

8.3 SYSTEM VOLTAGE PROFILES DURING SHUNT FAULTS

The trends of the sequence voltages for the various faults of Fig. 8.1 are illustrated in Fig. 8.2 for the power system at the top of the figure. Only the phase *a* sequence voltages are shown for the ideal case of $X_1 = X_2 = X_0$, an impossibility but simplifies the presentation without affecting the trends.

With the common assumption of no load, the system voltage is equal throughout the system, as indicated by the dashed lines. When a solid three-phase fault occurs, the voltage at the fault point becomes zero, but as indicated above, does not change in the source until the regulators act to change the generator fields. By this time the fault should have been cleared by protective relays. Thus the voltage profile is shown in (a).

For phase-to-phase faults (b), the positive-sequence voltage drops to half value ($Z_1 = Z_2$). This unbalance fault is the source of negative sequence and the V_2 drops are as shown, being zero in the generators.

For two-phase-to-ground faults (c) with $Z_1 = Z_2 = Z_0$, the positive-sequence voltage at the fault drops to one-third of V_1. The fault now generates both negative and zero sequences, which flows through the system, producing voltage drops as shown. V_2 becomes zero in the generators, while V_0 is zero at the grounded transformer neutral point.

The fault voltage for a phase-*a*-to-ground solid fault is zero and as phase *a* is documented in Fig. 8.2d, the sum of the positive, negative, and zero-voltage components at the fault add to zero. Thus the positive-sequence voltage drops to $\frac{2}{3}V_1$ when Z_1

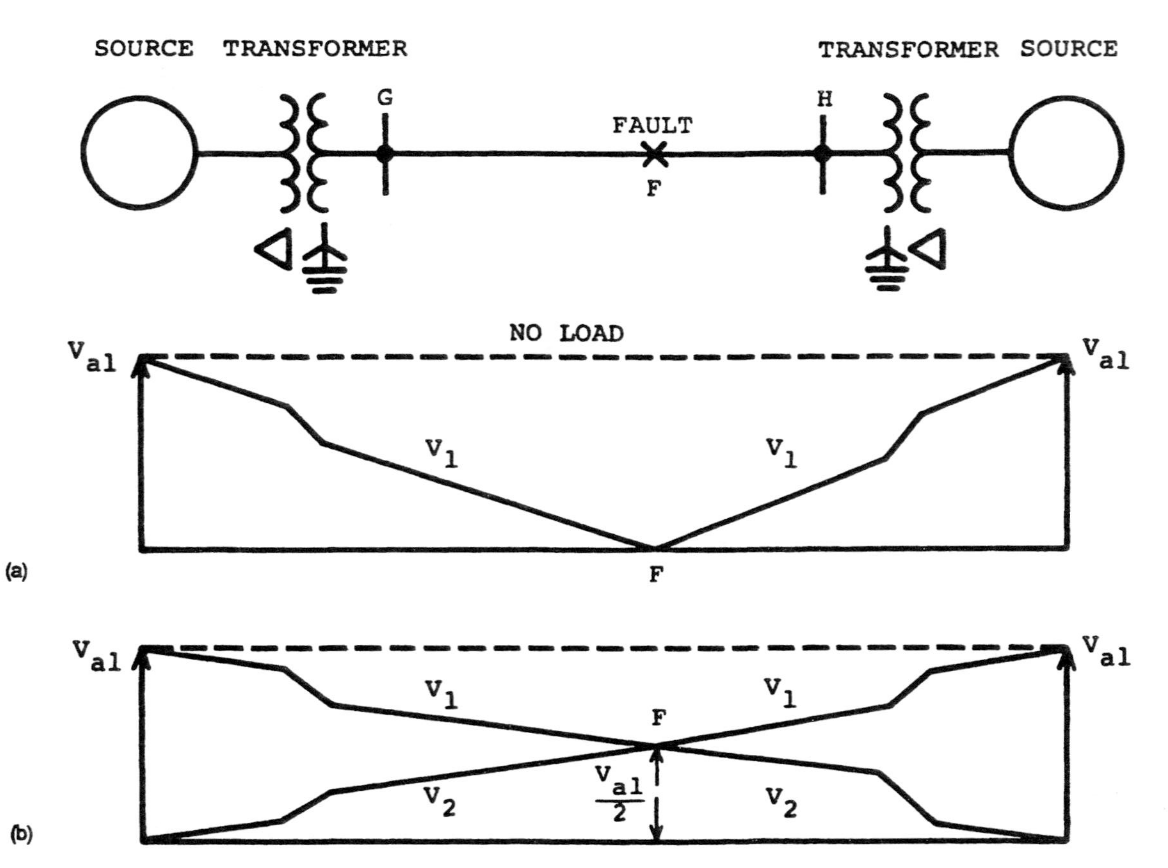
SOURCE TRANSFORMER
TRANSFORMER SOURCE
G
H
FAULT
F
NO LOAD
V_{a1}
V_{a1}
V_1
V_1
F
(a)
V_{a1}
V_{a1}
V_1
V_1
F
V_2
$\frac{V_{a1}}{2}$
V_2
(b)

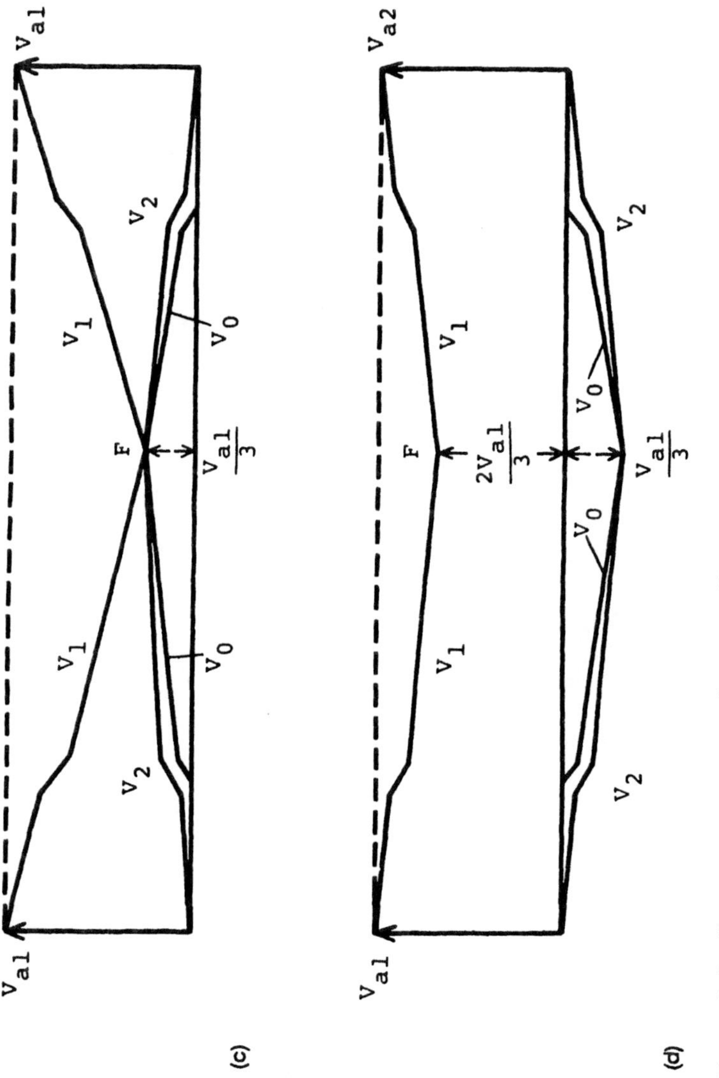

Figure 8.2 System sequence voltage profiles during shunt faults: (a) three-phase fault; (b) phase-to-phase (*bc*) fault; (c) two-phase-to-ground (*bc*-G) fault; (d) single-phase-to-ground (*a*-G) fault.

$= Z_2 = Z_0$ at the fault point, where $-\frac{1}{3}V_2$ and $-\frac{1}{3}V_0$ are generated. Again they drop to zero in the generator or source for the negative sequence and to zero at the grounded transformer bank neutral.

The fundamental illustrated in Fig. 8.2 is that *positive-sequence voltage is always maximum at the generators and minimum at the fault. The negative- and zero-sequence voltages are always maximum at the fault and minimum at the generator or grounded neutral.*

It is common to refer to the grounded-wye–delta or similar banks as "ground sources." This is really a misnomer, as the source of zero sequence is the unbalance, the ground fault. However, so designating these transformers as ground sources is practical, as by convention ground ($3I_0$) current flows up the grounded neutral, through the system, and down the fault into ground.

8.4 VOLTAGE AND CURRENT PHASORS FOR ALL COMBINATIONS OF THE FOUR SHUNT FAULTS

Another look at the sequence and total shunt fault phasors is presented in Figs. 8.3 and 8.4. The voltages and currents that are generated by the sources can be only positive sequence by design—nothing else. Yet the unbalanced faults require unbalanced quantities. How can this conflicting difference be resolved to satisfy both requirements: balanced quantities by the generators and unbalanced quantities at the faults? The resolution can be considered as the function of the negative-sequence quantities and for ground faults the zero-sequence quantities. This can be seen from Figs. 8.3 and 8.4. Considering the voltages of Fig. 8.3, positive-sequence voltage developed by the source or generator

Figure 8.3 Sequence voltages and the voltage at the fault point for the various fault types. Solid faults with $Z_1 = Z_2 = Z_0$ for simplicity. Magnitudes not to scale.

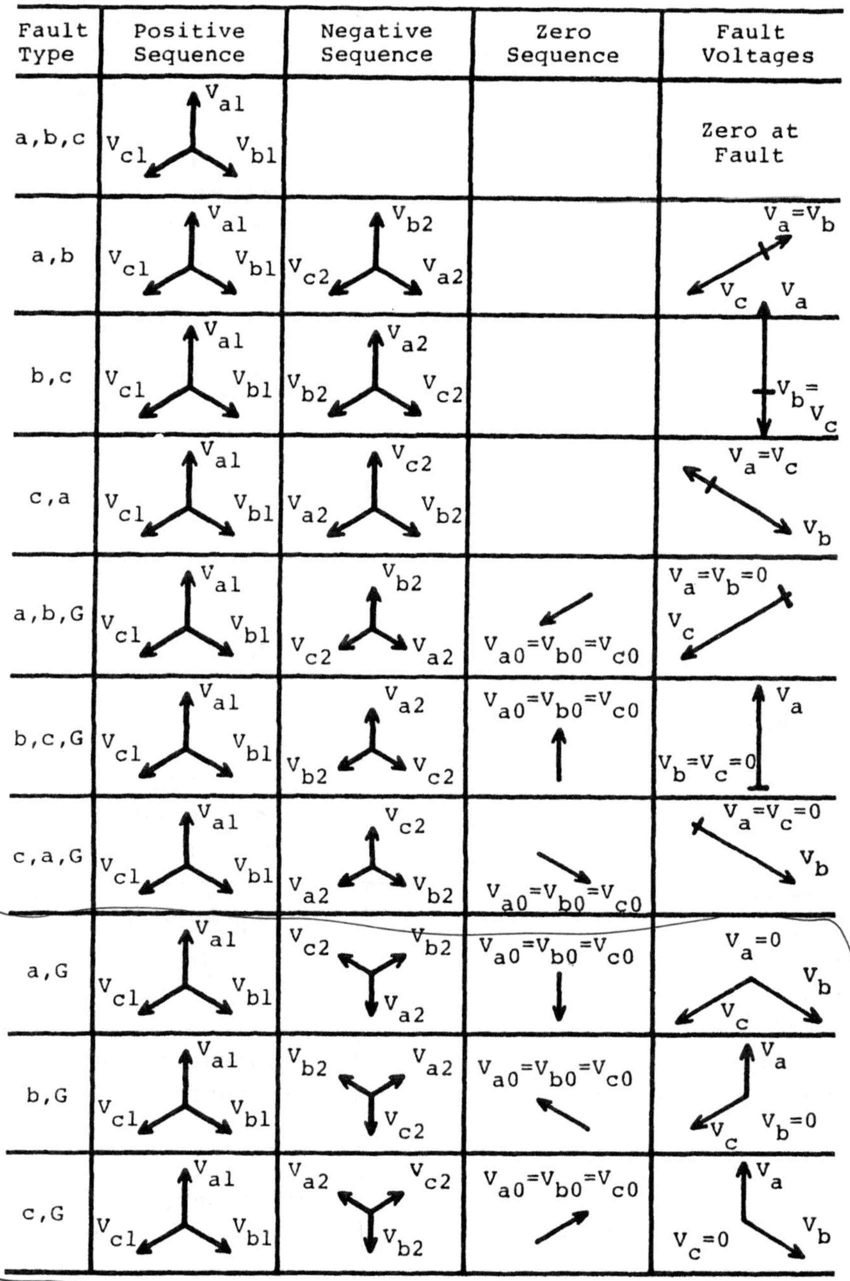

Fault Type	Positive Sequence	Negative Sequence	Zero Sequence	Fault Voltages
a,b,c	V_{a1} V_{c1} V_{b1}			Zero at Fault
a,b	V_{a1} V_{c1} V_{b1}	V_{b2} V_{c2} V_{a2}		$V_a=V_b$ V_c
b,c	V_{a1} V_{c1} V_{b1}	V_{a2} V_{b2} V_{c2}		V_a $V_b=V_c$
c,a	V_{a1} V_{c1} V_{b1}	V_{c2} V_{a2} V_{b2}		$V_a=V_c$ V_b
a,b,G	V_{a1} V_{c1} V_{b1}	V_{b2} V_{c2} V_{a2}	$V_{a0}=V_{b0}=V_{c0}$	$V_a=V_b=0$ V_c
b,c,G	V_{a1} V_{c1} V_{b1}	V_{a2} V_{b2} V_{c2}	$V_{a0}=V_{b0}=V_{c0}$	V_a $V_b=V_c=0$
c,a,G	V_{a1} V_{c1} V_{b1}	V_{c2} V_{a2} V_{b2}	$V_{a0}=V_{b0}=V_{c0}$	$V_a=V_c=0$ V_b
a,G	V_{a1} V_{c1} V_{b1}	V_{c2} V_{b2} V_{a2}	$V_{a0}=V_{b0}=V_{c0}$	$V_a=0$ V_b V_c
b,G	V_{a1} V_{c1} V_{b1}	V_{b2} V_{a2} V_{c2}	$V_{a0}=V_{b0}=V_{c0}$	V_a V_c $V_b=0$
c,G	V_{a1} V_{c1} V_{b1}	V_{a2} V_{c2} V_{b2}	$V_{a0}=V_{b0}=V_{c0}$	V_a $V_c=0$ V_b

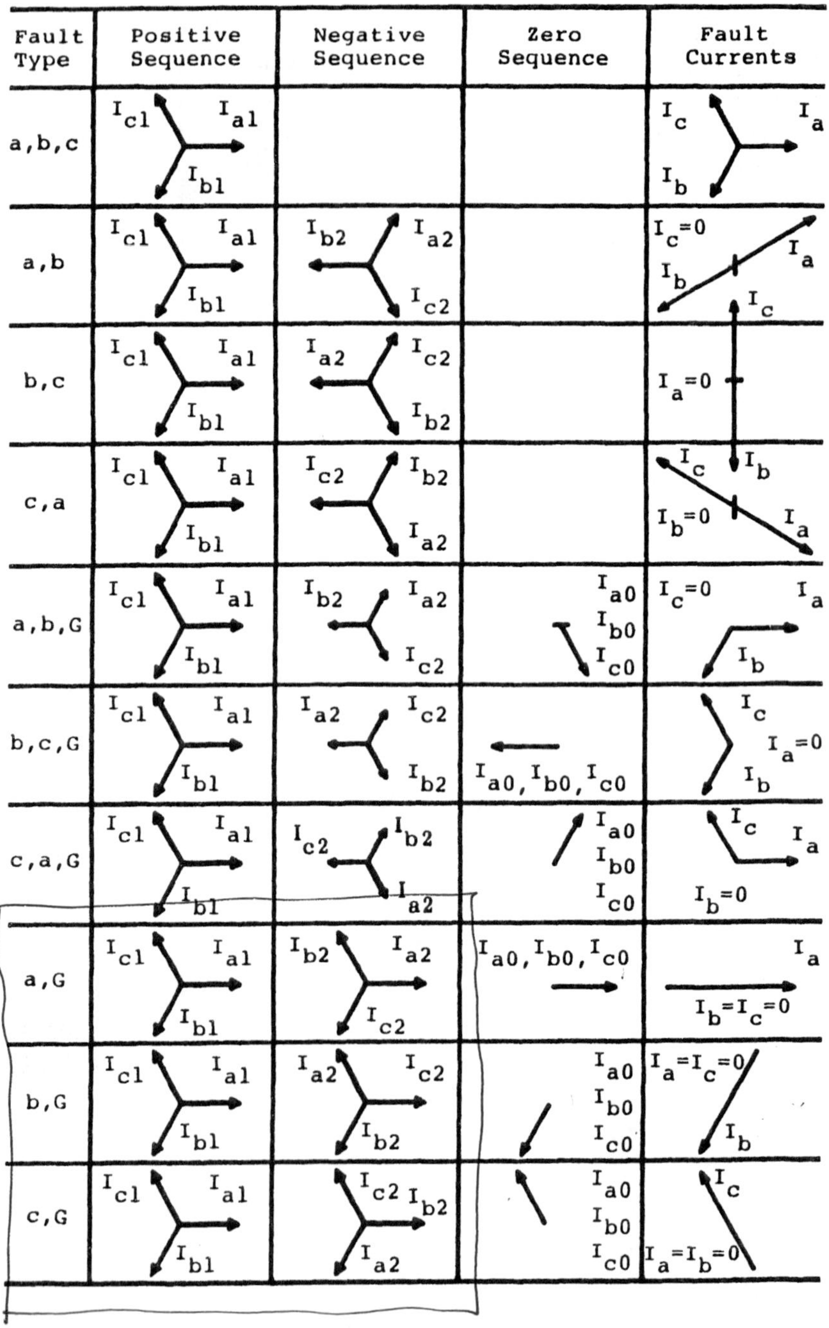

Fault Type	Positive Sequence	Negative Sequence	Zero Sequence	Fault Currents
a,b,c	I_{c1} I_{a1} I_{b1}			I_c I_a I_b
a,b	I_{c1} I_{a1} I_{b1}	I_{b2} I_{a2} I_{c2}		$I_c=0$ I_a I_b I_c
b,c	I_{c1} I_{a1} I_{b1}	I_{a2} I_{c2} I_{b2}		$I_a=0$
c,a	I_{c1} I_{a1} I_{b1}	I_{c2} I_{b2} I_{a2}		I_c I_b $I_b=0$ I_a
a,b,G	I_{c1} I_{a1} I_{b1}	I_{b2} I_{a2} I_{c2}	I_{a0} I_{b0} I_{c0}	$I_c=0$ I_a I_b
b,c,G	I_{c1} I_{a1} I_{b1}	I_{a2} I_{c2} I_{b2}	I_{a0}, I_{b0}, I_{c0}	I_c $I_a=0$ I_b
c,a,G	I_{c1} I_{a1} I_{b1}	I_{c2} I_{b2} I_{a2}	I_{a0} I_{b0} I_{c0}	I_c I_a $I_b=0$
a,G	I_{c1} I_{a1} I_{b1}	I_{b2} I_{a2} I_{c2}	I_{a0}, I_{b0}, I_{c0}	I_a $I_b=I_c=0$
b,G	I_{c1} I_{a1} I_{b1}	I_{a2} I_{c2} I_{b2}	I_{a0} I_{b0} I_{c0}	$I_a=I_c=0$ I_b
c,G	I_{c1} I_{a1} I_{b1}	I_{c2} I_{b2} I_{a2}	I_{a0} I_{b0} I_{c0}	I_c $I_a=I_b=0$

is the same for all faults. For three-phase faults no transition help is required, as these faults are symmetrical, hence no negative or zero sequences. For the phase-to-phase faults, negative sequence appears to provide the transition. Note that for the several combinations, *ab*, *bc*, and *ca* phases, the negative sequence is in different positions to provide the transition. The key is that for, say, an *ab* fault, phase *c* will be essentially normal, so V_{c1} and V_{c2} are basically in phase to provide this normal voltage. Correspondingly, for a *bc* fault, V_{a1} and V_{a2} are essentially in phase, and so on.

The two-phase-to-ground faults are similar; for *ab-G* faults, the uninvolved phase *c* quantities V_{c1}, V_{c2}, V_{c0} combine to provide the uncollapsed phase *c* voltage. In the figure these are shown in phase and half magnitude. In actual cases there will be slight variations, as the sequence impedances do not have the same magnitude or phase angle.

For single-phase-*a*-to-ground faults, the negative (V_{a2})- and zero (V_{a0})-sequence voltages add to cancel the positive-sequence V_{a1}, which will be zero at a solid fault. Correspondingly, for a phase *b* fault, V_{b2} and V_{b0} oppose V_{b1}, and similarly for the phase *c* fault.

The same concept applies to the sequence currents, as shown in Fig. 8.4. The positive-sequence currents are shown the same for all faults and for 90° lag (X-only system) with respect to the voltages of Fig. 8.3. These will vary depending on the system constants, but the concepts illustrated are valid. Again for three-phase faults, no transition help is required; hence there is no negative- or zero-sequence involvement.

For phase-to-phase faults negative sequence provides the necessary transition, with the unfaulted phase sequence currents in opposition to provide zero or low current. Thus for the *ab* fault, I_{c1} and I_{c2} are in opposition.

Figure 8.4 Sequence currents and the fault current for the various fault types. Solid faults with $Z_1 = Z_2 = Z_0$ for simplicity. Magnitudes not to scale.

Similarly for two-phase-to-ground faults; for an *ab-G* fault, I_{c2} and I_{c0} tend to cancel I_{c1}, and so on. For single-phase-to-ground faults the faulted phase components tend to add to provide a large fault current, since $I_{a1} + I_{a2} + I_{a0} = I_a$.

8.5 SUMMARY

A question often asked is: Are the sequence quantities real or only useful mathematical concepts? This has been debated for years, and in a sense they are both. Yes, they are real; positive sequence certainly since it is generated, sold, and consumed; zero sequence because it flows in the neutral and ground and deltas; and negative sequence because it can cause serious damage to rotating machines. Negative sequence, for example, cannot be measured directly by an ammeter or voltmeter. Networks are available and commonly used in protection to measure V_2 and I_2, but these are designed to solve the basic equations for those quantities.

In either case, symmetrical components analysis is an extremely valuable and powerful tool. Protection engineers automatically tend to think in its terms when evaluating and solving unbalanced situations in the power system.

It is important to remember always that any sequence quantity cannot exist in only one phase; this is a three-phase concept. If any sequence is in one phase, it *must* be in all three phases according to the fundamental definitions of Eqs. (4.2) through (4.4).

9

Transformer, Reactor, and Capacitor Characteristics

9.1 TRANSFORMER FUNDAMENTALS

Transformers are used in power systems to change the voltage levels for economics, safety, and convenience. There are many types, sizes, connections, and applications. The principal types are power, distribution, current, voltage, and auxiliary, primarily with two windings. However, there are also many three-winding and autotransformers. Sometimes they have taps for voltage regulation or application purposes.

For all transformers, the fundamental equation for the induced voltage is

$$V_{rms} = 4.44 fNAB_{max} \times 10^{-8} \qquad \text{volts} \tag{9.1}$$

where N is the number of turns, f the frequency in hertz, A the cross-sectional area of the iron core in square inches, and B_{max} the maximum flux density in lines per square inch.

The equivalent circuit of a two-winding transformer is shown in Fig. 9.1 either for ohms or per unit. An ideal transformer is

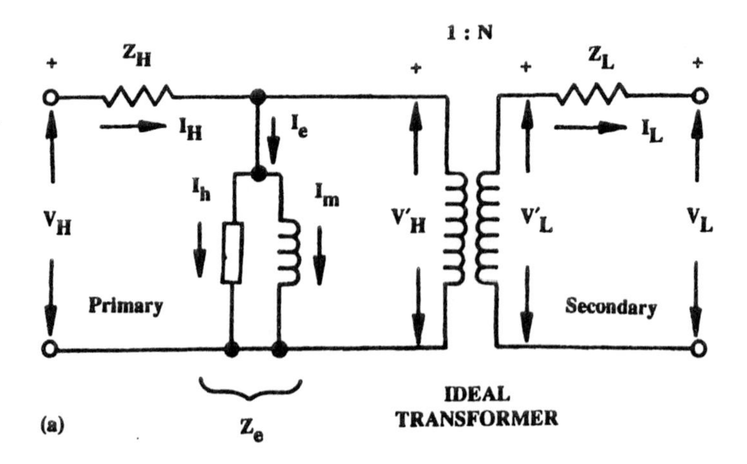

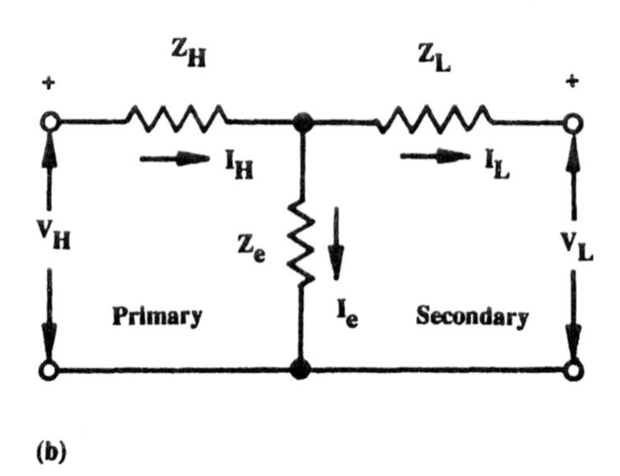

Figure 9.1 Equivalent circuit for a two-winding transformer. Values shown are (a) in ohms; (b) in per unit or percent.

required with ohms only for the purpose of changing the voltage level. This transformer has zero short-circuit impedance ($Z_H + Z_L = 0$) and infinite open-circuit impedance (Z_e infinite). This means no loss, no impedance drop, and no exciting current: a perfect transformer. The values to the left are on the primary voltage base, while those to the right are on the secondary voltage base.

With per unit (or percent) the ideal transformer is not necessary. The impedances are on the specified kVA (or MVA) base at either the primary or secondary voltage base, as covered in Chapter 2.

Z_H and Z_L are the primary and secondary leakage impedances, respectively, while Z_e is the magnetizing or exciting branch. With r_H and L_H as the resistance and self-inductance of the primary winding, r_L and L_L the resistance and self-inductance of the secondary winding, M the mutual inductance between the two windings, N_1 the primary turns, and N_2 the secondary turns, the following relations exists:

$$Z_H = r_H + j\omega\left(L_H - \frac{N_1}{N_2}M\right) \quad \text{primary ohms} \tag{9.2}$$

$$N = \frac{N_2}{N_1} \tag{9.3}$$

Since Z_H and Z_L cannot be segregated by test, it is assumed that

$$Z_H = \frac{1}{N_2}Z_L = \frac{1}{2}Z_{HL} \quad \text{primary ohms} \tag{9.4}$$

where Z_{HL} is the leakage impedance looking into the primary. Thus

$$Z_{HL} = Z_T = Z_H + \frac{1}{N^2}Z_L \quad \text{primary ohms} \tag{9.5}$$

Correspondingly, Z_{LH} is the leakage impedance looking into the secondary:

$$Z_{LH} = Z_T = Z_L + N^2Z_H = N^2Z_{HL} \quad \text{secondary ohms} \tag{9.6}$$

With per unit (percent):

$$Z_H = Z_L = \tfrac{1}{2}Z_T = \tfrac{1}{2}Z_{HL} = \tfrac{1}{2}Z_{LH} \quad \text{pu} \tag{9.7}$$

$Z_T(Z_{HL}, Z_{LH})$ impedance (reactance) normally is supplied by the manufacturer in percent on the rated kVA (MVA) and rated voltage of the windings. Z_{HL} or Z_{LH} can be determined by test. This is covered in Section 9.12. With taps the data are for the full winding. Impedances for the taps may or may not be available. Often, the taps only change a relative few turns, so impedance change is small. With multiple power ratings the percent impedance values are on the lowest rating (without fans, pumps, etc.). For three-phase transformer units, the nameplate values are on the three-phase rated kVA (MVA) and the kV phase-to-phase rated voltage.

For individual single-phase transformers, the transformer impedance is specified on the rated single-phase kVA (MVA) and the rated winding kV voltages. When three individual transformers are used in a three-phase system, the impedances should be on a three-phase kVA (MVA) and phase-to-phase kV base as indicated in Chapter 2.

When three individual transformers are connected in various wye or delta combinations in the power system, the individual nameplate percent impedance will be the Z_T leakage impedance but on the equivalent three-phase kVA (MVA) and the related kV phase-to-phase voltage. This is shown in more detail in Section 9.2. A typical phasor diagram with its associated circuit diagram is shown in Fig. 9.2 for a two-winding transformer.

9.2 EXAMPLE: IMPEDANCES OF SINGLE-PHASE TRANSFORMERS IN THREE-PHASE POWER SYSTEMS

Three single-phase transformers each have a nameplate rating of 20 MVA, 66.5 kV:13.8 kV, $X_T = 10\%$. Individually, their leakage reactance is

$$X_T = 0.10 \text{ pu on 20 MVA, 66.5 kV} \tag{9.8}$$

$$\textit{or} \quad \text{on 20 MVA, 13.8 kV}$$

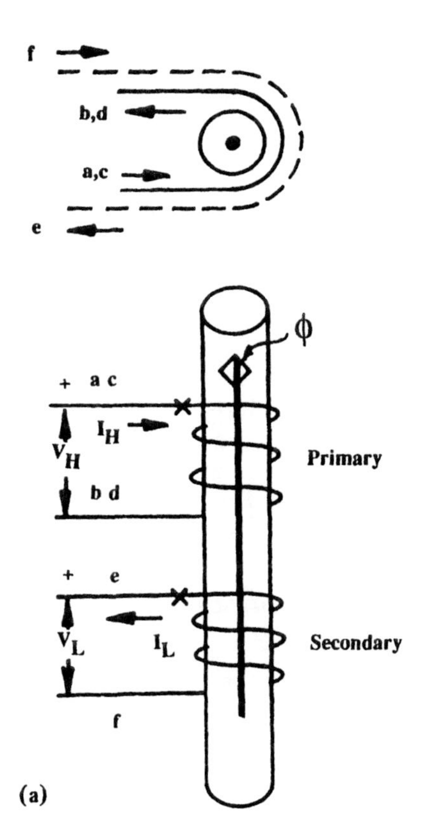

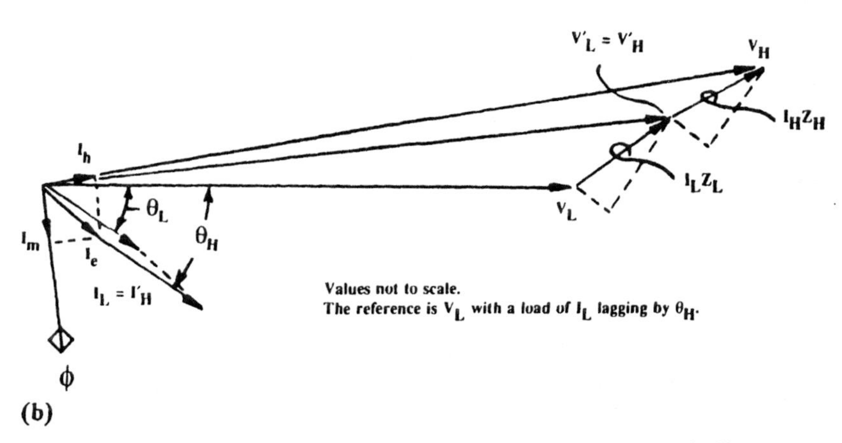

Figure 9.2 (a) Circuit and (b) phasor diagrams for a two-winding transformer. The turns ratio N is equal to 1 for simplification.

Converting to ohms [Eq. (2.17)], we have

$$X_{HL} = 66.5^2 \times \frac{0.10}{20} = 22.11\ \Omega \text{ at } 66.5\text{ kV} \tag{9.9}$$

or

$$X_{LH} = 13.8^2 \times \frac{0.10}{20} = 0.952\ \Omega \text{ at } 13.8\text{ kV} \tag{9.10}$$

Check: 0.952 Ω secondary × $(66.5/13.8)^2$ = 22.11 Ω primary.

Consider two possible applications of these individual transformers to a three-phase power system. These are intended to demonstrate the fundamentals and do not consider if the transformer windings are compatible or suitable for the system voltages shown.

9.2.1 Three Individual Transformers Connected 115-kV Wye and 13.8-kV Delta

As indicated, the bank impedance is

$$X_T = 0.10 \text{ pu on } 60\text{ MVA, } 115\text{ kV} \tag{9.11}$$

$$\textit{or} \quad \text{on } 60\text{ MVA, } 13.8\text{ kV}$$

This can be checked by converting the above to ohms using Eq. (2.17):

$$X_T = 115^2 \times \frac{0.10}{60} = 22.11\ \Omega \text{ primary} \tag{9.12}$$

or

$$X_T = 13.8^2 \times \frac{0.10}{60} = 0.317\ \Omega \text{ secondary} \tag{9.13}$$

Note that the individual transformer reactance equation (9.10) is 0.952 Ω. In this application 0.952 Ω is the reactance across the 13.8-kV phases. The equivalent phase to neutral value from

Eqs. (4.30) is 0.952/3 = 0.317 Ω as Eq. (9.13). This can be checked:

$$0.317\ \Omega\ \text{secondary} \times \left(\frac{115}{13.8}\right)^2 = 22.11\ \Omega\ \text{primary}$$

9.2.2 Three Individual Transformers Connected 66.5-kV Delta and 24-kV Wye

Now the transformer bank reactance is

$$X_T = 0.10\ \text{pu on 60 MVA, 66.5 kV} \qquad (9.14)$$
$$\textit{or}\quad \text{on 60 MVA, 24 kV}$$

Converting to ohms [Eq. (2.17)] yields

$$X_T = 66.5^2 \times \frac{0.10}{60} = 7.37\ \Omega\ \text{primary} \qquad (9.15)$$

The transformer reactance of 22.11 primary ohms [Eq. (9.9)] is now connected phase to phase. The equivalent on this line to neutral is $\frac{1}{3}$ or 7.37 primary ohms [Eqs. (4.30)].

On the wye side,

$$X_T = 24^2 \times \frac{0.10}{60} = 0.952\ \Omega\ \text{secondary} \qquad (9.16)$$

As a check:

$$0.952\ \Omega\ \text{secondary} \times \left(\frac{66.5}{24}\right)^2 = 7.37\ \Omega\ \text{primary}$$

9.3 POLARITY, STANDARD TERMINAL MARKING, AND PHASE SHIFTS

Transformer polarity along with its very useful applications for phasing and connecting has been covered in Section 3.5. The definitions apply equally to transformers with either additive or subtractive polarity (Fig. 3.4). Subtractive polarity is standard in the United States for all transformers except for single-phase transformers of 200 kVA and less with high-voltage ratings of 8660 V (winding voltage) and below. These smaller transformers have additive polarity.

The standard terminal markings for single-phase transformers are H_1 and H_2 for the high-voltage winding and x_1 and x_2 for the low-voltage winding.

For three-phase transformer banks the high-voltage terminals are H_1, H_2, and H_3 and the low-voltage x_1, x_2, and x_3. For transformers with more than two windings, the windings are marked H, x, y, z in order of decreasing voltage.

The ANSI standard angular displacement between the primary and secondary for three-phase transformer units (between H_1 and x_1, H_2 and x_2, H_3 and x_3) is (a) zero (in phase) for wye–wye or delta–delta banks, and (b) high-voltage side leading the low-voltage side by 30° for either wye–delta or delta–wye banks. The sequence phase shifts through wye–delta and delta–wye banks is discussed in Section 4.13.

9.4 TWO-WINDING TRANSFORMER BANKS: SEQUENCE IMPEDANCE AND CONNECTIONS

The positive- and negative-sequence impedances of a group of three transformers are always equal. In the absence of actual data, typical approximate reactance values for two-winding transformers for estimating purposes are [values on rated transformer kVA (MVA) and kV base]

Distribution	1–3%
Network	5%
Power through 69 kV	5–10%
92 kV through 115 kV	7–15%
138 kV through 161 kV	8–17%
196 kV through 230 kV	11–21%

The dc resistance varies from 0.35 to 1.0%.

The zero-sequence impedance of a group of three transformers is the same as the positive- and negative-sequence impedances or is infinite except in some types of construction, such as the three-phase core types. For these transformer banks the zero-sequence reactance generally is 90% to almost 100% of the positive-sequence reactance. This results because there is no or very limited return for the zero-sequence exciting flux except via the insulation medium and the transformer tank. The result is that the flux linkages with the zero-sequence exciting current is low and on the order of 50 to 300%, compared to several thousand percent for the positive sequence. Transformer banks of this type can be considered as a three-winding bank where the effect of the restricted magnetic circuit is replaced by a fictitious delta tertiary of very high reactance. In this manner the net zero-sequence reactance can be reduced slightly as required.

The sequence impedances and connections for two winding banks are shown in Fig. 4.9. It will be noted that the positive- and negative-sequence connections and impedance are the same and independent of the transformer connections. The zero-sequence connections are determined by whether the three equal and in-phase zero-sequence currents can flow.

9.5 THREE-WINDING TRANSFORMER BANKS

Power transformers with three windings on the same core are encountered frequently in power systems. Each winding usually is at a different voltage level, with at least one set connected in delta as a tertiary for third harmonic suppression and system grounding. Two common designations have been used over the years for the three windings; (1) H, M, L (high, medium, and low voltage) and (2) H, T, L (high, tertiary, and low voltage).

As for two-winding transformers, the leakage impedance can be measured between pairs of windings, so that

$$Z_{HM} = Z_H + Z_M \tag{9.17}$$

This is the impedance looking into the *H* winding with the *M* winding short circuited and the *L* winding open:

$$Z_{HL} = Z_H + Z_L \tag{9.18}$$

This is the impedance looking into the *H* winding with the *L* winding short circuited and the *M* winding open:

$$Z_{ML} = Z_M + Z_L \tag{9.19}$$

This is the impedance looking into the *M* winding with the *L* winding short circuited and the *H* winding open.

The three values are provided by the manufacturer normally or can be determined by test. The manufacturer usually gives these in percent on the kVA (MVA), kV rating of the two windings involved so that they are not on a common base. The *first* step is to convert to a common base, usually the system base.

On a common base the individual branches of an equivalent wye network that represents the transfer of power (fault current, etc.) between the terminals of the three windings are

$$Z_H = \tfrac{1}{2}(Z_{HM} + Z_{HL} - Z_{ML}) \tag{9.20}$$

$$Z_M = \tfrac{1}{2}(Z_{HM} + Z_{ML} - Z_{HL}) \tag{9.21}$$

$$Z_L = \tfrac{1}{2}(Z_{ML} + Z_{HL} - Z_{HM}) \tag{9.22}$$

Any of these values can be zero or negative and they should be dealt with in the calculations in that form. The wye is a mathematical equivalent between the three terminals and is valid only for loads, faults, and so on, connected at or outside the terminals. It is important to recognize that the wye point of the equivalent is a fictitious point and does not represent the system neutral or any other point in the transformer.

The clue to remembering these formulas is that when solving for any equivalent winding impedance, it is one-half the sum of the two, which include the same letter minus the third one, which does not. After solving for these three branches, always recheck by adding any two of them, which should total to the corresponding value of Eqs. (9.17), (9.18), or (9.19).

9.6 THREE-WINDING TRANSFORMERS: SEQUENCE IMPEDANCE AND CONNECTIONS

The sequence impedances and connections for three-winding transformer banks are shown in Fig. 4.10. Again the positive- and negative-sequence values and connections are the same and independent of the bank connections.

The zero-sequence impedances are either the same or infinite, depending on whether the bank connections will permit zero-sequence current to flow. If slightly different zero-sequence impedance values result, this is because of the transformer construction as outlined under the two-winding transformers.

9.7 EXAMPLE: THREE-WINDING TRANSFORMER EQUIVALENT

Consider the 161-kV wye, 115-kV wye, 13.8-kV delta transformer shown in Fig. 9.3. The winding reactances are

$$\begin{aligned} X_{HM} &= 10\% \text{ on } 30 \text{ MVA} \\ X_{HL} &= 6\% \text{ on } 10 \text{ MVA} \\ X_{ML} &= 14\% \text{ on } 15 \text{ MVA} \end{aligned} \tag{9.23}$$

First convert to a common base; on a 100 MVA

$$\begin{aligned} X_{HM} &= \frac{100}{30}(10\%) = 33.33\% \\ X_{HL} &= \frac{100}{10}(6\%) = 60.00\% \\ X_{ML} &= \frac{100}{15}(14\%) = 93.33\% \end{aligned} \tag{9.24}$$

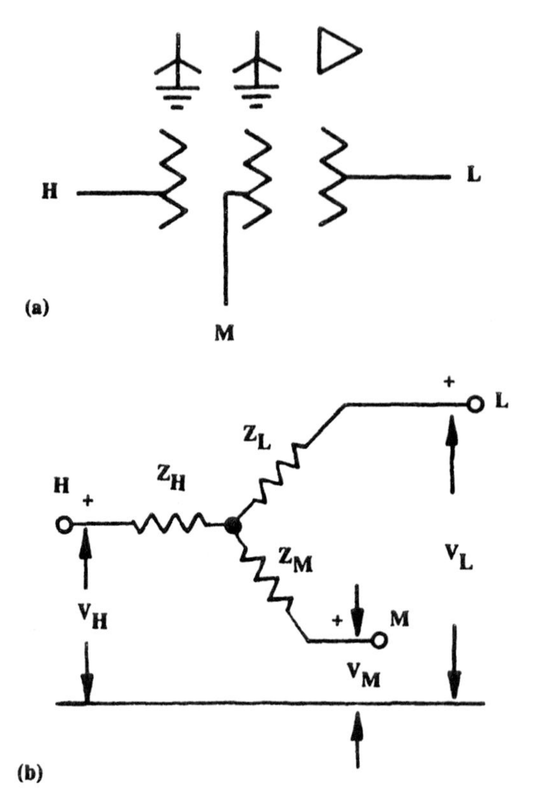

Figure 9.3 Three-winding transformer bank and its wye equivalent circuit.

From Eqs. (9.17) through (9.19),

$$\begin{aligned} X_H &= \tfrac{1}{2}(33.33 + 60.00 - 93.33) = 0 \\ X_M &= \tfrac{1}{2}(33.33 + 93.33 - 60.00) = 33.33\% \qquad (9.25) \\ X_L &= \tfrac{1}{2}(60.00 + 93.33 - 33.33) = 60.00\% \end{aligned}$$

This equivalent wye circuit is connected into the sequence networks as shown in Fig. 4.10 and in the example of Fig. 7.14.

9.8 EXAMPLE: THREE-WINDING TRANSFORMER FAULT CALCULATION

Consider the system of Fig. 9.4 with a three-winding transformer at bus H. Typical constants are shown on the bases indicated. The first step is to convert them to a common base. Usually the reactances between the three windings X_{HM}, X_{HL}, and X_{ML} are given on different MVA or kVA ratings but have been previously converted to a 150-MVA base. For this example, a 100-MVA base at the various voltages indicated has been selected.

The positive sequence network is illustrated in Fig. 9.5. Since there are no synchronous equipment or faults being considered on the three-winding transformer L winding, the three-winding X_{HM} is the only one involved.

If there were synchronous equipment connected to the L winding, or if 13.2-kV faults or load were involved, then the equivalent circuit shown and derived from Eqs. (9.20) through (9.22) would be required.

For a fault at bus G, the right-hand reactances ($j0.18147 + j0.03667 + j0.03 = j0.2481$) are parallel with the left-hand re-

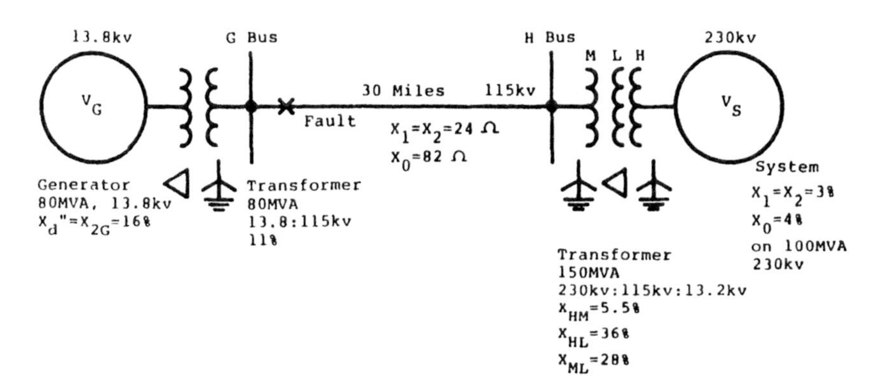

Figure 9.4 Power system example for fault calculations of Section 9.8.

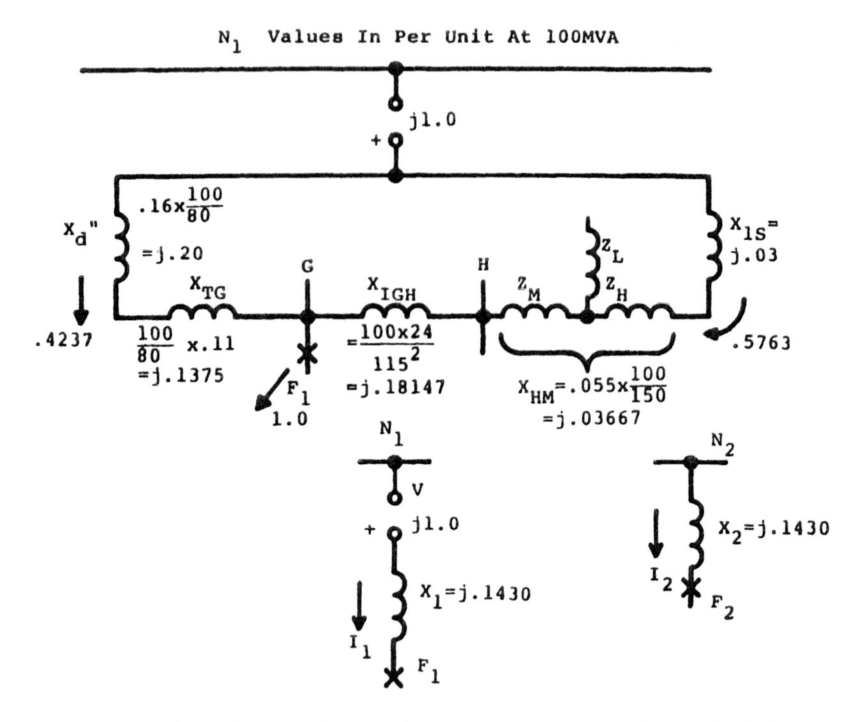

Figure 9.5 Positive- and negative-sequence networks and their reduction to a single impedance for a fault at bus *G* in the power system of Fig. 9.4.

actances ($j0.20 + j0.1375 = j0.3375$). Thus

$$X_1 = \frac{\overset{(0.5763)}{0.3375} \times \overset{(0.4237)}{0.2481}}{0.5856} = j0.1430 \text{ pu}$$

The numbers in parentheses represent the division of 0.3375/0.5856 = 0.5763 and 0.2481/0.5856 = 0.4237. They provide a partial check, as 0.5763 + 0.4237 must equal 1.0, and are the distribution factors indicating the per unit current flowing on either side of a 1.0 per unit fault. The distribution factors are shown in Fig. 9.5. Thus for faults at bus G and assuming the positive and negative sequences reactances are equal,

$$X_1 = X_2 = j0.1430 \quad \text{on 100 MVA base}$$

The zero sequence network for the system of Fig. 9.4 is shown in Fig. 9.6. Again the reactance values must be converted to the common base as the positive and negative sequence values or 100 MVA in this example. The three-winding transformer equivalents from Eqs. 9.20, 9.21, and 9.22 are important as the transformer is connected as shown per Fig. 4.10b with $Z_{NH} = Z_{NM} = 0$ since the bank neutrals are solidly grounded. These conversions and equivalents for the three-winding bank are

$$X_{HM} = 0.055 \times \frac{100}{150} = 0.03667 \text{ pu}$$

$$X_{HL} = 0.360 \times \frac{100}{150} = 0.2400 \text{ pu}$$

$$X_{ML} = 0.280 \times \frac{100}{150} = 0.18667 \text{ pu}$$

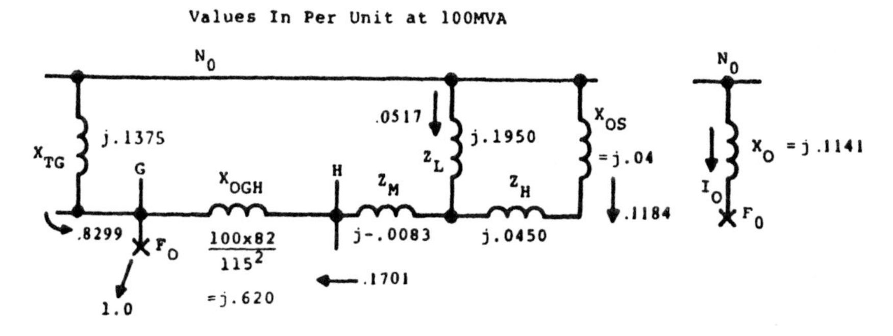

Figure 9.6 Zero-sequence network and its reduction to a single impedance for a fault at bus G in the power system of Fig. 9.4.

and from Eq. (9.20) through (9.22),

$$X_H = \tfrac{1}{2}(0.03667 + 0.2400 - 0.18667) = 0.0450 \text{ pu}$$

$$X_M = \tfrac{1}{2}(0.03667 + 0.18667 - 0.240) = -0.00833 \text{ pu}$$

$$X_L = \tfrac{1}{2}(0.2400 + 0.18667 - 0.03667) = 0.1950 \text{ pu}$$

This network is reduced for a fault at bus G by first paralleling $X_{0S} + X_H$ with X_L and then adding X_M and X_{0GH};

$$\frac{\overset{(0.6964}{0.1950} \times \overset{(0.3036)}{0.0850}}{0.280} = \begin{array}{rl} j0.0592 & \\ -j0.0083 & (Z_M) \\ j0.620 & (X_{0GH}) \\ \hline j0.6709 & \end{array}$$

This is the right-hand branch. Paralleling with the left-hand branch,

$$X_0 = \frac{\overset{(0.8299)}{0.6709} \times \overset{(0.1701)}{0.1375}}{0.8084} = j0.1141 \text{ pu at 100 MVA}$$

The values (0.8299) and (0.1701) shown add to 1.0 as a check and provide the current distribution on either side of the bus G fault, as shown on the zero-sequence network. The distribution factor 0.1701 for the right side is further divided up by $0.6964 \times 0.1701 = 0.1184$ pu in the 230-kV system neutral, and $0.3036 \times 0.1701 = 0.0517$ pu in the three-winding transformer H neutral winding. These are shown on the zero-sequence network.

9.8.1 Three-Phase Fault at Bus *G*

For this fault,

$$I_1 = I_{aF} = \frac{j1.0}{j0.143} = 6.993 \text{ pu}$$

$$= 6.993 \frac{100{,}000}{\sqrt{3} \times 115} = 3510.8 \text{ A at 115 kV}$$

The divisions of current from the left (I_{aG}) and right (I_{aH}) are

$$I_{aG} = 0.4237 \times 6.993 = 2.963 \text{ pu} = 1487.56 \text{ A at } 115 \text{ kV}$$
$$I_{aH} = 0.5763 \times 6.993 = 4.030 \text{ pu} = 2023.24 \text{ A at } 115 \text{ kV}$$

9.8.2 Single-Phase-to-Ground Fault at Bus *G*

For this fault,

$$I_1 = I_2 = I_0 = \frac{j1.0}{j(0.143 + 0.143 + 0.1141)} = 2.50 \text{ pu}$$
$$I_{aF} = 3 \times 2.5 = 7.5 \text{ pu} \quad \text{at } 100 \text{ MVA}$$
$$= 7.5 \times \frac{100{,}000}{\sqrt{3} \times 115} = 3764.4 \text{ A} \quad \text{at } 115 \text{ kV}$$

Normally, the $3I_0$ currents are documented in the system, as these are used to operate the ground relays. As an aid to understanding, these are illustrated in Fig. 9.7 with the phase currents. Equations (4.5) through (4.7) provide the three phase currents. Since $X_1 = X_2$, so that $I_1 = I_2$, these reduce to $I_b = I_c = -I_1 + I_0$ for the phase b and c currents, since $a + a^2 = -1$. The currents shown are determined by adding $I_1 + I_2 + I_0$ for I_a, $-I_1 + I_0$ for I_b and I_c, and $3I_0$ for the neutral currents.

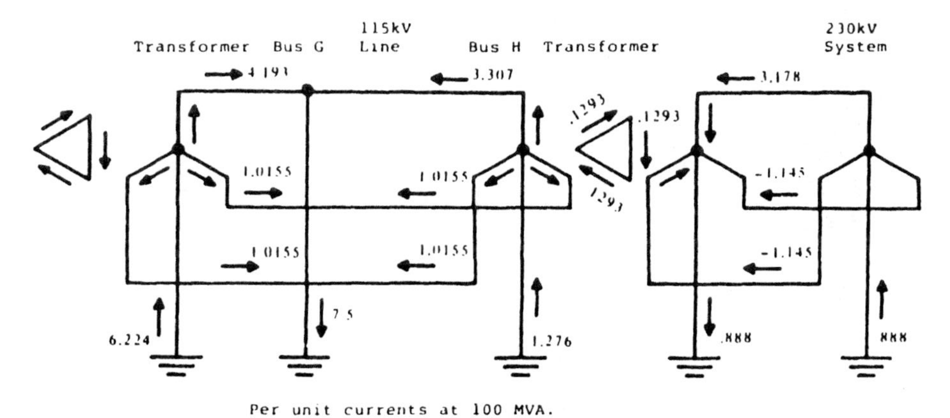

Figure 9.7 Phase and $3I_0$ current distribution for a single-phase-to-ground fault at bus G in Fig. 9.4. 1 pu current = 502 A at 115 kV, 251 A at 230 kV.

In the 115-kV system the sum of the two neutral currents is equal and opposite to the current in the fault. In the 230-kV system the current up the neutral equals the current down the other neutral.

The calculations assumed no load, so prefault, all currents in the system were zero. With the fault involving phase *a* only, it will be observed that current flows in the *b* and *c* phases. This is because the distribution factors in the zero-sequence network are different from the positive- and negative-sequence distribution factors. On a radial system where positive-, negative-, and zero-sequence currents flow only from one source and in the same direction, the distribution factors in all three networks will be 1.0, although the zero-sequence impedances are different from the positive-sequence impedances. Then $I_b = I_c = -I_1 + I_0$ above becomes zero, and fault current only flows in the faulted phase. In this type $I_a = 3I_0$ throughout the system for a single-phase-to-ground fault.

9.9 AUTOTRANSFORMERS

Autotransformers are very widely used, especially at the higher voltages. Most of these are three-winding types, with the third winding used as a tertiary. The determination of the equivalent circuit and the sequence interconnections are the same as for the three-winding transformer. The connections are shown in Fig. 4.10d. The positive- and negative-sequence impedances are equal. The zero-sequence impedance is usually equal to the positive-sequence impedance or infinite where zero sequence cannot flow.

The major difference between three-winding transformers and autotransformers is in the determination of the actual currents flowing in the autotransformer windings. This is illustrated best by a typical example.

9.10 EXAMPLE: AUTOTRANSFORMER FAULT CALCULATION

A 138-kV/230-kV autotransformer with a 11-kV tertiary is connected to a 138-kV power system as shown in Fig. 9.8. The

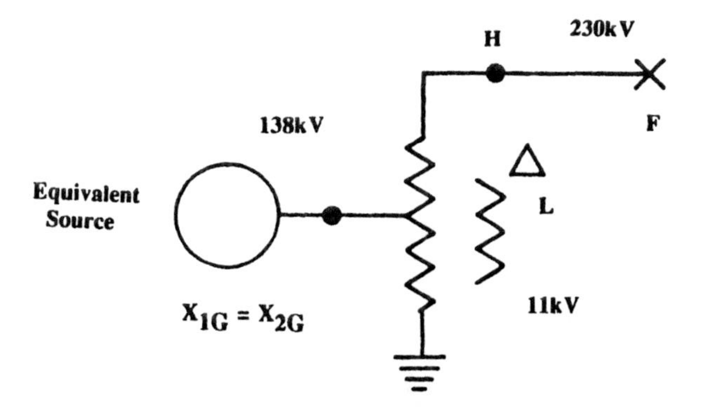

X_{HM} = 9.5% on 37,500 KVA
X_{ML} = 9.2% on 11,000 KVA
X_{HL} = 14.0% on 11,000 KVA
$X_{1G} = X_{2G}$ = 20% on 37,500 KVA
X_{0G} = 10% on 37,500 KVA

Figure 9.8 Typical example of an autotransformer system.

various reactance values required for calculating a ground fault at the 230-kV terminals are shown in the diagram. The 230-kV system has been omitted for simplification.

The phase-*a*-to-ground fault calculations are as follows: First, change all reactance values to a common base; 37.5 MVA was selected as a convenient base. The transformer reactance values become

$$X_{HM} = 9.5\%$$

$$X_{ML} = \frac{37.5}{11}(9.2) = 31.36\%$$

$$X_{HL} = \frac{37.5}{11}(14) = 47.73\%$$

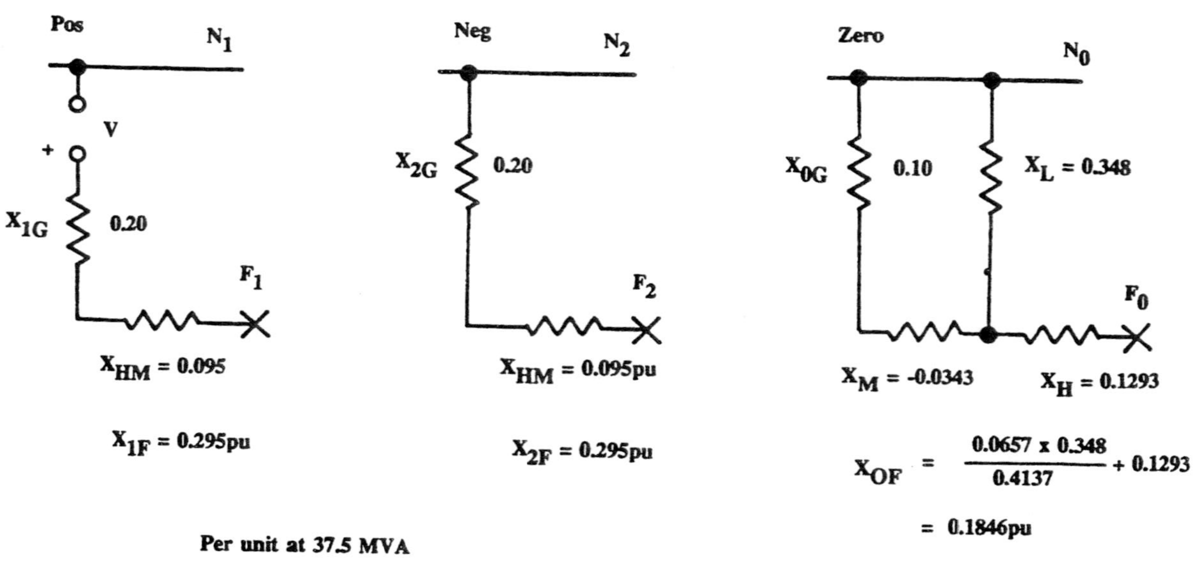

Figure 9.9 The three sequence networks for the system of Fig. 9.8.

From Eqs. (9.25),

$$X_H = \tfrac{1}{2}(9.5 + 47.73 - 31.36) = 12.93\%$$

$$X_M = \tfrac{1}{2}(9.5 + 31.36 - 47.73) = -3.43\%$$

$$X_L = \tfrac{1}{2}(31.36 + 47.73 - 9.5) = 34.80\%$$

At this point check to see that the sums of X_H, X_M, and X_L add up to X_{HM}, X_{ML}, and X_{HL} before continuing. For this system the three sequence networks and the reactance values to the fault are as shown in Fig. 9.9. For the line-to-ground fault,

$$I_{1F} = I_{2F} = I_{0F} = \frac{1}{0.7746} = 1.29 \text{ pu}$$

$$I_{aF} = 3 \times 1.29 = 3.87 \text{ pu}$$

$$I_{aF} = 3.87\left(\frac{37{,}500}{\sqrt{3}\ 230}\right) = 364.57 \text{ A at } 230 \text{ kV} \tag{5.2}$$

Now convert the currents to amperes at the proper voltage bases, as in Fig. 9.10. From these sequence currents the phase and neutral currents are calculated using Eqs. (4.5) through (4.7) as in Fig. 9.11. The 11-kV I_0 is the current inside the delta, so the transformation ratio is $230/\sqrt{3}$ to 11.

For the autotransformer it is to be noted that currents at one voltage level are *combined directly* with currents at another voltage level. In the example 364.57 A at 230 kV are subtracted from 575.46 A at 138 kV to give 210.89 A in the transformer common winding. These are real amperes but without a convenient base. It is not amperes at 138 kV or at 230 kV. Thus for autotransformers, per unit is very difficult; amperes, as outlined above, are far easier. Where per unit must be used, it is necessary to establish bases for the per unit values based on turns ratio of the parts—very tedious and difficult!

This example indicates that current flows down the autotransformer neutral. For high-side faults this neutral current can flow

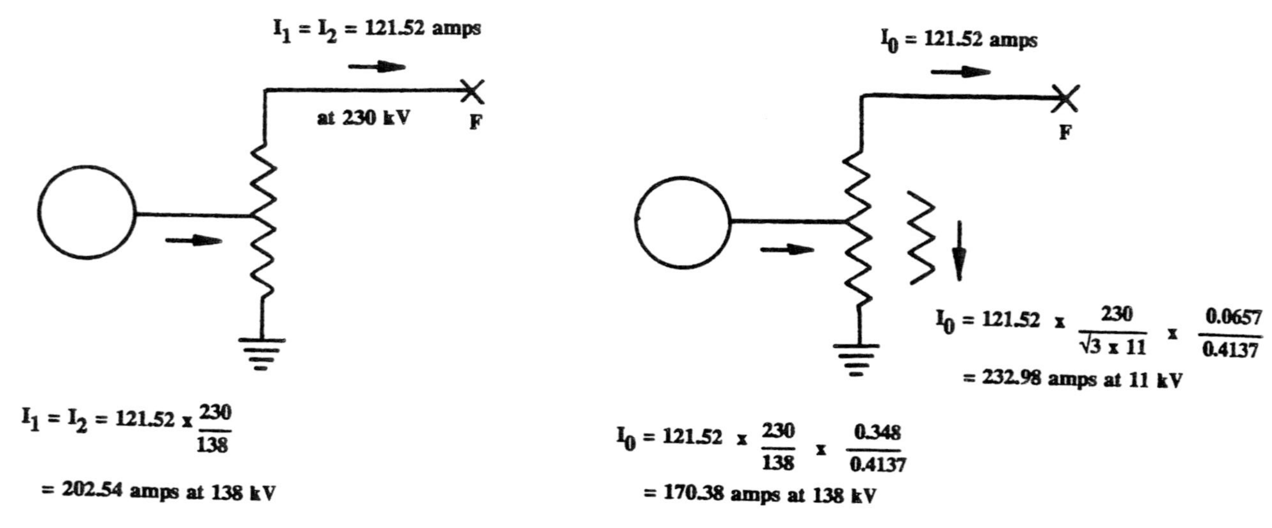

Figure 9.10 Sequence currents flowing in the system of Fig. 9.8.

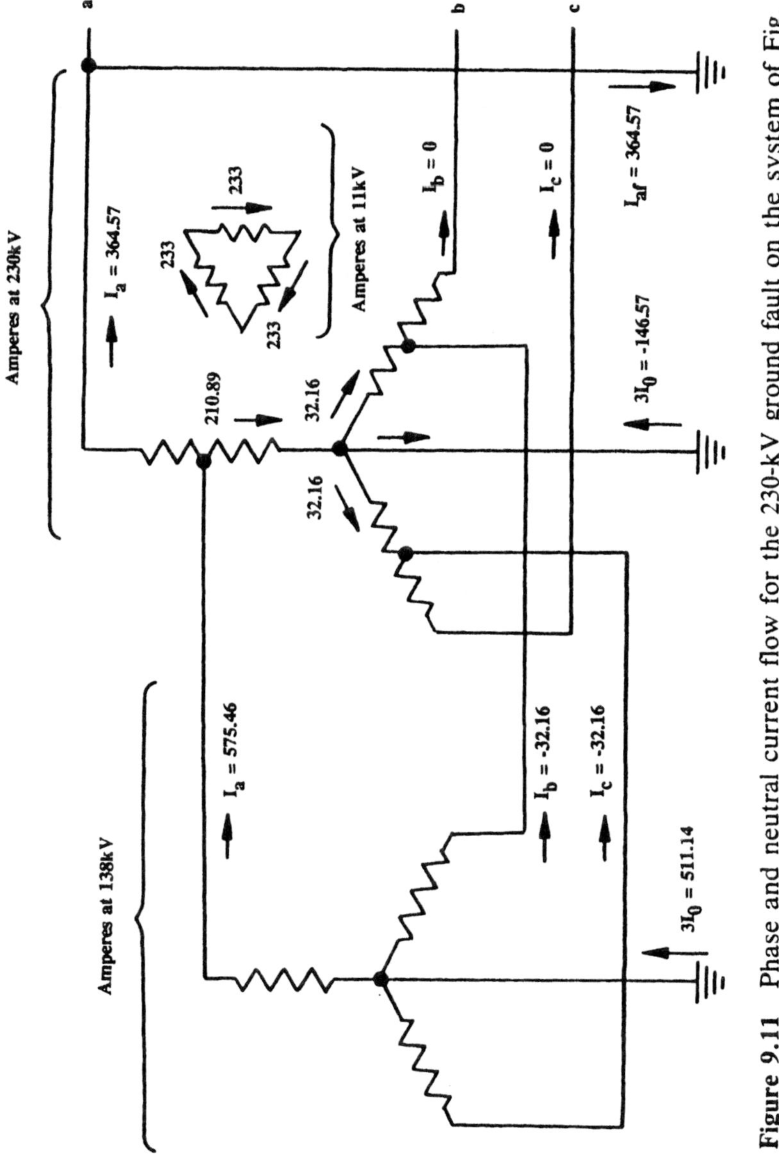

Figure 9.11 Phase and neutral current flow for the 230-kV ground fault on the system of Fig. 9.8.

up, down, or be zero, depending on the characteristics of the transformer and the connected low-voltage system. For faults on the low side of the auto, current will always be up the neutral, as conventional in two-winding banks except for the simultaneous faults discussed in Chapter 7.

As in the example one branch of the equivalent wye may be negative. Generally, this is in the medium-voltage leg but can be in the high-voltage leg. With a negative leg in a small MVA autotransformer connected into a very solidly grounded system, it is possible that the total of this negative leg and the zero-sequence connected system impedance may be negative. When transferred to the high-system base, the negative leg of the small autotransformer amplified this value where it can be greater than the zero low-system zero sequence. The net result is to reverse the zero sequence in the tertiary delta.

The direction of the current flow in the tertiary can be a question. An ampere-turn check will provide the values as well as check the direction of the current flow. In the example, assume that the 230-kV winding is 1 pu turn. Then the related per unit turns are as in Fig. 9.12. The ampere-turns in the main windings should be equal and opposite to the ampere-turns in the tertiary.

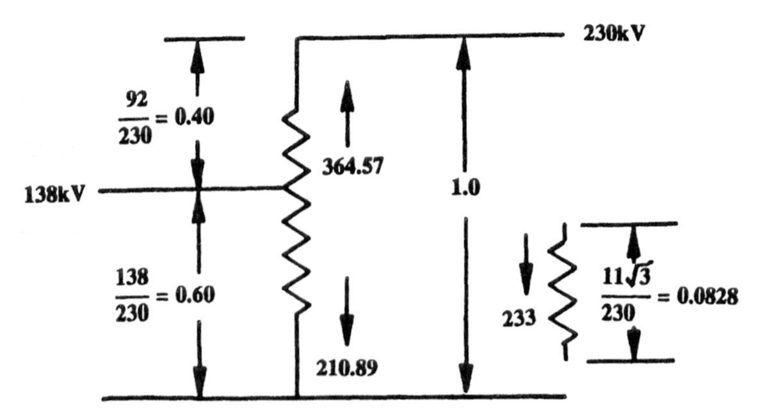

Figure 9.12 Ampere-turn check to determine current flow in the autotransformer tertiary windings.

Thus comparing ampere-turns up in the 138/230-kV windings with the ampere-turns down in the tertiary 11-kV winding.

Phase *a*:

$$\text{ampere-turns up } (364.57 \times 0.4 - 210.89 \times 0.6) = 19.3$$
$$\text{opposite ampere-turns down } (233 \times 0.0828) = 19.3$$

Phases *b* and *c*:

$$\text{ampere-turns out } (32.16 \times 0.6) = 19.3$$
$$\text{opposite ampere-turns } (233 \times 0.0828) = 19.3$$

Thus the directions shown in Fig. 9.11 are correct.

9.11 UNGROUNDED AUTOTRANSFORMERS WITH TERTIARY AND GROUNDED AUTOTRANSFORMERS WITHOUT TERTIARY

These two types are not too common. Their documentation is shown in Figs. 9.13 and 9.14.

9.12 TEST MEASUREMENTS FOR TRANSFORMER IMPEDANCE

When the leakage impedance of a transformer is not known it can be measured as from either the primary or secondary winding as follows: Short one winding and apply a voltage to pass rated current in the other. From Fig. 9.1b the impedance measured is

$$Z_T = \frac{V_H}{I_H} = Z_H + \frac{Z_L Z_e}{Z_L + Z_e} \tag{9.26}$$

Since Z_e unsaturated is very large compared to Z_L, the term $Z_L Z_e / Z_L + Z_e$ approaches and is approximately equal to Z_L, so that

$$Z_T = \frac{V_H}{I_H} = Z_{HL} = Z_H + Z_L \tag{9.27}$$

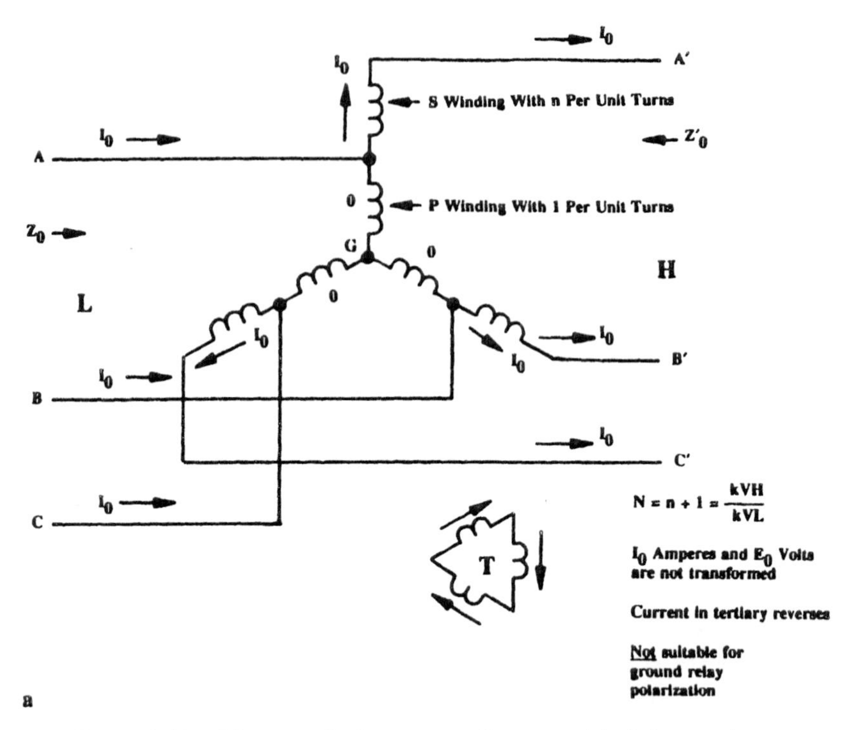

Figure 9.13 Ungrounded autotransformer with delta tertiary.

For the 20-MVA, 66.5-kV/13.8-kV, 10% transformer (Section 9.2), the rated primary current is

$$I_H = \frac{20,000}{66.5} = 300.75 \text{ A}$$

With 10% reactance, $V_w = 0.10 \times 66,500 = 6650$ V, so

$$Z_T = \frac{6650}{300.75} = 22.11 \ \Omega \text{ primary}$$

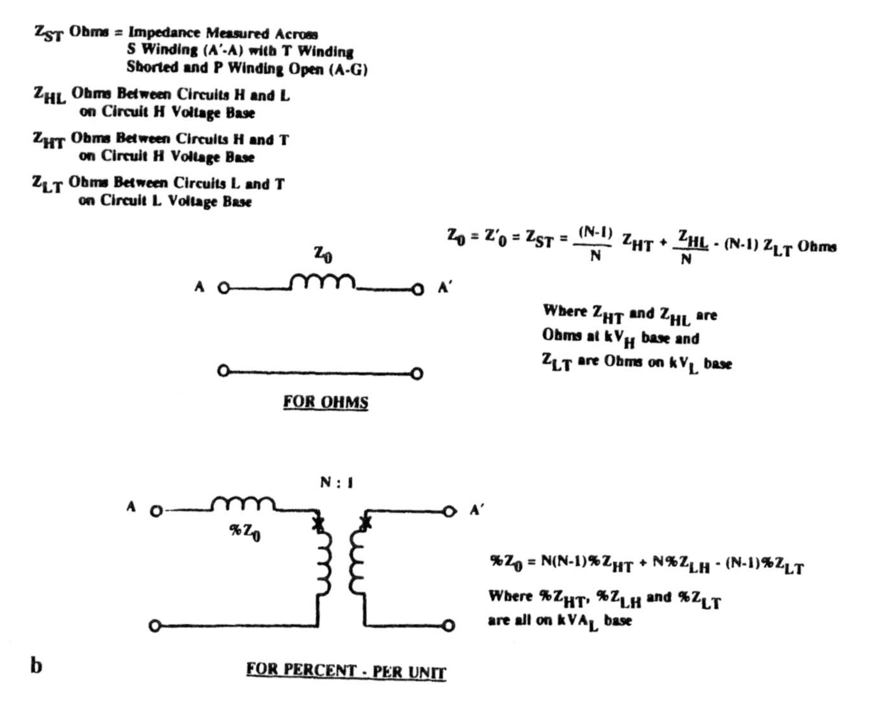

This checks the Eq. (9.9) determination. Where impedance is desired, a wattmeter can be used to measure the I^2r resistance portion.

On the test side (either primary as in the example above, or secondary), the base impedance is

$$Z_B = \frac{V_R}{I_R} \qquad \text{ohms at } V_R \tag{9.28}$$

a) I_0 Flowing Low to High

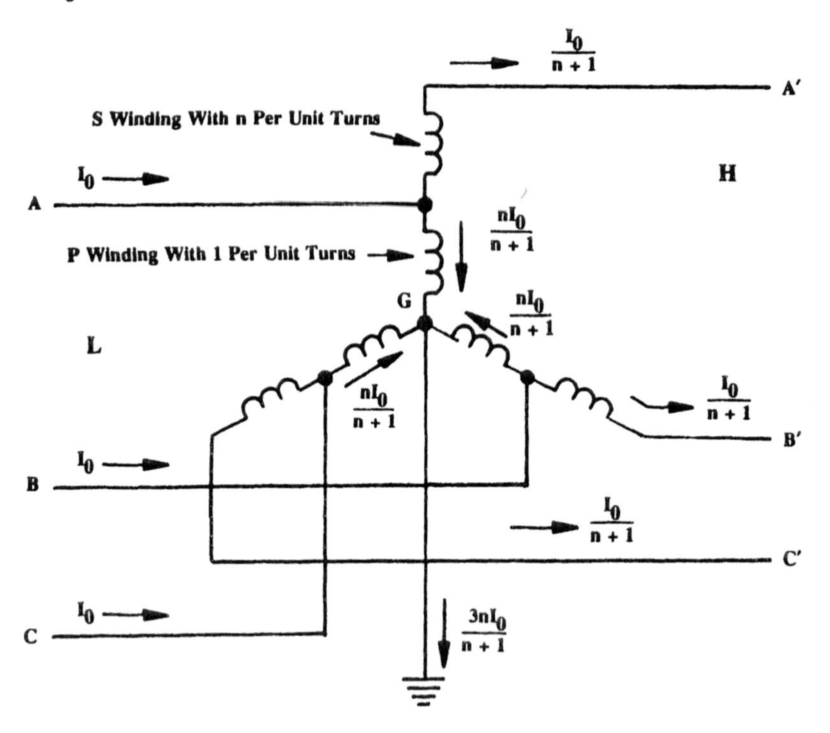

b) I_0 Flowing High to Low

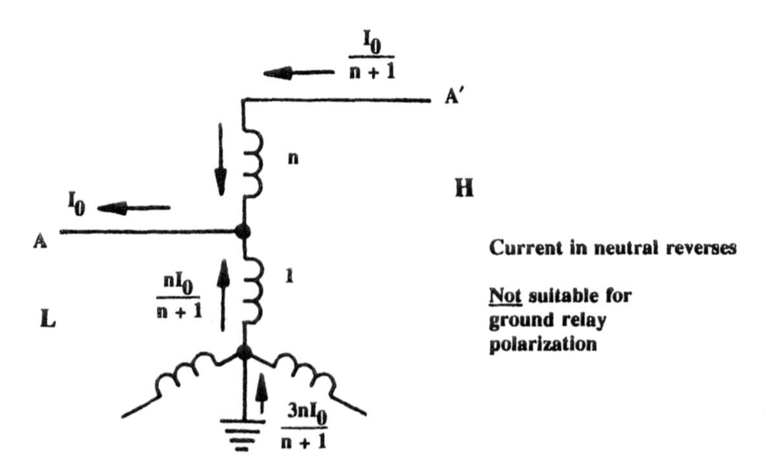

Figure 9.14 Ungrounded autotransformer without delta tertiary.

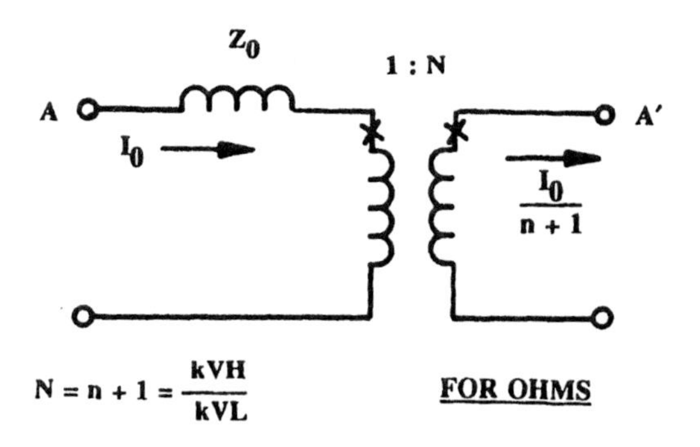

$\%Z_0 = \%Z_{LH}$

A A'

FOR PERCENT - PER UNIT

Z_{PS} Ohms = Impedance Measured Across P Winding (A-G) with S Winding Shorted (A'-A)

Z_{SP} Ohms = Impedance Measured Across S Winding (A'-A) with P Winding Shorted (A-G)

$$Z_{SP} = (N - 1)^2 Z_{PS} = n^2 Z_{PS}$$

$$Z_0 \text{ Ohms} = \frac{(N - 1)^2}{N^2} Z_{PS} = Z_{LH}$$

(Ohms between circuits L and H on circuit L voltage base)

where V_R and I_R are the rated voltage and current. Then the per unit impedance from Eq. (2.1) is

$$Z_T = \frac{Z_T \text{ in ohms}}{Z_B \text{ in ohms}} = \frac{V_W I_R}{I_R V_R} = \frac{V_W}{V_R} \quad \text{pu} \tag{9.29}$$

For the example,

$$Z_B = \frac{66{,}500}{300.75} = 221.11\ \Omega$$

$$Z_T = \frac{21.11}{221.11} = 0.10 \text{ pu} = 10\%$$

9.13 DETERMINATION OF THE EQUIVALENT ZERO-SEQUENCE IMPEDANCES FOR THREE-WINDING THREE-PHASE TRANSFORMERS WHERE THE TERTIARY DELTA WINDING IS NOT AVAILABLE

The following tests and calculations will provide the star equivalent zero-sequence impedances (Fig. 9.3). The three tests required are shown in Fig. 9.15.

From (a): $$\frac{V_{WH}}{I_{WH}} = Z_{0HL} = Z_{0H} + Z_{0L} \tag{9.30}$$

From (b): $$\frac{V_{WM}}{I_{WM}} = Z_{0ML} = Z_{0M} + Z_{0L} \tag{9.31}$$

From (c): $$\frac{V_{WP}}{I_{WP}} = Z'_{0H} = Z_{0H} + \frac{Z_{0L}Z_{0M}}{Z_{0L} + Z_{0M}} \tag{9.32}$$

From Eq. (9.30): $$Z_{0H} = Z_{0HL} - Z_{0L} \tag{9.33}$$

From Eq. (9.31): $$Z_{0M} = Z_{0ML} - Z_{0L} \tag{9.34}$$

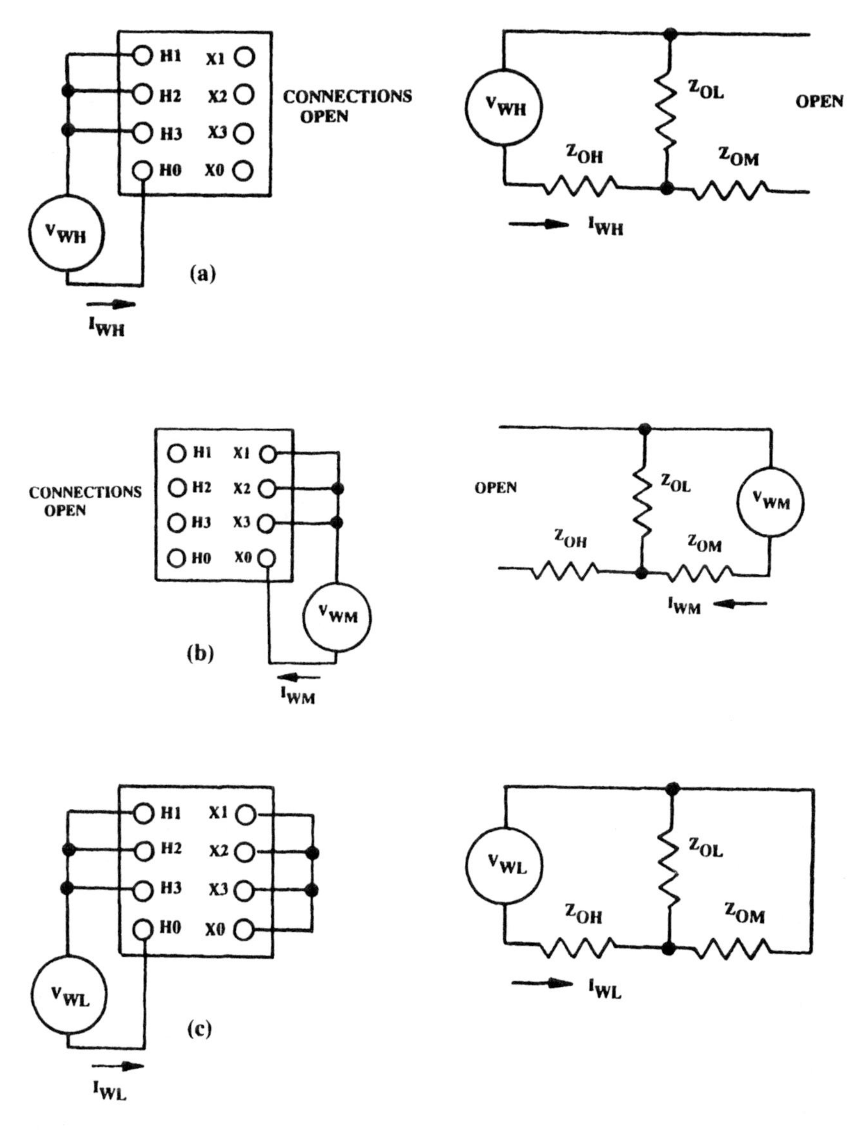

Figure 9.15 Tests for determining equivalent transformer impedances where the tertiary winding L is not available. (a) Test on H side with M open; (b) test on M side with H open; (c) test on H side with M shorted.

Substitute Eqs. (9.33) and (9.34) in Eq. (9.32):

$$Z'_{0H} = Z_{0HL} - Z_{0L} + \frac{Z_{0L}(Z_{0ML} - Z_{0L})}{Z_{0L} + Z_{0ML} - Z_{0L}} \tag{9.35}$$

$$Z'_{0H} = Z_{0HL} - Z_{0L} + \frac{Z_{0ML}Z_{0L} - Z_{0L}^2}{Z_{0ML}}$$

Multiplying both sides of Eq. (9.35) by Z_{0ML}:

$$Z'_{0H}Z_{0ML} = Z_{0HL}Z_{0ML} - Z_{0L}Z_{0ML} + Z_{0L}Z_{0ML} - Z_{0L}^2$$

$$Z_{0L}^2 = Z_{0ML}(Z_{0HL} - Z'_{0H})$$

$$Z_{0L} = \sqrt{Z_{0ML}(Z_{0HL} - Z'_{0H})} \tag{9.36}$$

Then Z_{0H} and Z_{0M} can be calculated from Eqs. (9.33) and (9.34).

9.14 DISTRIBUTION TRANSFORMERS WITH TAPPED SECONDARY

This type of transformer is utilized to supply three-phase and single-phase secondary loads simultaneously. Typical secondary voltages are 240/120 V on the secondary. The equivalent circuit, shown in Fig. 9.16, is complicated because the loads on the secondary may not be equal and usually one of the secondary midpoints is grounded.

The impedance from primary to one-half of the secondary differs from the impedance primary to total secondary. Thus, in the equivalent circuit the leakage impedance components appear in the primary (H) circuit and in each of the secondary (X) circuits. In Fig. 9.16:

$$Z_T = Z_{HX1-3} \tag{9.37}$$

(impedance looking into the H_1–H_2 winding with X_1, X_2, and X_3 shorted)

$$Z_{T2} = Z_{HX1-2} \tag{9.38}$$

(impedance looking into the H_1–H_2 winding with X_1 and X_2 shorted with X_3 open)

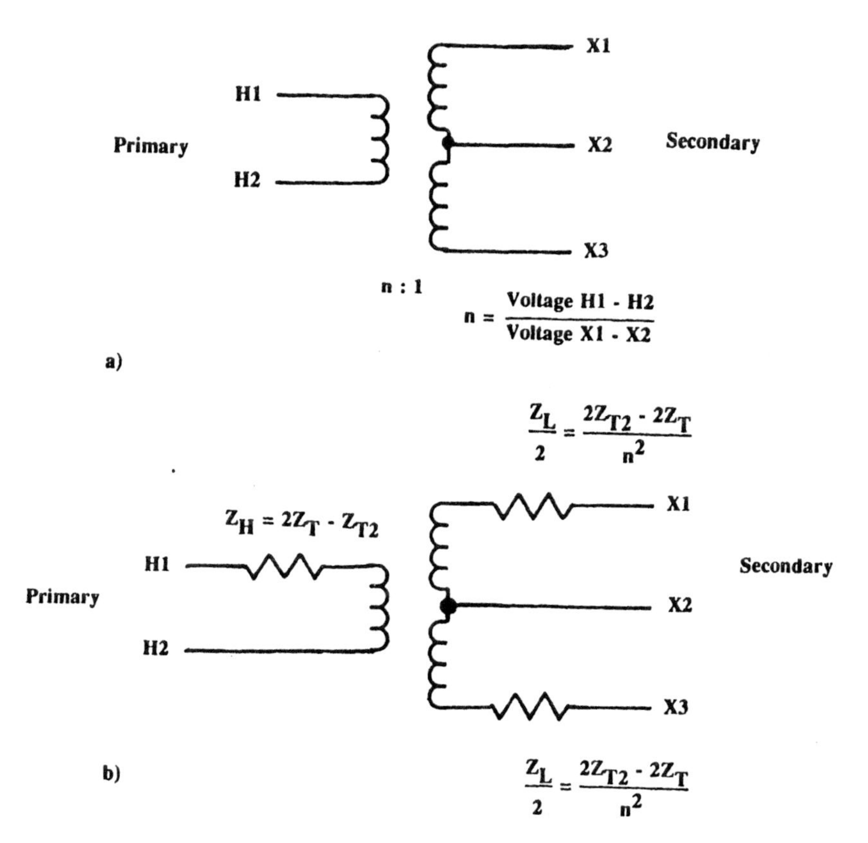

Figure 9.16 A distribution-type transformer with secondary mid-tap. (a) The mid-tapped transformer diagram; (b) the equivalent circuit.

These impedances can be in per unit (percent) or in ohms at the respective primary or secondary voltage. At around 480 V and below, the resistance of the windings and associated circuits can be significant and so should be considered in any calculations.

Where the value of Z_{T2} is not specified or known, the Westinghouse *Distribution Systems* handbook, (p. 223, eq. 14) indicates it can be approximated as

$$Z_{T2} = Z_{HX1-2} = 1.5R_{HX1-3} + j1.2X_{HX1-3} \quad (9.39)$$

Consider an example of a single-phase 25-kVA, 2.4-kV to volt transformer with $Z_T = 0.013 + j0.018$ per unit. With Z_{T2} unknown, use the approximation of Eq. 9.39;

$$Z_{T2} = 1.5(0.013) + j1.2(0.018)$$
$$= 0.0195 + j0.0216 \text{ pu at 25 kVA, } \quad 2.4 \text{ kV, or } 120 \text{ V} \tag{9.40}$$

Then, the Fig. 9.12 equivalents are

$$\begin{aligned} 2Z_T - Z_{T2} = &\quad 0.026 + j0.036 \\ &-0.0195 - j0.0216 \\ \hline &\quad 0.0065 + j0.0144 \text{ pu at 25 kVA} \end{aligned} \tag{9.41}$$

or

$$\begin{aligned} 2Z_T - Z_{T2} &= 1000(2.4)^2 Z_{pu}/25 \\ &= 1.498 + j3.318 \text{ ohms at 2.4 kV} \end{aligned} \tag{9.42}$$

and

$$\begin{aligned} 2Z_{T2} - 2Z_T = &\quad 0.038 + j0.0432 \\ &-0.026 - j0.036 \\ \hline 2Z_{T2} - 2Z_T = &\quad 0.013 + j0.0072 \text{ pu at 25 kVA,} \\ &\quad 2.4 \text{ kV, or 120 V} \end{aligned} \tag{9.43}$$

or

$$\begin{aligned} 2Z_{T2} - 2Z_T &= 1000(0.12)^2 Z_{pu}/25 \\ &= 0.00749 + j0.00415 \text{ ohms at 120 V} \end{aligned} \tag{9.44}$$

As a check, express Z_T and Z_{T2} in primary ohms:

$$Z_T = \begin{matrix} 1000(2.4)^2 Z_{Tpu}/25 \\ 3.0 + j4.15 \text{ ohms at 2.4 kV} \end{matrix} \tag{9.45}$$

$$Z_{T2} = 4.5 + j4.98 \text{ ohms at 2.4 kV} \tag{9.46}$$

Then

$$2Z_T - Z_{T2} = 1.5 + j3.32 \text{ ohms at } 2.4 \text{ kV as above} \quad (9.47)$$

$$2Z_{T2} - 2Z_T = 3.0 + j1.66 \text{ ohms at } 2.4 \text{ kV} \quad (9.48)$$

$$(2Z_{T2} - 2Z_T)/n^2 = 0.00749 + j0.00415 \text{ ohm at } 120 \text{ V} \quad (9.49)$$

as above where $n = 2400/120 = 20$.

Examples for the calculation of secondary ground faults with one mid-tap grounded are in Chap. 6.

9.15 ZIG-ZAG CONNECTED TRANSFORMERS

The zig-zag connected transformer, Fig. 9.17, is basically three 1:1 ratio single-phase transformers connected to pass zero se-

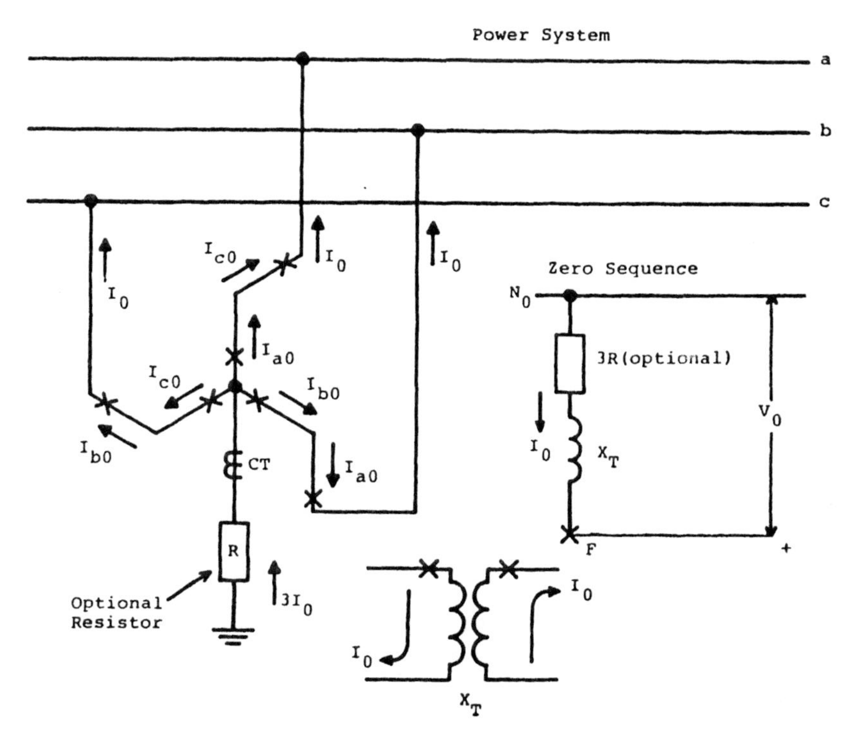

Figure 9.17 The zig-zag transformer for low impedance grounding.

quence current only. With the transformer polarities shown and since $I_{a0} = I_{b0} = I_{c0}$, zero sequence current can flow. The positive and negative sequence currents cannot, since $I_{a1} \neq I_{b1} \neq I_{c1}$ and $I_{a2} \neq I_{b2} \neq I_{c2}$. The impedance of the zero sequence path is the leakage impedance of the transformer X_T as shown.

With line to neutral voltage of 1 per unit, the voltage across each transformer winding is 0.866 per unit. The zig-zag transformer grounding is essentially reactance grounding as the transformer resistance is very low. If the zig-zag X_T is too low for the desired fault limiting, a resistor or reactor can be used as shown.

9.16 REACTORS

Reactors connected in shunt are used to reduce the capacitance of long transmission lines, and in series to limit fault currents. For the series type, the rating of a typical reactor might be 5% reactance at 100 A at 46 kV. Thus the voltage drop across the reactor is

$$0.05 \times \frac{46{,}000}{\sqrt{3}} = 1327.91 \text{ V}$$

and

$$X = \frac{1327.91}{100} = 13.28 \ \Omega \qquad (9.37)$$

The reactor kVA rating is 1327.91 × 100/1000 = 132.79 kVA.

This is a single-phase rating for the individual reactor and indicated the kVA drop through the reactor. However, this kVA rating cannot be used in per unit or percent fault calculations. It cannot be compared with a transformer kVA rating, which is a system kVA and not the drop across the transformer.

For system calculations, the reactor has 5% reactance at

$$\frac{46}{\sqrt{3}} \times 100 = 2655.81 \text{ kVA single-phase} \qquad (9.38)$$

or

$$46 \times \sqrt{3} \times 100 = 7967.43 \text{ kVA three-phase} \qquad (9.39)$$

As a check, 5% X at 7967.43 kVA, 46 kV from Eq. (2.18) is

$$\frac{10 \times 46^2 \times 5}{7967.43} = 13.28\ \Omega \text{ at } 46 \text{ kV}$$

as in Eq. (9.37).

For the shunt reactors, the percent (per unit) reactance should be specified as Eq. (9.39). As these reactors normally are used to compensate for capacitance or control transients, and so on, their values are quite large. Connected as shunt devices similar to loads, generally they will have minimum impact on fault currents (Section 6.3, assumption 3).

9.17 CAPACITORS

Shunt capacitors are widely used in distribution systems to regulate voltage and reactive power flow. With high values of $-X_C$ to the low values of impedance that determine fault current, shunt capacitors generally are neglected (Section 6.3, assumption 3).

Series capacitors are used in long EHV transmission lines to reduce the total impedance across the line for lower var requirements and improved system stability. Thus their $-X_C + X_L$ of the line must be considered in fault calculations.

10

Generator and Motor Characteristics

10.1 INTRODUCTION

Generators, synchronous machines, are the source of power for the electric system. They also supply positive sequence to faults and unbalances. A sudden increase in the generator current such as occurs during a fault results in a transient symmetrical current and may cause a transient dc component. A typical three-phase fault at the terminals of a generator operating at no load initially and with constant excitation is illustrated in Fig. 10.1. Phase *a* is symmetrical; phases *b* and *c* are not. The intent of this chapter is to review briefly and provide a basic understanding of the various generator constants involved in symmetrical components.

10.2 TRANSIENTS IN RESISTANCE–INDUCTANCE SERIES CIRCUITS

A review of the basic fundamentals of transients will aid in understanding of the asymmetrical currents that can occur. In the

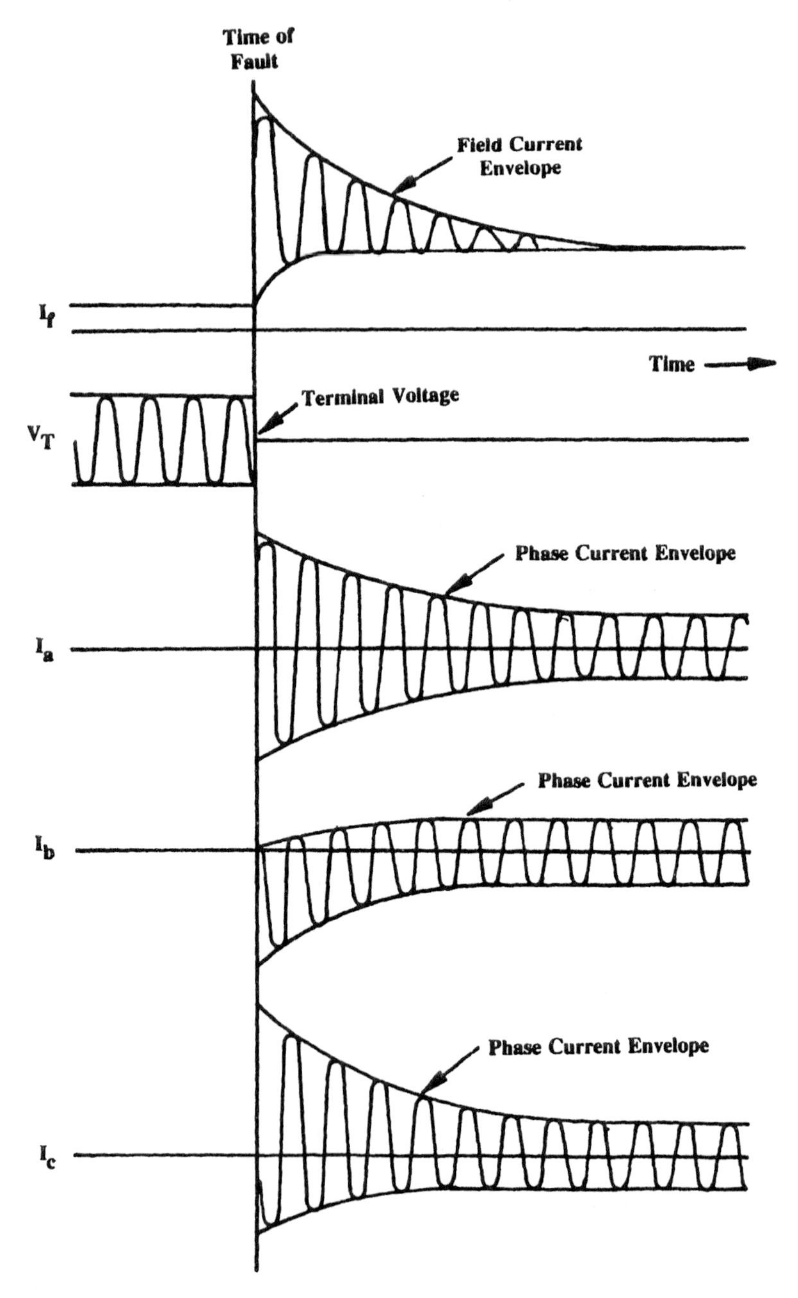

Figure 10.1 Typical three-phase fault at the generator terminals operating at no load and with constant excitation.

series circuit of Fig. 10.2, and neglecting any harmonics and, for the moment, the ac decrement, we obtain

$$v = V_M \sin(\omega t + \theta) = Ri + L\frac{di}{dt} \tag{10.1}$$

This is a first-order differential equation whose solution can be shown to be

$$i = \underbrace{-\frac{V_M}{Z}\sin\left(\theta - \tan^{-1}\frac{X}{R}\right)e^{-Rt/L}}_{\text{Dc component}} + \underbrace{\frac{V_M}{Z}\sin\left(\omega t + \theta - \tan^{-1}\frac{X}{R}\right)}_{\text{Ac component}} \tag{10.2}$$

The maximum transient will occur when switch S is closed at or near zero on the voltage wave. In a highly inductive circuit the current at this time should be at or near maximum value rather than zero. Hence the dc component occurs such that the

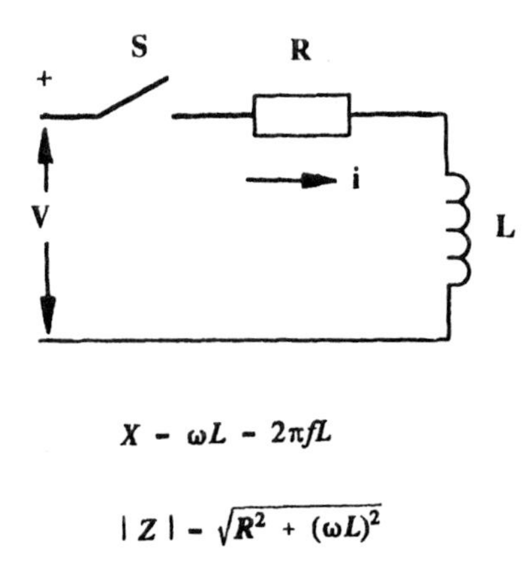

Figure 10.2 A series inductance circuit.

initial sum is at or near zero. For this case Eq. (10.2) simplifies to

$$i = \underbrace{-I_M e^{-t/T}}_{\text{Dc component}} + \underbrace{I_M \cos \omega t}_{\text{Ac component}} \qquad (10.3)$$

where $I_M = V_M/Z$ and the time constant $T = L/R$. For an offset above the X axis, the sign of the two components of Eqs. (10.2) and (10.3) are reversed. The two components add to give the maximum asymmetrical current as illustrated in Fig. 10.3.

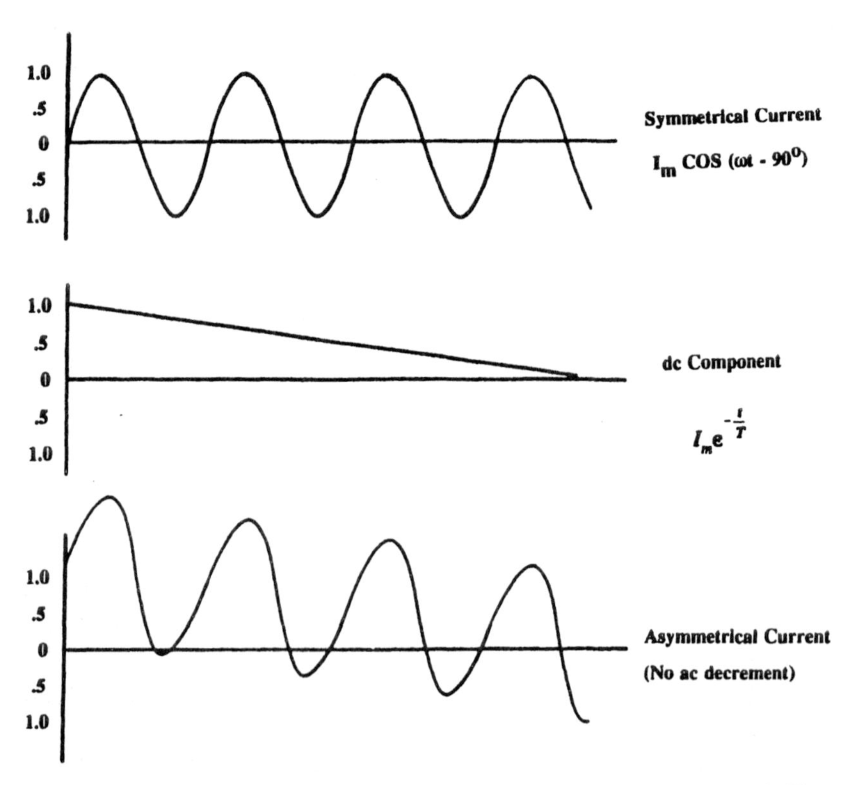

Figure 10.3 Maximum asymmetrical current when switch S of Fig. 10.2 is closed when V is at or near maximum value.

With the dc component established to provide proper phase relations between the voltage and current following any sudden change, its function is finished. The dc component then decays as a function of the inductance and resistance in the circuit as defined by the time constant T.

10.3 TRANSIENT GENERATOR CURRENTS

A sudden change in voltage and current in a generator such as occurs during faults produces transients similar to those described above. Thus any of the armature currents may be divided into two components:

1. A symmetrical ac component (the associated component in the field is a dc current).
2. A dc component (the associated component in the field and damper windings is an ac current).

One way to visualize generator operation is to consider it as a transformer with one winding stationary and the other winding rotating at synchronous speed. Thus dc in the field produces ac in the armature, and vice versa. Hence dc in the armature produces ac in the field. This very simplified analogy can be helpful in understanding negative-sequence currents, as will be discussed later.

10.3.1 Symmetrical AC Component of Armature Current

In symmetrical components, the principal concern is with the symmetrical component and its associated constants. These are used to obtain the symmetrical fault currents for various faults and unbalances. The maximum asymmetrical currents can be determined from these symmetrical values when desired.

Fault studies do not include the dc component and are seldom required for the application and setting of protective relays. If necessary, as for circuit breaker applications, various factors are available from the standards, manufacturers, or other sources to account for the peak values that can occur.

For synchronous machines, the symmetrical ac component can be resolved into three distinct components:

1. The subtransient component [the double-prime (″) values]
2. The transient component [the single-prime (′) values]
3. The steady-state component

These are shown on an envelope of the symmetrical ac current of Fig. 10.4.

Subtransient Component

The air-gap flux at the first instant of a sudden change is prevented from changing to any great extent by the presence of damper windings or other closed rotor circuits for eddy currents. Currents are induced in these circuits which tend to maintain the air-gap flux resulting in an increase of the initial value of the current, i''_d. This is shown as "oc" in Fig. 10.4. For a three-phase fault at the terminals,

$$i''_d = \frac{v''}{X''_d}; \text{ or at no load, } = \frac{v_{\text{rated}}}{X''_d} \tag{10.4}$$

The subtransient reactance (X''_d) approaches the armature leakage reactance but is higher as a result of the damper windings, and so on. The subtransient current i''_d does not last long as the damper windings have a relatively high resistance so the subtransient time constant T''_d is low. Typical values are in the order of 0.01 to 0.05 s.

Transient Component

By plotting the excess of current over the substained or steady-state current on semilog (current log and time ordinary scale), the envelope of the current wave will be a straight line except for the subtransient first few cycles. Projecting this

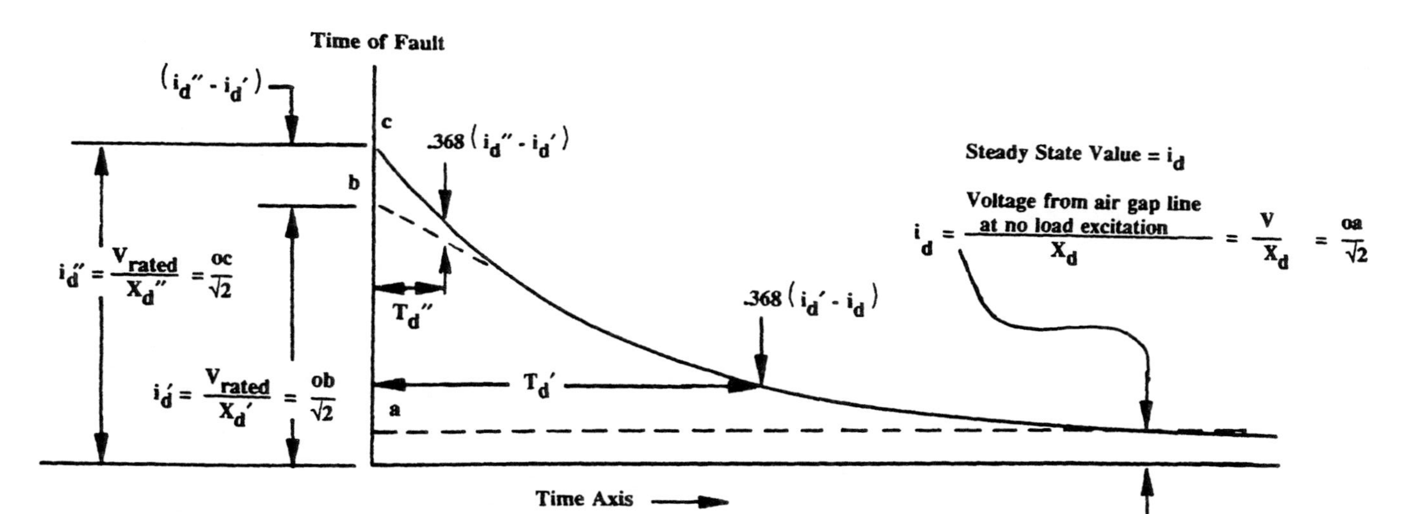

Figure 10.4 Component of the symmetrical ac current of a synchronous machine at no load where $V_{rated} = V'' = V' = V$. i''_d, i'_d, and i_d are rms values.

straight line back to the time origin and adding the steady-state value determines the transient component, i'_dor 0*b* in Fig. 10.4.

$$i'_d = \frac{v'}{X'_d}; \text{ or at rated voltage, } = \frac{v_{\text{rated}}}{X'_d} \tag{10.5}$$

When the machine is short circuited, the currents suddenly appearing in the armature windings act to demagnetize the field and decease the flux linkages with the field winding. The term *flux linkages* is the summation or integral of all the elements of flux multiplied by the fraction of the total current linked by each. Now flux linkages with any inductive circuit cannot be changed instantly, as this would require an infinitely large voltage but tend to remain constant. Thus when the fault occurs, the flux linkages with each circuit on the rotor and armature remain constant for the first instant. This results in a current induced in the field winding to annul the demagnetizing effect of the armature current.

The increase in field current increases the field flux linkages and decreases the air-gap flux. Thus the transient reactance includes the effect of both armature and field leakages and is higher than the armature leakage reactance.

The induced field current has no voltage to substain it, so it decays exponentially. There is always constant proportionality between the ac current in the armature and the dc component in the field both during steady-state and transient conditions. The transient short-circuit time constant (T'_d) varies typically from 0.35 to 3.3 s.

Steady-State Component

The transient component eventually decays to the steady-state current (i_d) if the fault has not been cleared long before by the protective relays, as it should be. Assuming no change of the field current by the excitation system during the fault, the load prior to the fault will determine the steady-state component.

With no load before the fault,

$$i_d = \frac{\text{air-gap line voltage corresponding to no-load excitation}}{X_d} \qquad (10.6)$$

At short circuit, the demagnetizing effect of the large fault current occurs along the direct axis and the flux density within the machine falls below the saturation point. Thus for the fault X_d is the unsaturated direct-axis reactance.

With reference to Fig. 10.5:

I_{fg} = field current at normal open-circuit voltage on the air-gap line

$I_{fs} = OP$ = field current necessary to circulate full load current under short-circuit conditions with terminal voltage equal zero

$V_{PQ} = PQ$ = line-to-neutral voltage when the short circuit is removed, maintaining field current equal to I_{fs}

The unsaturated synchronous reactance per phase is

$$X_d = \frac{V_{PQ}}{I_{\text{rated}}} \quad \text{ohms} \qquad (10.7)$$

$$X_d = \frac{V_{PG}}{V_T} = \frac{I_{fs}}{I_{fg}} \qquad (10.8)$$

Total Symmetrical AC Component of the Synchronous Generator Armature Current

The total symmetrical ac current for a three-phase short circuit is

$$i_{\text{ac}} = (i''_d - i'_d)e^{-t/T''_d} + (i'_d - i_d)e^{-t/T'_d} + i_d \qquad (10.9)$$

These quantities may be either rms or instantaneous values. The

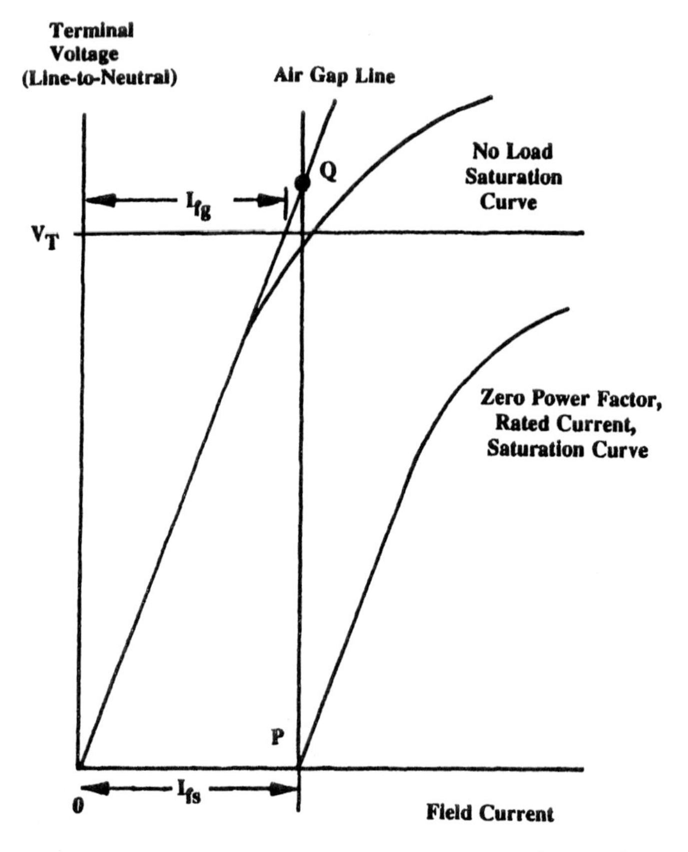

Figure 10.5 Typical characteristic curves of a synchronous generator.

currents in all three phases are equal and 120° displaced for the three-phase fault.

10.3.2 DC Component of Armature Current

The dc component does not occur uniformly in all three phases or consistently with repeated faults. Its initial magnitude depends on the time of the fault with respect to the voltage, type of fault, system constants, and load conditions prior to the fault. In a highly reactive system (typical of large generators), its maximum value is when the fault occurs at or near zero voltage. Thus

$$i_{dc(max)} = \sqrt{2}\,\frac{v''}{X''_d} = \sqrt{2}\,i''_d; \quad \text{or at no load,} = \sqrt{2}\,\frac{v_{rated}}{X''_d} \tag{10.10}$$

Typical cases are shown in Figs. 10.1 and 10.3.

The dc component decays exponentially as defined by T_a, the armature short-circuit or dc time constant. X_2, as discussed in the next section, is sort of an average reactance of the armature with the field short circuited. Thus where r_a is the dc resistance of the armature,

$$T_a = \frac{X_2}{2\pi f r_a} \tag{10.11}$$

10.4 NEGATIVE-SEQUENCE COMPONENT

Unbalances in the three-phase power system produce negative-sequence voltages and currents as discussed previously. When negative-sequence currents flow in the synchronous generator armature they produce a magnetic field in the air gap that rotates at synchronous speed in a direction opposite to that resulting from the normal motion of the field structure. As a result, double-frequency current flows in the dc field, damper windings, rotor wedges, and so on. This can be visualized by the relative motion of three-phase currents that rotate counterclockwise but with opposite phase sequence from that produced by the synchronous rotation (i.e., negative sequence a, c, b versus positive sequence a, b, c) across the air gap between the fixed winding and the rotating winding. Thus during an unbalance in the power system, $2f$ current can be seen in the dc field.

The subtransient reactances can be measured by blocking the rotor with the field winding shorted and applying a single-phase voltage across any two terminals. As the position of the rotor is changed, the measured reactance per phase will vary considerably if the machine has salient poles without dampers and very

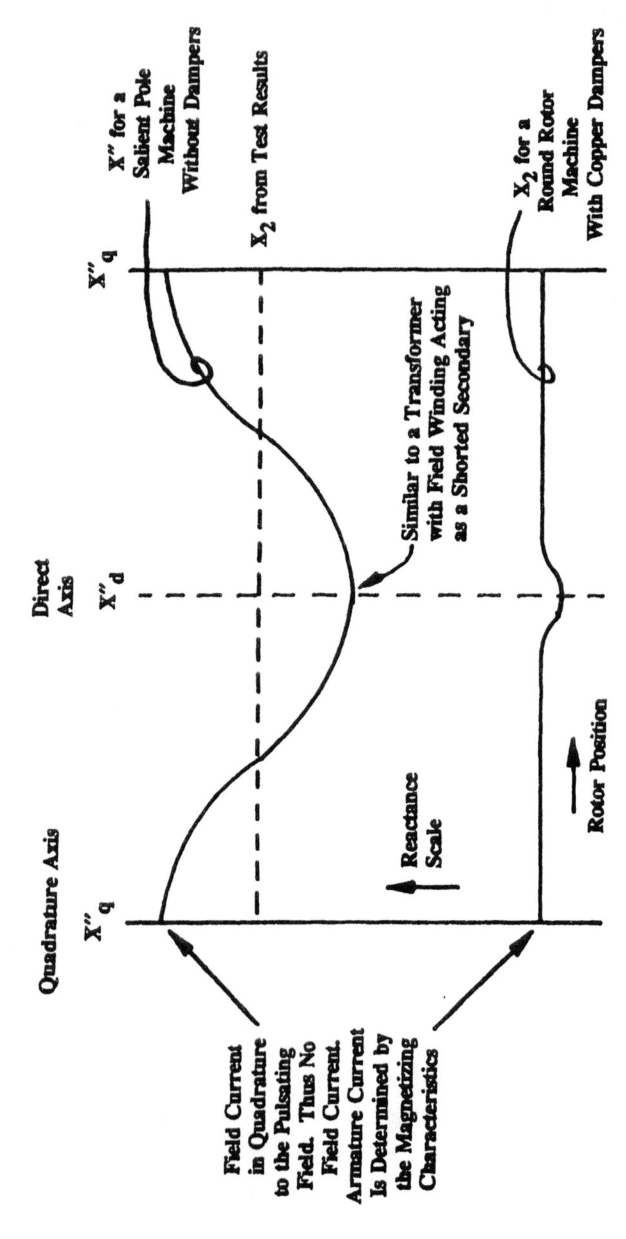

Figure 10.6 Typical subtransient reactance variations with machine rotor positions.

little if damper winding exists or if the machine has a round rotor. Typical variations are shown in Fig. 10.6.

For negative-sequence currents a similar phenomenon exists except that the rotor is at $2f$ with relation to the field set up by the applied voltage. There are a number of ways to approximate the negative-sequence reactance. A common approach is to consider the negative-sequence reactance as a mean between the direct axis X''_d and quadrature axis X''_q values, or

$$X_2 = \tfrac{1}{2}(X''_d + X''_q) \tag{10.12}$$

10.5 ZERO-SEQUENCE COMPONENT

The zero-sequence reactance of synchronous machines is quite variable, depending largely on pitch and breath factors of the armature winding. Generally, the values of X_0 are much smaller than the X_1 and X_2 values.

10.6 TOTAL RMS ARMATURE CURRENT

At any instant of time the total rms current is

$$i_{\text{rms}} = \sqrt{i_{\text{dc}}^2 + i_{\text{ac}}^2} \tag{10.13}$$

At no load and assuming no decrement or decrease in either the dc or ac components,

$$\begin{aligned} i_{d\ \text{rms}} &= \sqrt{\left(\frac{\sqrt{2}\ v_{\text{rated}}}{X''_d}\right)^2 + \left(\frac{v_{\text{rated}}}{X''_d}\right)^2} \\ &= \sqrt{3}\ \frac{v_{\text{rated}}}{X''_d} = \sqrt{3}\ i''_{d\ \text{rms}} \end{aligned} \tag{10.14}$$

With reference to Fig. 10.3, the maximum asymmetrical current occurs at the first positive peak of the symmetrical current. At that time both the dc and the ac will have decayed slightly. This is recognized by generally using 1.6 instead of $\sqrt{3}$ or 1.732 in Eq. (10.14). At 5 kV and below, 1.5 may be used instead of 1.6 and at 600 V or less, 1.25 instead of 1.6.

10.7 ROTATING MACHINE REACTANCE FACTORS FOR FAULT CALCULATIONS

In the selection of interruption devices such as circuit breakers, various factors may be applicable. These should be checked with the equipment manufacturer and all applicable standards.

One of these factors is the momentary duty, which is the current that the device must be capable of carrying without damage immediately after a fault. This was discussed in Section 10.6. A second factor is the interrupting duty, the current at the instant the device opens to interrupt the fault current. This current will depend on the time constants and operating speed of the device.

10.8 TIME CONSTANTS FOR VARIOUS FAULTS

The time constant of any circuit always is

$$T = \frac{\sum L}{\sum R} = \frac{\sum X}{\sum \omega R} \quad \text{seconds} \tag{10.15}$$

The values and equations in the various sections of this chapter are for three-phase faults. For any other type of fault or unbalance, the sequence networks and their interconnections will determine the time constant per Eq. (10.15). Thus for phase-to-phase and phase-to-ground faults:

For phase-to-phase faults:

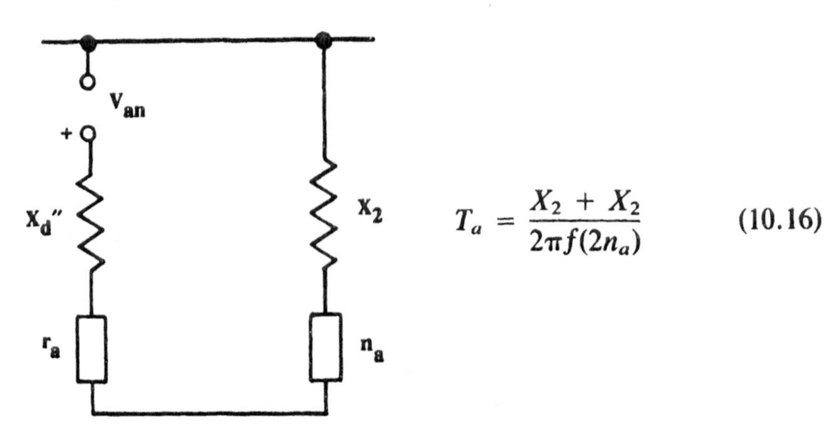

$$T_a = \frac{X_2 + X_2}{2\pi f(2n_a)} \tag{10.16}$$

For phase-to-ground faults:

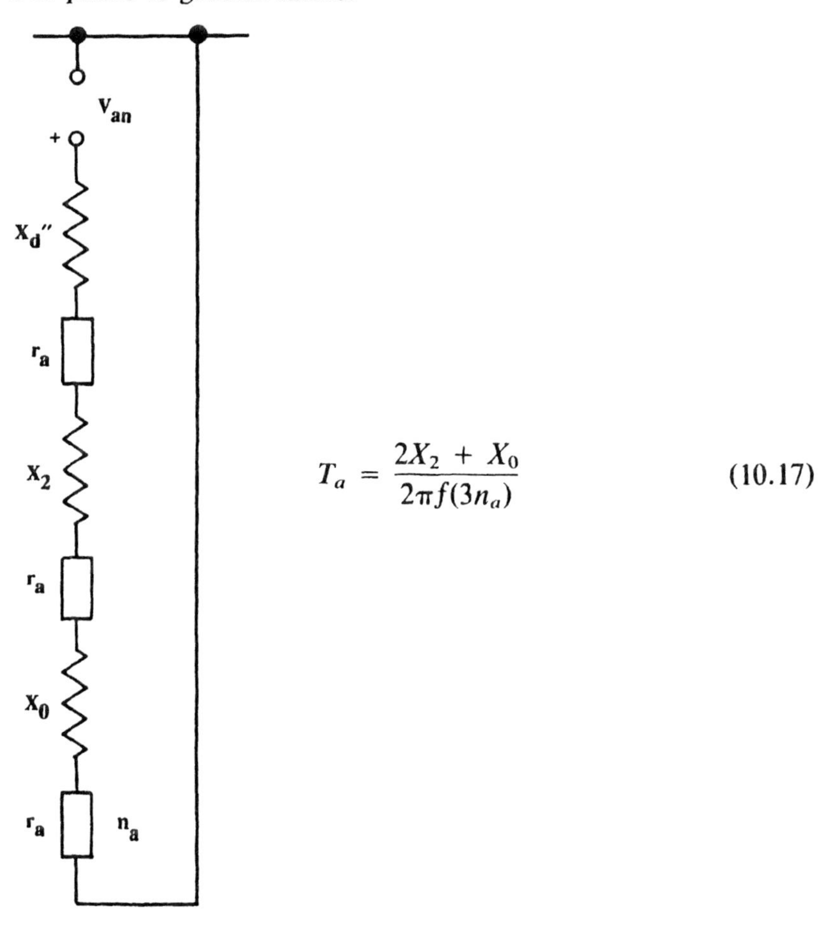

$$T_a = \frac{2X_2 + X_0}{2\pi f(3n_a)} \tag{10.17}$$

Note that X_2 is used for X''_d when calculating the dc time constant as explained above. These networks are for a fault at the machine terminals with no external circuits connected. External connected circuits would provide parallel paths and the time constant would be a function of the total X and total R from the neutral buses to the fault point in each network and the networks interconnected as for the fault calculation.

By way of review, typical dc time constants (in seconds) of 60-Hz electrical equipment arc:

	Average	Range
Two-pole turbine generators	0.13	0.04–0.24
Four-pole turbine generators	0.20	0.15–0.35
Salient pole generator and motors (with dampers)	0.15	0.03–0.25
Salient pole generator (no dampers)	0.30	0.10–0.50
Condensers	0.17	0.10–0.30

Typical time constants for power transformers are:

$$\text{5–10 MVA: } \frac{X}{R} = 10,\ T_a = 0.0265$$

$$\text{20–50 MVA: } \frac{X}{R} = 20,\ T_a = 0.053$$

$$\text{50–200 MVA: } \frac{X}{R} = 30,\ T_a = 0.0795$$

$$\text{200–500 MVA: } \frac{X}{R} = 45,\ T_a = 0.119$$

$$\text{500 MVA: } \frac{X}{R} = 50,\ T_a = 0.133$$

Typical time constants for transmission lines are given by

$$T_a = \frac{X_2}{377R} = \frac{\tan\theta}{377}$$

where θ is the line angle.

θ	T_a
60°	0.0046
70°	0.0073
75°	0.0099
80°	0.015
85°	0.0304
86°	0.038
87°	0.051

10.9 INDUCTION MACHINES

Equivalent diagrams for an induction machine are shown in Fig. 10.7 along with typical per unit resistance and reactance values on the motor kVA and kV. Where the symmetrical starting or locked rotor current is known in per unit,

$$jX''_d = \frac{1}{I_{\text{starting}}} \quad \text{pu} \tag{10.18}$$

or with the typical values given in Fig. 10.7,

$$I_{\text{starting}} = \frac{1}{0.15} = 6.67 \text{ pu} \tag{10.19}$$

Since the shunt jX_M is high relative to the other impedances in the motor equivalent circuit (Fig. 10.7b), the input equivalent impedances (Z_{M1} and Z_{M2}) are essentially equal to X''_d when the motor is stalled ($S = 1$).

Using the typical values of Fig. 10.7b,

$$\text{Stalled } (S = 1)\text{:} \quad Z_{M1} = Z_{M2} = 0.144\angle 82.39° \text{ pu} \tag{10.20}$$

$$\text{Running } (S = 0.01)\text{:} \quad Z_{M1} = 0.927\angle 25.87° \text{ pu} \tag{10.21}$$

$$Z_{M2} = 0.144\angle 84.19° \text{ pu} \tag{10.22}$$

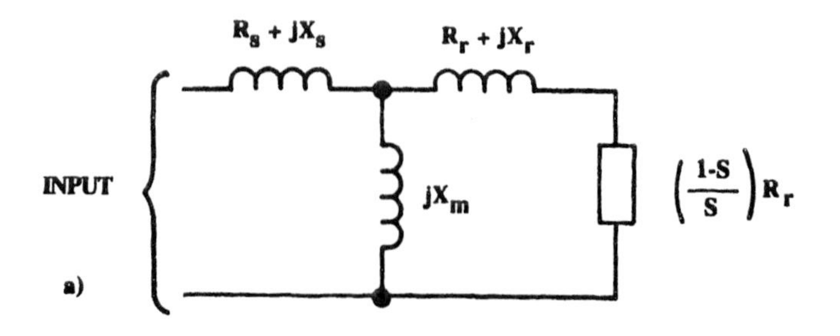

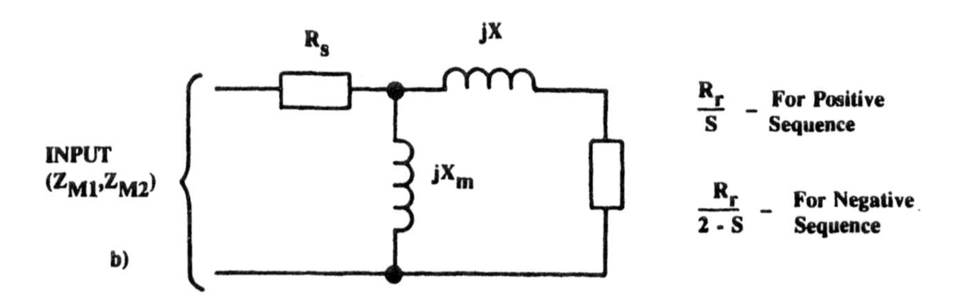

R_s = 0.01 pu = Stator Resistance

jX_s = Stator Leakage Reactance at Rated Frequency

R_r = 0.01 pu = Rotor Resistance

jX_r = Rotor Leakage Reactance at Rated Frequency

JX_m = j3.0 pu = Shunt Exciting Reactance

$jX = jX_s + jX_r = jX_d'' = j0.15$ pu

Values shown are typical for an induction motor and are per unit on the motor kVA and kV base

$$S = \frac{\text{Synchronous RPM - Rotor RPM}}{\text{Synchronous RPM}} = \begin{matrix}\text{1.0 for stalled condition} \\ \text{0+ for running conditions}\end{matrix}$$

Figure 10.7 Equivalent-circuit diagrams for induction motors: (a) equivalent diagram for an induction motor; (b) simplified equivalent induction motor diagram.

Thus from stalled to running the positive-sequence impedance changes from approximately 0.15 pu to around 0.9 to 1.0 pu, while the negative-sequence impedance remains essentially constant at approximately 0.15 pu. These values are based on the rated motor kVA and kV, which is roughly equal to the motor horsepower (hp). Actually,

$$\mathrm{kVA}_{\mathrm{rated}} = \frac{(\text{horsepower})(0.746)}{(\text{efficiency})(\text{power factor})} \tag{10.23}$$

The values will vary with each individual motor so that the manufacturer or test data should be used when available. In the absence of specific data, the typical values above are good approximations and quite useful in most cases.

10.10 SUMMARY

The characteristics of positive-sequence sources, synchronous machines, and their reaction to negative- and zero-sequence currents have been discussed briefly. The principal reactances are called *constants*. They are provided by the manufacturer and used as constants in the sequence networks to determine currents and voltages during various unbalances and faults. Beyond the scope of this chapter they are used to determine operation during various load conditions and during transient performance (stability studies).

Operating normally around rated voltage the permeability of the iron in the magnetic structures varies with voltage so that the reactances change depending on the operating prefault voltage and field excitation. As a result, the positive-sequence direct-axis reactances have several forms.

As a general reference the following values are identified:

$$X''_d = X''_{dv} = X''_d\ (\text{rated voltage}) = X''_d\ (\text{saturated}) \tag{10.24}$$

$$X''_{di} = X''_d\ (\text{rated current}) = X''_d\ (\text{unsaturated}) \tag{10.25}$$

$$X'_d = X'_{dv} = X'_d\ (\text{rated voltage}) = X'_d\ (\text{saturated}) \tag{10.26}$$

$$X'_{di} = X'_d \text{ (rated current)} = X'_d \text{ (unsaturated)} \quad (10.27)$$

$$X_{dv} = X_d \text{ (rated voltage)} = X_d \text{ (saturated)} \quad (10.28)$$

$$X_d = X_{di} = X_d \text{ (rated current)} = X_d \text{ (unsaturated)} \quad (10.29)$$

The rated voltage (v) values are measured when the excitation produces rated voltage at no load before the short circuit. The rated current (i) values are measured when the excitation is reduced to produce from no load a transient component of short-circuit current equal to rated current.

In symmetrical components, the rated voltage X''_d, X'_d, and X_d are the values normally used in the positive-sequence network and for fault and unbalance calculations. The other reactances, including the quadrature-axis (X_q) values, are involved in steady-state and stability studies and calculations. The X_2 and X_0 constants were outlined in earlier sections.

For system protection fault studies the almost universal practice is to use the subtransient (X''_d) reactance for the rotating machines in positive-sequence networks. This provides a maximum value of fault current useful for high-speed relaying.

Although slower-speed protection (time overcurrent relays) may operate after the subtransient reactance has decayed into the transient reactance period, the general practice is to use X''_d. In many cases the time overcurrent protection is remote from the generators so that the difference in fault current between the subtransient and transient is negligible. This effect is illustrated in Fig. 10.8. Cases (A) and (B) are the most common situations where the use of X''_d has a negligible effect on the protection. Here the higher system Z_S tends to negate the source decrement effects.

For case (C) X'_d may be used. There are special programs to account for the decremental decay in fault current with time for setting slower-speed protective relays, but these tend to be difficult and tedious and may not provide significant advantages.

The decrease in fault current level with time should not cause protective relay coordination problems unless the time-current characteristics of various devices used are significantly different.

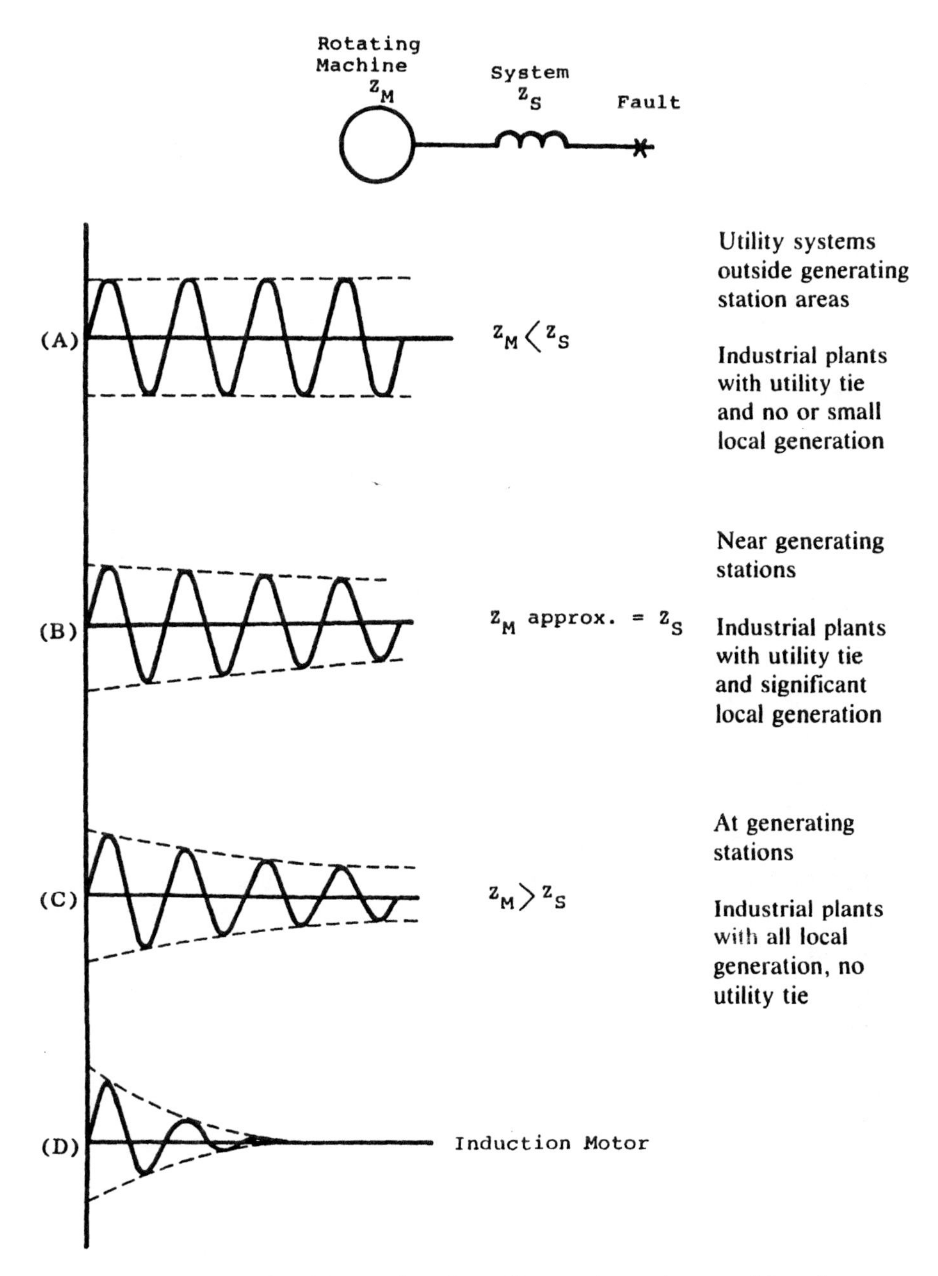

Figure 10.8 Guide illustrating the effects of rotating machine decrements on the symmetrical fault current.

Where Z_M predominates, the fault levels tend to be high and well above the maximum load current. The practice of setting the protection as sensitively as possible but not operating on maximum load (phase devices) should provide good protection sensitivity in the transient reactance period when using X''_d.

At a rated load of 1 pu, $V_{rated} = 1.0$ pu and X_d (saturated) = 1.0 pu. However, for sustained three-phase faults that fail to be cleared by the protection the X_d (unsaturated) value is used. Thus the sustained three-phase fault current will be less than maximum load unless the regulator system changes the excitation. By raising the voltage, the sustained fault current is increased.

Normally, the protective system will have removed the fault long before reaching the synchronous reactance area and before the regulating system can effect any change. Hence the regulating system is not considered in fault calculations.

In most cases induction motors are not considered sources of fault current for protection purposes. As shown in Fig. 10.8, case (D), their short-time feedback of current to a fault is low relative to the main source of power to operate the motors.

However, under ANSI/IEEE standards, induction motors must be considered in circuit breaker applications.

APPENDIX: TYPICAL CONSTANTS OF THREE-PHASE SYNCHRONOUS MACHINES

Values are typical value/range of values. Resistances and reactances are in per unit of the rated MVA (kVA) and voltage of the unit. Time constants are in seconds.

Appendix

Turbine generators	Two-pole: Conventional cooled	Two-pole: Conductor cooled	Four-pole: Conventional cooled	Four-pole: Conductor cooled
X_d (unsat)	1.65/1.0–1.75	1.85/1.5–2.25	1.65/1.0–1.75	1.85/1.5–2.25
X_q rated current	1.61/0.96–1.71	1.81/1.46–2.21	1.61/.96–1.71	1.81/1.46–2.21
X_d' rated voltage	0.17/0.12–0.25	0.28/0.20–0.35	0.25/0.2–0.3	0.35/0.25–0.45
X_d'' rated voltage	0.12/0.08–0.18	0.22/0.15–0.28	0.16/0.12–0.20	0.28/0.20–0.32
X_2 rated current	$= X_d''$	$= X_d''$	$= X_d''$	$= X_d''$
X_o rated current[1]				
x_p Potier reactance	0.07–0.17	0.2–0.45	0.12–0.24	0.25–0.45
r_2[2]	0.025–0.04	0.025–0.04	0.03–0.045	0.03–0.045
r_1[3]	0.004–0.011	0.001–0.008	0.003–0.008	0.001–0.008
r_a[3]	0.001–0.007	0.001–0.005	0.001–0.005	0.001–0.005
T_{do}'	5	5	8	6
T_d'	0.6	0.75	1.0	1.2
T_d''	0.035	0.035	0.035	0.035
T_a	0.13–0.45	0.2–0.55	0.2–0.4	0.25–0.55
H	2.5–3.5	2.5–3.5	3–4	3–4

Salient Pole Generators and Motors:

With dampers—$X_d' = 0.37/0.25–0.5$, $X_d'' = 0.24/0.13–0.32$, $X_2 = X_2''$

Without dampers—$X_d' = 0.35/0.25–0.5$, $X_d'' = 0.32/0.20–0.5$, $X_2 = 0.4/0.3–0.45$.

Synchronous condensers:

$X_d' = 0.40$, $X_d'' = 0.25$, $X_2 = 0.24$.

Notes: (1) X_o varies so critically with armature winding pitch that an average value can hardly be given. Variation is from 0.1 to 0.7 of X_d''; (2) r_2 varies with damper resistance; (3) r_1 and r_a vary with machine rating.

11

Overhead Line Characteristics: Inductive Impedance

11.1 INTRODUCTION

Line characteristics are not available from manufacturers except for the individual conductors that make up the three-phase circuit. Thus it is necessary to determine them using these conductor values, the configuration of the conductors, and the phases that make up the line circuit, and for ground involvement, the nature of the zero-sequence return path(s). In this chapter we review the fundamentals and their application to determine the positive-, negative-, and zero-sequence reactive impedances for overhead line circuits. In later chapters we cover capactive impedance and cable characteristics.

11.2 REACTANCE OF OVERHEAD CONDUCTORS

Overhead or aerial lines are assumed to be long, straight conductors for the purpose of calculating the reactance. The fact that the conductors actually sag from pole to pole or tower to tower, and seldom run in a straight line between terminals, introduces errors. Fortunately, these are small and negligible.

First, consider single phase, or the inductance (L) of a long, straight cylindrical conductor (1) with a circular cross section of radius r and a return conductor (2) separated at distance D_{12}. The inductance can be shown to be

$$L = \left(2 \ln \frac{D_{12}}{r} + \frac{u'}{2}\right) 10^{-7} \qquad \text{henries/meter} \tag{11.1}$$

Here ln is the natural logarithm and u' is the permeability. For nonmagnetic conductors, $u' = 1$. Reactance is $2\pi fL$, where f is the frequency. Equation (11.1) can be converted to reactance in ohms per kilometer or miles using common logarithms (log), where ln = 2.30358 log and 1 mile = 1609 m. Thus

$$X = 2\pi f \left(461 \log \frac{D_{12}}{r} + 50\right) \times 10^{-6} \qquad \text{ohms/kilometer} \tag{11.2}$$

$$X = \pi f \left(741 \log \frac{D_{12}}{r} + 80\right) \times 10^{-6} \qquad \text{ohms/mile} \tag{11.3}$$

At 60-Hz frequency, Eq. (11.3) becomes

$$X = 0.2794 \log \frac{D_{12}}{r} + 0.0302 \ \Omega/\text{mi at 60 Hz} \tag{11.4}$$

The inductance value of 0.0302 results from the flux within the conductor linking the current, assuming uniform current density (no skin effect). To make the equation more general so that skin effect, nonsymmetrical conductors, and so on, can be dealt with, this 0.0302 constant can be replaced with

$$0.2794 \log 10^{K} \tag{11.5}$$

Now adding the line conductor resistance r_a, the line impedance can be expressed from Eqs. (11.4) and (11.5) as

$$Z = r_a + j\left(0.2794 \log \frac{10^K}{r} + 0.2794 \log D_{12}\right) \quad \text{ohms/mile at 60 Hz} \quad (11.6)$$

This leads to two parallel approaches to determine line impedance. One is to use the conductor(s) GMR and separation GMD values, and the other the X_a and X_d constants. These are described in the following section.

11.3 GMR AND GMD VALUES

GMR is the geometric mean radius of a conductor (or a group of conductors, as will be developed later). The GMR is the radius of an infinitesimally thin circular tube that has the same internal reactance as the actual conductor. For the conductor above,

$$\text{GMR} = \frac{r}{10^K} \quad (11.7)$$

This is a very useful value, as it can include skin effect, the effects of stranding (multiple wires in a conductor), noncircular conductors, multiple conductors for a phase wire, the three phases combined for zero sequence, and so on. These combinations are discussed later. For the homogeneous round single conductor neglecting the skin effect, GMR = r. With skin effect included, GMR = $0.799r$. For stranded conductors and various combinations of different materials, such as ACSR (aluminum cable steel reinforced), the GMR will vary from about $0.7r$ to $0.85r$. These GMR values usually are given as part of the cable characteristics and in conductor tables.

The general equation for GMR of a group of parallel conductors (1), (2), (3), . . . , (n) is: D is the distance between the conductors

$$\mathrm{GMR}_n = \sqrt[n^2]{\begin{array}{c}(\mathrm{GMR}_1 D_{12} D_{13} \cdots D_{1n})(\mathrm{GMR}_2 D_{21} D_{23} \cdots D_{2n}) \\ \cdots(\mathrm{GMR}_n D_{n1} D_{n2} D_{n3} \cdots D_{n(n-1)})\end{array}} \tag{11.8}$$

The distance and radius values must be in the same units.

GMD is the geometrical mean distance or separation between the conductor(s) and the return conductor(s). The individual distances are measured between the conductor centers. For the single-phase line above,

$$\mathrm{GMD} = D_{12} \tag{11.9}$$

The general equation for the GMD between two areas, or between wires of a conductor (1), (2), (3), . . . , (n) and its return or associated conductors (1′), (2′), (3′), . . . , (m) with D as the distance, is

$$\mathrm{GMD}_{mn} = \sqrt[nm]{\begin{array}{c}(D_{11'} D_{12'} D_{13'}) \cdots D_{1m})(D_{21'} D_{22'} D_{23'} \cdots D_{2m}) \\ \cdots(D_{n1'} D_{n2'} D_{n3'} \cdots D_{nm})\end{array}} \tag{11.10}$$

These general equations for GMR and GMD will be useful in later discussions.

Now Eq. (11.6) can be written in several useful forms:

$$\begin{aligned} Z &= r_a + j0.2794 \log \frac{\mathrm{GMD}}{\mathrm{GMR}} \quad \text{ohms/mile at 60 Hz} \\ Z &= r_a + j0.2328 \log \frac{\mathrm{GMD}}{\mathrm{GMR}} \quad \text{ohms/mile at 50 Hz} \\ Z &= r_a + j0.1736 \log \frac{\mathrm{GMD}}{\mathrm{GMR}} \quad \text{ohms/km at 60 Hz} \\ Z &= r_a + j0.1447 \log \frac{\mathrm{GMD}}{\mathrm{GMR}} \quad \text{ohms/km at 50 Hz} \end{aligned} \tag{11.11}$$

The GMR is sometimes called the self-GMD or GMD_s, and the GMD is called the mutual GMD or GMD_m. These are extremely useful in line impedance calculations for (1) reducing multiconductor cables or conductors to an equivalent single conductor, and (2) to obtain equivalent separations to the return conductors.

The conductors or strands to be combined in the equivalent must be paralleled, such as in stranded cable, bundled conductors, or in the three-phase circuit for zero-sequence current flow.

11.4 THE X_a AND X_d LINE CONSTANTS

This method of determining line impedance is quite convenient where manufacturers or reference books provide data in this form. Equation (11.6) can be rewritten as

$$Z = r_a + j(X_a + X_d) \qquad \text{ohms/mile at 60 Hz} \tag{11.12}$$

where

$$X_a = 0.2794 \log \frac{10^K}{r} \quad \text{and} \quad X_d = 0.2794 \log D_{12} \tag{11.13}$$

or in general terms,

$$X_a = 0.2794 \log \frac{1}{\text{GMR}} \quad \text{and} \quad X_d = 0.2794 \log \text{GMD} \tag{11.14}$$

These are ohms/mile at 60 Hz. See Eq. (11.11) for other units.

Now the X_a component is the reactance within the conductor and external up to a distance of 1 unit of the unit of r (e.g., 1 ft, where r is in feet). Thus X_a is independent of the location of the return conductor(s). As indicated, line conductor characteristic tables usually give the values of r and X_a for each size and type of conductor.

The X_d component is only a function of the separation of the return conductor(s) and so is independent of the conductor itself. It is the reactance from 1 unit to the return conductor(s) in the same units: where feet are used, the reactance from 1 ft to the distance in feet to the return conductor(s). One reference table gives X_d values for different separations, and can be used for any type of conductors.

11.5 POSITIVE- AND NEGATIVE-SEQUENCE IMPEDANCE

For a three-phase line where the three conductors are identical, the return path for any phase is the other two phases. This is because the positive- and negative-sequence currents always add to zero at the neutral.

Unless these return paths are always equidistant, the flux linkages and the inductances of each of the three phases will be unequal and nonsymmetrical impedances result. Equilateral spacing or transposing of the three conductors is necessary for symmetrical line impedances. A complete transposition of a three-phase line requires that each conductor occupy the position of each of the other two for one-third of the total line distance. This transposition is illustrated in Fig. 11.1.

Equilateral spacing results in $D_1 = D_2 = D_3$. In many cases $D_1 \neq D_2 \neq D_3$, and so as shown in the three diagrams, conductor A is moved to the other three positions for one-third of the total length; similarly for the other two conductors. Many lines do not have equilateral spacing and are not transposed except possibly at the substations where the positions of the phases on several lines are interchanged. The unbalanced error and the interference with adjacent shielded telephone cables usually is small and negligible. However, on lines on the order of 100 miles in length, transpositions probably will be necessary.

All of the following discussions assume equilateral spaced or transposed lines and hence symmetrical impedances in the three phases. Thus from Eq. (11.10), the GMD separation values for several cases are as shown in Table 11.1.

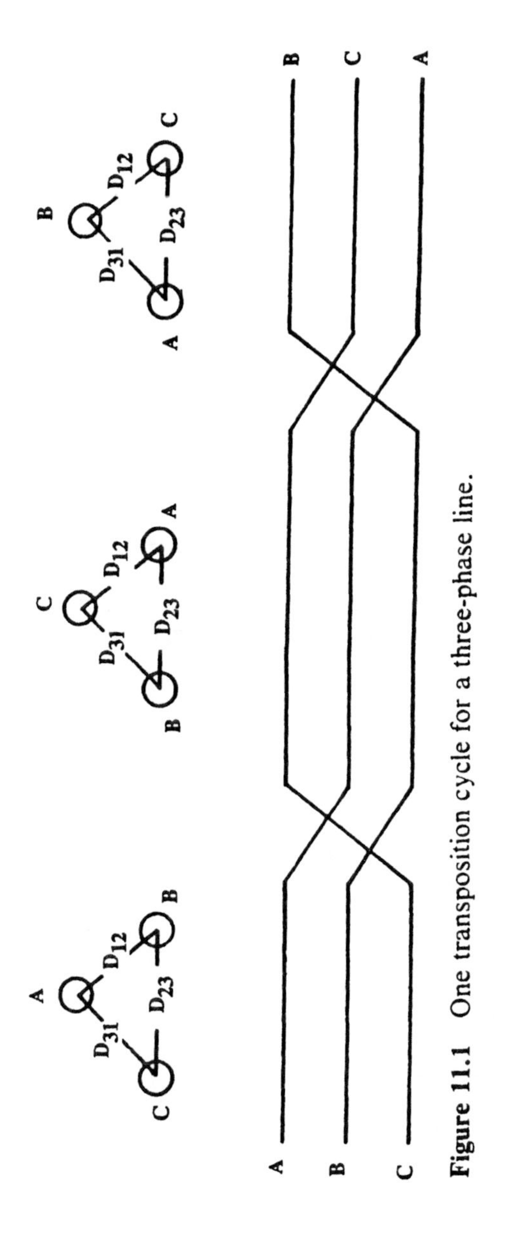

Figure 11.1 One transposition cycle for a three-phase line.

Table 11.1

Between	GMD separation
Two conductors, 1 and 2	D_{12}
Three conductors 1, 2, and 3	$\sqrt[3]{D_{12}D_{23}D_{31}}$
Group 1, 2, 3 and group 4, 5, 6	$\sqrt[9]{D_{14}D_{15}D_{16}D_{24}D_{25}D_{26}D_{34}D_{35}D_{36}}$

11.6 EXAMPLE

Consider a three-phase line that has a flat or horizontal spacing of 10 ft between phases A and B, and 12 ft between phases B and C (not usual; unequal spacings used for illustration). The conductors are 4/0 ACSR. Determine Z_1 and Z_2.

Either Eq. (11.11) or (11.12) can be used. In either case the resistance and reactance of the conductors must be obtained from manufacturer's data or tables available in many reference books. Characteristics for a number of cables are given in the Appendix. The reactance value may be X_a or the conductor GMR, from which X_a can be determined from Eq. (11.14). Thus from Table A.5, for the 4/0 ACSR (Penguin) conductors: r_a = 0.563 and X_a = 0.553, both ohms/mile at 60 Hz. The resistance value is for 50°C (current approximately 75% capacity).

Table A.5 does not give the GMR but it can be calculated if desired:

$$\text{GMR} = \frac{1}{10^{X_a 0.2794}} = 0.01049 \text{ ft}$$

For the spacings, D_{12} = 10 ft, D_{23} = 12 ft, and D_{31} = 22 ft, Table 11.1 can be used to obtain the equivalent spacing or GMD.

$$\text{GMD}_{\text{gr. of 3}} = \sqrt[3]{10 \times 12 \times 22} = 13.82 \text{ ft}$$

Then

$$X_d = 0.2794 \log 13.82 = 0.3187\ \Omega/\text{mi at 60 Hz}$$

Alternatively,

$$X_d = (\tfrac{1}{3})0.2794(\log D_{12} + \log D_{23} + \log D_{31})$$

or

$$X_d = (\tfrac{1}{3})X_d \text{ for } 10 \text{ ft} + X_d \text{ for } 12 \text{ ft} + X_d \text{ for } 22 \text{ ft})$$

from the tables. Thus

$$\begin{aligned} Z_1 = Z_2 &= 0.563 + j(0.553 + 0.319) \\ &= 0.563 + j0.872 = 1.04\underline{/57.15^\circ}\ \Omega/\text{mi at 60 Hz} \end{aligned}$$

In the absence of exact data to calculate the positive- and negative-sequence reactance, a value of 0.8 Ω/mi can be used for good approximation for single-type conductors. For bundled conductors a value of 0.6 Ω/mi can be used. The average of many different lines of various spacings and conditions will fall close to these values.

11.7 LINES WITH BUNDLED CONDUCTORS

To reduce the impedance many HV (high-voltage) and EHV (extrahigh-voltage) lines use several conductors in parallel for each of the three phases. These are known as *bundled conductor lines*. Their impedance can be determined as for a single conductor line by reducing the several conductors per phase to a single equivalent conductor using GMD–GMR techniques. This analysis assumes complete transpositions of the conductors in each phase, which is practical. From the general equations (11.8) and (11.10), commonly encountered cases can be obtained.

11.7.1 Two Bundled Conductors

The conductors have the same conductor GMR as in Fig. 11.2: The phase a conductors are $a1$–$a2$, phase b conductors $b1$–$b2$,

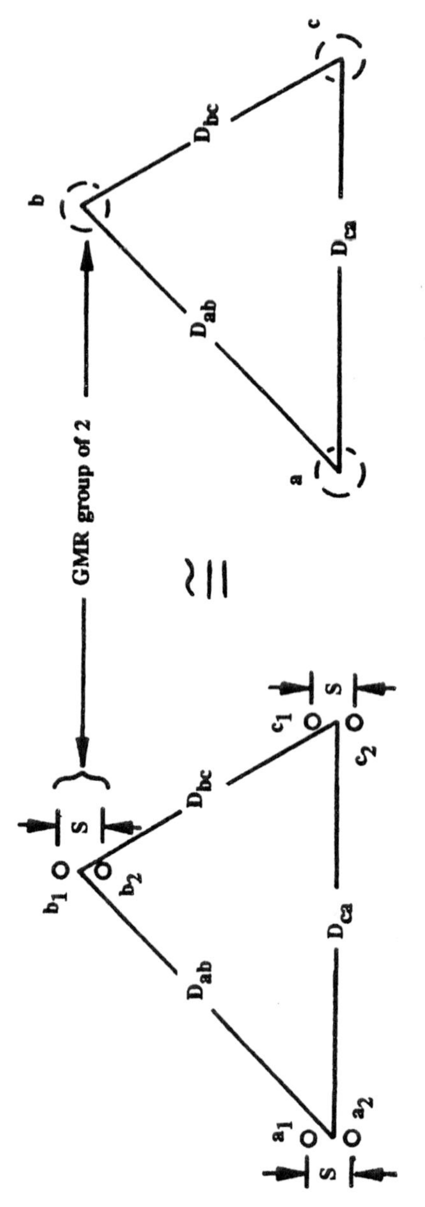

Figure 11.2 A three-phase line with two bundled conductors each with the same GMR.

and phase c conductors are $c1-c2$. S is the distance between the two-phase conductors. Then where GMR is the GMR of each conductor,

$$\begin{aligned}
\text{GMR}_{\text{gr. of 2}} &= \sqrt[4]{\text{GMR}^2 \cdot S^2} = \sqrt{\text{GMR} \cdot S} \\
\text{GMD}_{\text{sep. of } a \text{ and } b} &= \sqrt[4]{(D_{a1b1})(D_{a1b2})(D_{a2b1})(D_{a2b2})} \qquad (11.15) \\
&= D_{ab} \quad \text{approximately}
\end{aligned}$$

Similar expressions can be written for the separation of b and c equal approximately D_{bc} and for c and a equal approximately D_{ca}. Then

$$\text{GMD}_{\text{sep. of 3}} = \sqrt[3]{D_{ab}D_{bc}D_{ca}} \quad \text{approximately} \qquad (11.16)$$

From Eq. (11.11),

$$\begin{aligned}
Z_1 &= Z_2 \\
&= \frac{1}{2} r_a + j.2794 \log \frac{\sqrt[3]{D_{ab}D_{bc}D_{ca}}}{\sqrt{\text{GMR} \cdot S}} \quad \text{ohms/mile at 60 Hz}
\end{aligned}$$

$$(11.17)$$

This is the general equation using GMR and GMD values. This can be transformed into an equivalent expression using X_a, X_d, and X_s, defined as

$$X_a = 0.2794 \log \frac{1}{\text{GMR}} \qquad (11.14)$$

$$X_d = 0.2794 \log \sqrt[3]{D_{ab}D_{bc}D_{ca}} \qquad (11.18)$$

$$X_s = 0.2794 \log S \qquad (11.19)$$

$$Z_1 = Z_2 = \tfrac{1}{2} r_a + j\tfrac{1}{2}(X_a - X_s) + X_d \quad \text{ohms/mile at 60 Hz}$$

$$(11.20)$$

Either Eq. (11.17) or (11.20) can be used, as convenient.

11.7.2 Three Bundled Conductors

The conductors have the same conductor GMR as in Fig. 11.3: The phase a conductors are $a1$, $a2$, $a3$; phase b conductors, $b1$, $b2$, $b3$; and phase c conductors, $c1$, $c2$, $c3$. Now the GMR from Eq. (11.8) and the GMD separation between phases a and b from Eq. (11.10) are

$$\mathrm{GMR}_{\mathrm{gr.\ of\ 3}} = \sqrt[3]{\mathrm{GMR}\cdot S^2} \tag{11.21}$$

$$\begin{aligned}\mathrm{GMD}_{\mathrm{sep.\ gr.\ 3\ to\ gr.\ 3}} &= \sqrt[9]{\begin{matrix}(D_{a1b1})(D_{a1b2})(D_{a1b3})(D_{a2b1})(D_{a2b2}) \\ (D_{ab3})(D_{a3b1})(D_{a3b2})(D_{a3b3})\end{matrix}} \\ &= D_{ab} \quad \text{approximately}\end{aligned} \tag{11.22}$$

Corresponding GMD separations can be written for the separations of group b from group c = D_{bc} approximately, and group c from group a = D_{ca} approximately. Then the impedance of

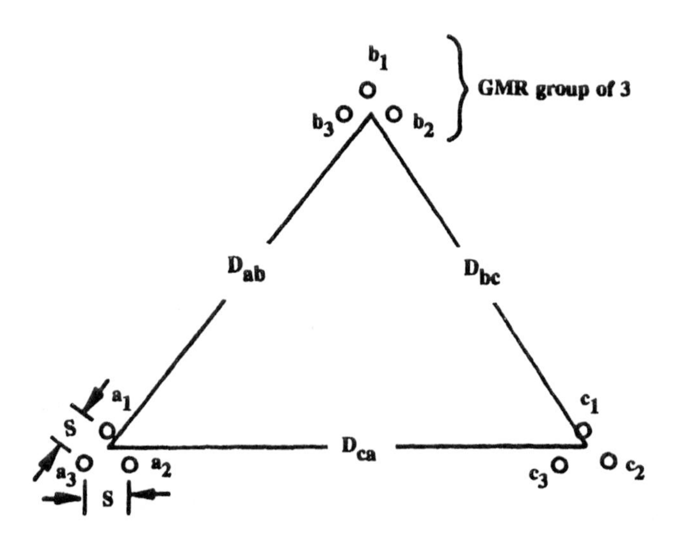

Figure 11.3 A three-phase line with three bundled conductors each with the same GMR.

any phase is similar to the equations above:

$$Z_1 = Z_2 = \frac{1}{3} r_a + j0.2794 \log \frac{\sqrt[3]{D_{ab} D_{bc} D_{ca}}}{\sqrt[3]{\mathrm{GMR} \cdot S^2}} \quad \text{ohms/mile at 60 Hz} \tag{11.23}$$

This general equation can be expressed in terms of the X_a, X_d, and X_s values of Eqs. (11.14), (11.18), and (11.19):

$$Z_1 = Z_2 = \tfrac{1}{3} r_a + j\tfrac{1}{3}(X_a - 2X_s) + X_d \quad \text{ohms/mile at 60 Hz} \tag{11.24}$$

11.7.3 Four Bundled Conductors

The conductors have the same conductor GMR as in Fig. 11.4. The phase a conductors are $a1$, $a2$, $a3$, $a4$; the phase b conductors are $b1$, $b2$, $b3$, $b4$; and the phase c conductors are $c1$, $c2$, $c3$, $c4$. With the four conductors in a square array separated by distance S, the distance between $a1$ and $a2$ is S, between $a1$ and $a3$ is $\sqrt{2}\,S$, and between $a1$ and $a4$ is S. Now the GMR from Eq. (11.8) and the GMD separation between phases a and b from Eq. (11.10) are

$$\mathrm{GMR}_{\text{gr. of 4}} = \sqrt[16]{\begin{matrix}(\mathrm{GMR}_{a1})(S)(S)(\sqrt{2}S)(\mathrm{GMR}_{a2})(S)(S)(\sqrt{2}S) \\ (\mathrm{GMR}_{a3})(S)(S)(\sqrt{2}S)(\mathrm{GMR}_{a4})(S)(S)(\sqrt{2}S)\end{matrix}} \tag{11.25}$$

This reduces to

$$\mathrm{GMR}_{\text{gr. of 4}} = \sqrt[4]{(\mathrm{GMR})(S^2)(\sqrt{2}S)} \tag{11.26}$$

$$\mathrm{GMD}_{\text{sep. gr. 4 to gr. 4}} = \sqrt[16]{\begin{matrix}(D_{a1b1})(D_{a1b2})(D_{a1b3})(D_{a1b4})(D_{a2b1})(D_{a2b2}) \\ (D_{a2b3})(D_{a2b4})(D_{a3b1})(D_{a3b2})(D_{a3b3}) \\ (D_{a3b4})(D_{a4b1})(D_{a4b2})(D_{a4b3})(D_{a4b4})\end{matrix}} \tag{11.27}$$

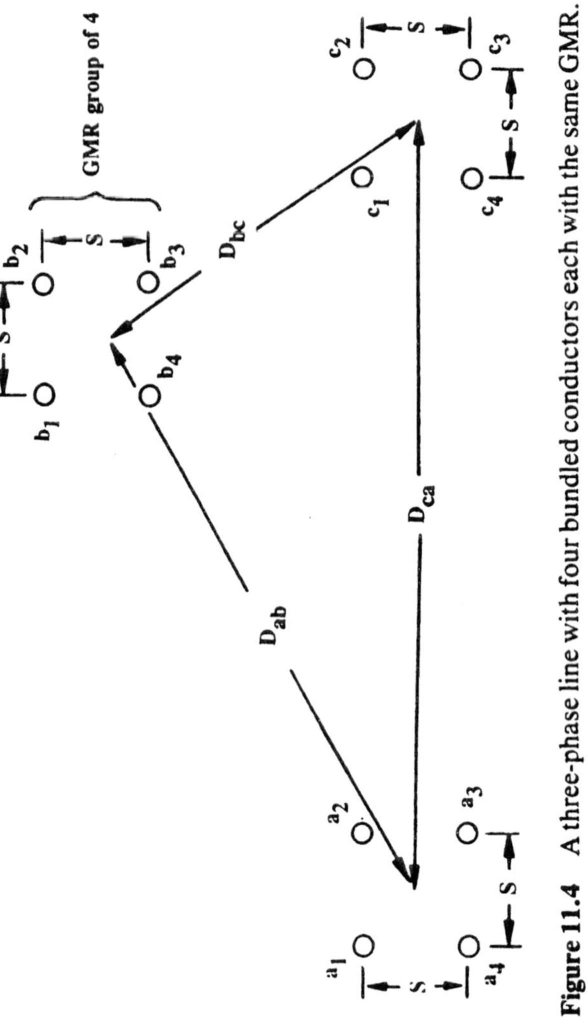

Figure 11.4 A three-phase line with four bundled conductors each with the same GMR.

For the configuration of Fig. 11.4, the equation will reduce to

$$\text{GMD}_{\text{sep. gr. 4 to gr. 4}} = \sqrt[4]{(\text{GMR})(S^2)(\sqrt{2}\,S)}$$
$$= D_{ab} \quad \text{approximately} \qquad (11.28)$$

Similar expressions can be written for the separation of b and c $= D_{bc}$ approximately and the separation of c and a $= D_{ca}$ approximately. Then from Eq. (11.11) the impedance of each phase is

$$Z_1 = Z_2$$
$$= \frac{1}{4} r_a + j0.2794 \log \frac{\sqrt[3]{D_{ab}D_{bc}D_{ca}}}{\sqrt[4]{\text{GMR}\cdot S^2\cdot\sqrt{2}S}}$$
$$\text{ohms/mile at 60 Hz} \qquad (11.29)$$

This general equation can be expressed in terms of X_a, X_d, and X_s of Eqs. (11.14), (11.18), and (11.19):

$$Z_1 = Z_2 = \tfrac{1}{4}r_a + j\tfrac{1}{4}(X_a - 3X_s)$$
$$+ X_d - 0.0105\ \Omega/\text{mi at 60 Hz} \qquad (11.30)$$

The constant of 0.0105 results from the square root of 2 in the conductor separation values.

11.8 ZERO-SEQUENCE IMPEDANCE

The zero-sequence currents divide equally among the three conductors of a single-phase line and have a common return through the earth and, if used, the ground wires. This return current through the earth tends to follow the path of the line rather than taking any shortcuts that might exist.

The earth is a conductor of enormous dimensions and nonuniform conductivity. Thus the earth current distribution is not uniform. To calculate the impedance of the conductors with earth return, it is necessary to know the distribution of the current returning in the earth.

Many engineers have attacked this problem using different assumptions and methods. Of all of these, the work of J. R. Carson of the Bell Telephone Laboratories is generally accepted as the best. This work was published in 1926. Carson started with the following assumptions:

1. The conductors are parallel to the ground.
2. The earth is a solid with a plane surface, infinite in extent, and of uniform conductivity.

Now while these are very far from true in power lines, in general, the results check experimental results rather closely. On short lines some error is introduced because of end effects, but again the error is not great.

Carson's formulas are quite complicated, but fortunately, they can be simplified with little error for power lines. These simplified formulas are:

The self-impedance Z_{11} of a conductor or group of conductors (1) with earth return is

$$Z_{11} = r_a + 1.588f \times 10^{-3} + j4.657f \times 10^{-3} \log \frac{D_e}{\text{GMR}} \quad \text{ohms/mile} \tag{11.31}$$

The mutual impedance Z_{12} between two parallel conductors or groups of conductors (1) and (2) with a common earth return is

$$Z_{12} = 1.588f \times 10^{-3} + j4.657f \times 10^{-3} \log \frac{D_e}{\text{GMD}_{12}} \quad \text{ohms/mile} \tag{11.32}$$

The log is to the base 10, as before. The circuit and nomenclature are shown in Fig. 11.5. D_e is the equivalent depth of earth return, a mathematical fiction, as the actual ground current flows at or near the surface under the conductors.

$$D_e = 2160 \sqrt{\frac{p}{f}} \tag{11.33}$$

where p is the earth resistivity in meter-ohms and f is the fre-

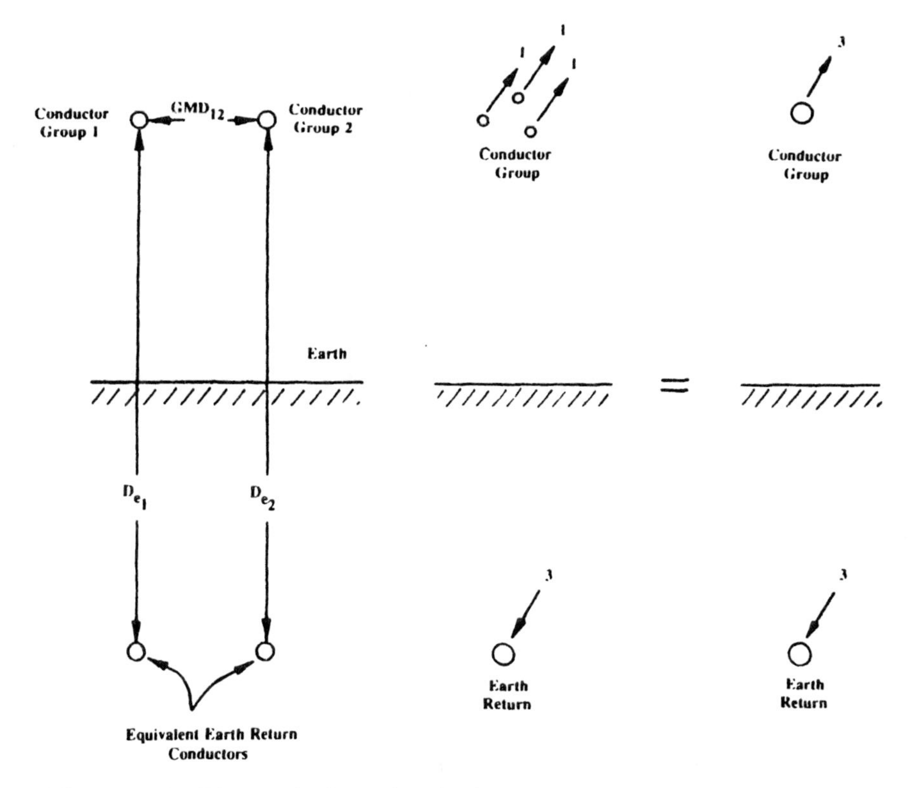

Figure 11.5 The equivalent circuits for the mutual impedance between two parallel lines.

quency. The distance to the equivalent earth plane is $\frac{1}{2}D_e$ assuming infinite conductivity. The values for various soils are stated in Table 11.2.

One of the reasons why these equations provide good cor-

Table 11.2

Soil	p/f	D_e at 60 Hz (ft)	Log D_e
Damp earth[a]	100	2800	3.45
Dry earth	1000	8840	3.95
Seawater	1	280	2.45

[a]The average of a large number of determinations and commonly used in the absence of specific data.

relation with actual values is the effect of the logarithm. This can be seen by the relative low log D_e value range compared to the very wide difference between the equivalent depth of earth return values. Carson's equations are for single-phase line circuits and must be modified for zero-sequence applications.

Unit zero-sequence current consists of 1 pu current flowing in each phase circuit and 3 pu current flowing in the earth or return. By replacing the three phases by one equivalent conductor, 3 pu current flows in this equivalent. This is illustrated in Fig. 11.5. However, only 1 pu current flows in the zero-sequence networks by fundamental definition. The same voltage drop results if 1 pu current flows in 3 pu impedance as 3 pu current in 1 pu impedance. Thus it is necessary to multiply Carson's basic equations (11.31) and (11.32) by 3.

$$Z_{011} = Z_0 = 3Z_{11} \qquad (11.34)$$
$$Z_{012} = Z_{0m} = 3Z_{12}$$

With r as the resistance of each single-phase conductor, $r/3$ is the resistance of the equivalent conductor. Thus $3r/3 = r$ for the zero-sequence values. Therefore, Carson's equations for zero sequence are

$$Z_0 = r_a + 0.0047f + j0.01397f \log \frac{D_e}{\text{GMR}} \quad \text{ohms/mile}$$

$$Z_0 = r_a + 0.286 + j0.8382 \log \frac{D_e}{\text{GMR}} \quad \text{ohms/mile at 60 Hz}$$

$$Z_0 = r_a + 0.238 + j0.6985 \log \frac{D_e}{\text{GMR}} \quad \text{ohms/mile at 50 Hz}$$

$$Z_0 = r_a + 0.178 + j0.5208 \log \frac{D_e}{\text{GMR}} \quad \text{ohms/km at 60 Hz}$$

$$Z_0 = r_a + 0.148 + j0.4340 \log \frac{D_e}{\text{GMR}} \quad \text{ohms/km at 50 Hz}$$

(11.35)

$$Z_{0m} = 0.0047f + j0.01379f \log \frac{D_e}{\text{GMD}} \qquad \text{ohms/mile}$$

$$Z_{0m} = 0.286 + j0.8382 \log \frac{D_e}{\text{GMD}} \qquad \text{ohms/mile at 60 Hz}$$

$$Z_{0m} = 0.238 + j0.6985 \log \frac{D_e}{\text{GMD}} \qquad \text{ohms/mile at 50 Hz}$$

$$Z_{0m} = 0.178 + j0.5208 \log \frac{D_e}{\text{GMD}} \qquad \text{ohms/km at 60 Hz}$$

$$Z_{0m} = 0.148 + j0.4340 \log \frac{D_e}{\text{GMD}} \qquad \text{ohms/km at 50 Hz}$$

(11.36)

11.9 ZERO-SEQUENCE IMPEDANCES OF VARIOUS LINES

11.9.1 Transposed Single Three-Phase Lines, No Ground Wires

For three conductors a, b, c, each with GMR, Eq. (11.35) becomes, at 60 Hz,

$$Z_0 = r_a + 0.286 + j0.8382 \log \frac{D_e}{\text{GMR}_{\text{gr. 3}}}$$

$$= r_a + 0.286 + j0.8382 \log \frac{D_e}{\sqrt[9]{\text{GMR}^3 D_{ab}^2 D_{bc}^2 D_{ca}^2}} \qquad (11.37)$$

$$Z_{0m} = 0$$

Expanding Eq. (11.37) yields

$$Z_0 = r_a + 0.286 + j0.8382 \left(\log D_e - \tfrac{3}{9} \log \text{GMR} - \tfrac{2}{9} \log D_{ab} D_{bc} D_{ca}\right)$$

$$= r_a + 0.286 + j0.8382 \log D_e + j0.2794 \log \frac{1}{\text{GMR}} - j0.2794(2) \log \sqrt[3]{D_{ab} D_{bc} D_{ca}}$$

Defining and substituting the following values in ohms/mile at 60 Hz,

r_a = resistance of a single conductor of GMR

r_e = 0.286, the resistance of the ground return

$X_e = j0.8382 \log D_e$

$X_a = j0.2794 \log \frac{1}{\text{GMR}}$ [same as Eq. (11.14)]

$X_d = j0.2794 \log \sqrt[3]{D_{ab}D_{bc}D_{ca}}$ [same as Eq. (11.16)]

Eq. (11.37) becomes

$$Z_0 = r_a + r_e + j(X_a + X_e - 2X_d) \tag{11.38}$$

It is important to note that the positive- and negative-sequence impedances Z_1 and Z_2 are the self-impedance of each phase from one end to the other, a one-way impedance. In contrast, the zero-sequence impedance is a loop impedance, out and back, or the sum of the phases and ground return impedances.

From Eqs. (11.12) and (11.38),

$$Z_0 = Z_1 + r_e + j(X_e - 3X_d) \tag{11.39}$$

Here Z_1 is the resistance and reactance of the conductor for one unit of current or of the equivalent conductor with three units of current (equivalent conductor impedance of $\frac{1}{3} \times 3$ units of current = 1). r_e is the resistance of the earth return for 3 units of current. $(X_e - 3X_d)$ is the reactance of the earth return from the earth to the GMD distance for 3 units of current. Since X_d is for 1 unit of current, it must be multiplied by 3 for 3 units of current.

11.9.2 Lines with Ground Wires

Consideration of the general case provides the method of approach for specific cases. The $\text{GMR}_{gr.\ a}$ of the line circuit in Fig. 11.6 can represent any type of three-phase line configurations;

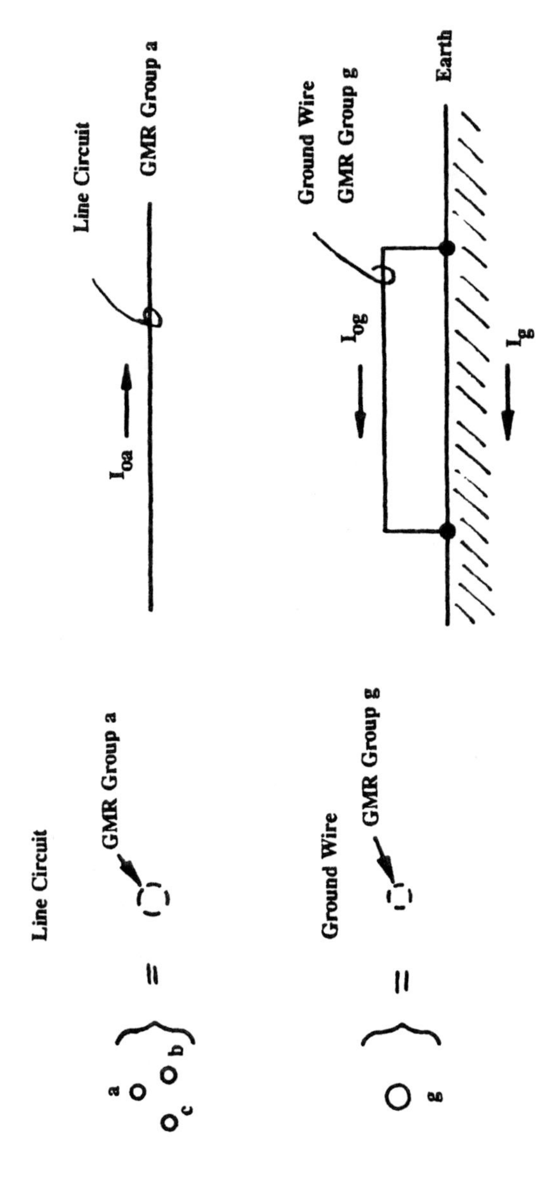

Figure 11.6 The circuit for zero-sequence current return in the ground wire and earth.

either single or double circuit, and the $GMR_{gr.\ g}$, any configuration of ground wires that are in parallel with the line circuit. The ground wire(s) provide a parallel path with the earth for the zero-sequence currents so that the return current divides between the earth and the ground wire(s), as shown.

The ground wire circuit is tied to the earth at each pole or tower. In practice these connections to earth will have some impedance (mostly resistance), known as tower footing resistance. Good design practice attempts to keep these values low, and there are many in parallel to help. In zero-sequence line impedance calculations, the tower footing resistance is assumed zero, which is practical for fault calculations.

From Fig. 11.6 general equations can be written:

The drop across the three-phase line circuit is

$$V_{0a} = Z_{0a}I_{0a} - Z_{0m}I_{0g} \tag{11.40}$$

where Z_{0a} is the three-phase line zero-sequence self-impedance from Eq. (11.35) or (11.38) and Z_{0m} is the zero-sequence mutual impedance between the line group and ground wire group from Eq. (11.36). Expanding and collecting gives us

$$\begin{aligned} V_{0a} &= Z_{0a}I_{0a} - Z_{0m}I_{0a} + Z_{0m}I_{0a} - Z_{0m}I_{0g} \\ &= (Z_{0a} - Z_{0m})I_{0a} + Z_{0m}(I_{0a} - I_{0g}) \\ &= Z'_{0a}I_{0a} + Z_{0m}I_g \end{aligned} \tag{11.41}$$

where $Z'_{0a} = Z_{0a} - Z_{0m}$ and $I_g = I_{0a} - I_{0g}$.

The drop across the ground wire circuit is

$$V_{0g} = Z_{0g}I_{0g} - Z_{0m}I_{0a} = 0$$

with the ground wires grounded at both ends. Expanding this yields

$$\begin{aligned} V_{0g} &= Z_{0g}I_{0g} - Z_{0m}I_{0g} + Z_{0m}I_{0g} - Z_{0m}I_{0a} \\ &= (Z_{0g} - Z_{0m})I_{0g} - Z_{0m}(I_{0a} - I_{0g}) \\ &= Z'_{0g}I_{0g} - Z_{0m}I_g = 0 \end{aligned} \tag{11.42}$$

In Eqs. (11.41) and (11.42) the various quantities are:

$Z'_{0a} = Z_{0a} - Z_{0m}$ = leakage impedance of the conductors

$Z'_{0g} = Z_{0g} - Z_{0m}$ = leakage impedance of the ground wires

Z_{0m} = mutual impedance between the conductors and the ground wire(s)

I_g = current in the earth

From these equations an equivalent circuit can be set up to represent the zero-sequence impedance of the conductors and the ground wire(s) as shown in Fig. 11.7. The total zero-sequence impedance of the line with earth and ground wire(s) is

$$Z_0 = Z'_{0a} + \frac{Z'_{0g}Z_{0m}}{Z'_{0g} + Z_{0m}} \tag{11.43}$$

The key to the application of Eq. (11.43) is proper interpretation of the GMR and GMD values in Eqs. (11.35) and (11.36) that are involved in Eq. (11.43).

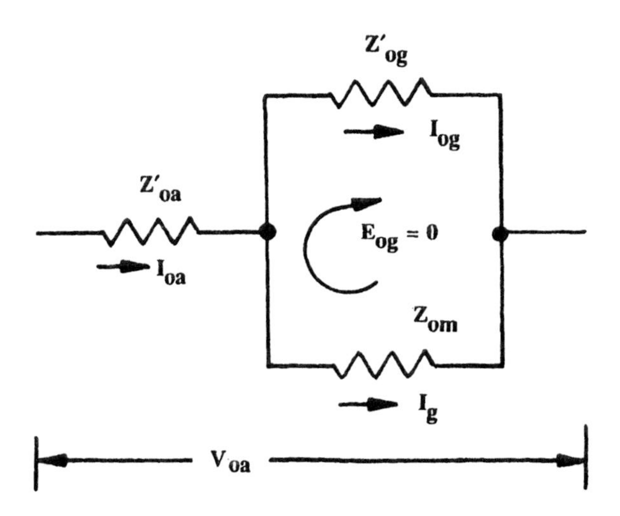

Figure 11.7 The equivalent circuit for the zero sequence current return in the ground wire and earth.

11.9.3 Transposed Single Three-Phase Lines, One Ground Wire

All impedances are ohms/mile at 60 Hz. For the three-phase line circuit (a), Z_{0a} is the same as Eq. (11.37), or

$$Z_{0a} = r_a + 0.286 + j0.8382 \log \frac{D_e}{\text{GMR}_{a(\text{gr. of } 3)}}$$

From Eq. (11.36),

$$Z_{0m} = 0.286 + j0.8382 \log \frac{D_e}{\text{GMD}_{\text{gr. 3 to gr. 1}}} \tag{11.44}$$

Thus

$$Z'_{0a} = r + j0.8382 \log \frac{\text{GMD}_{\text{gr. 3 to gr. 1}}}{\text{GMR}_{a(\text{gr. of } 3)}} \tag{11.45}$$

For the ground wire (g) from Eq. (11.35),

$$Z_{0g} = 3r_g + 0.286 + j0.8382 \log \frac{D_e}{\text{GMR}_{g(\text{gr. of } 1)}}$$

Thus

$$Z'_{0g} = 3r_g + j0.8382 \log \frac{\text{GMD}_{\text{gr. 3 to gr. 1}}}{\text{GMR}_{g(\text{gr. of } 1)}} \tag{11.46}$$

Equations (11.44) through (11.46) can be converted into forms useful with table data, so that

$$Z'_{0a} = r_a + j(X_a + 3X_{d(a \text{ to } g)} - 2X_d) \tag{11.47}$$

$$Z'_{0g} = 3r_g + j(3X_g + 3X_{d(a \text{ to } g)}) \tag{11.48}$$

$$Z_{0m} = r_e + j(X_e - 3X_{d(a \text{ to } g)} \tag{11.49}$$

In the above, r_a and r_g are the resistance of one phase and the single ground conductor, respectively. r_e is the ground return resistance (0.286), as defined previously. X_a and X_g are the reactances per Eq. (11.14), where GMR for X_a is the GMR for a group of 3 [as in Eq. (11.37)], and for X_g is the GMR of the single ground wire. $X_e = j0.8382 \log D_e$, as defined previously. X_d is Eq. (11.16) for the GMD separation of the three-phase conductors, while $X_{d(a \text{ to } g)}$ is the GMD separation between the group of three-phase conductors and the single ground wire.

11.9.4 Example: Single Three-Phase Line, One Ground Wire

This circuit is shown in Fig. 11.8. The conductors and ground wire are 397,000 CM ACSR (30 × 7) with a GMR = 0.0278 ft. $r_a = r_g = 0.259\ \Omega$/mile at 60 Hz. Thus

$$\begin{aligned} \text{GMR}_{a(\text{gr. of } 3)} &= \sqrt[9]{(0.0278)^3 \times (18)^2 \times (18)^2 \times (36)^2} \\ &= 2.43 \text{ ft} \\ \text{GMD}_{\text{gr. } 3 \text{ to gr. } 1} &= \sqrt[3]{28.8 \times 13.45 \times 13.45} \\ &= 17.34 \text{ ft} \end{aligned}$$

From Eq. (11.45),

$$\begin{aligned} Z'_{0a} &= 0.259 + j0.8382 \log \frac{17.34}{2.43} \\ &= 0.259 + j0.715 = 0.760 \text{ at } 70.09° \end{aligned}$$

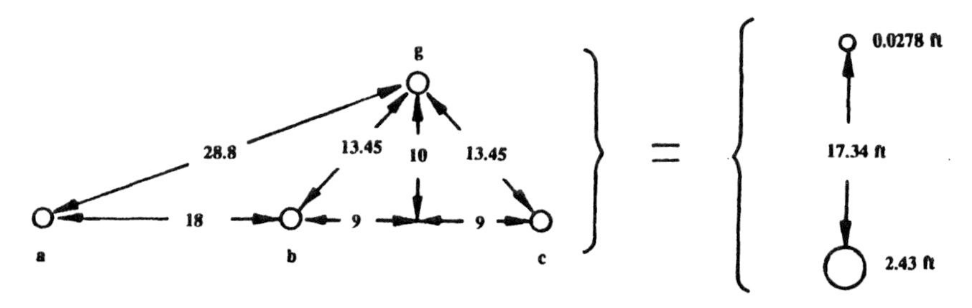

Figure 11.8 A typical example of a single three-phase line with one ground wire.

From Eq. (11.46),

$$Z'_{0g} = 3(0.259) + j0.8382 \log \frac{17.43}{0.0278}$$
$$= 0.777 + j2.345 = 2.47 \text{ at } 71.67°$$

From Eq. (11.44),

$$Z_{0m} = 0.286 + j0.8383 \log \frac{2800}{17.34}$$
$$= 0.286 + j1.851 = 1.873 \text{ at } 81.22°$$

Then from Eq. (11.43) the zero-sequence impedance of this circuit is

$$Z_0 = 0.259 + j0.715 + \frac{(2.47\angle 71.67°)(1.873\angle 81.22°)}{1.063 + j4.196 = 4.329\angle 75.78°}$$

$$Z_0 = 0.498 + j1.757 = 1.826 \text{ at } 74.18° \qquad \text{ohms/mile at 60 Hz}$$

From Fig. 11.8 equivalent circuit, the percent of the current returning in the ground wire will be

percent in ground wire

$$= 100 \frac{Z_{0m}}{Z'_{0g} + Z_{0m}} = 100 \left(\frac{1.873}{4.329}\right) = 43.3\%$$

11.9.5 Transposed Single Three-Phase Lines, Two Ground Wires

This is similar to case of Section 11.9.3 with modifications in the GMR and GMD values. a, b, and c are the three phases, $g1$, $g2$ the two ground wires.

$$\text{GMR}_{a(\text{gr. of }3)} = \sqrt[9]{(\text{GMR}_a)^3(D_{ab})^2(D_{bc})^2(D_{ca})^2} \qquad (11.50)$$

$$\text{GMD}_{\text{gr. 3 to gr. 2}} = \sqrt[6]{(D_{ag1})(D_{ag2})(D_{bg1})(D_{bg2})(D_{cg1})(D_{cg2})} \qquad (11.51)$$

$$\text{GMR}_{g(\text{gr. of 2})} = \sqrt{(\text{GMR}_g)(D_{g1g2})} \tag{11.52}$$

Similar to before, the values of Z'_{0a}, Z'_{0g}, and Z_{0m} are

$$Z'_{0a} = r_a + j0.8382 \log \frac{\text{GMD}_{\text{gr. 3 to gr. 2}}}{\text{GMR}_{a(\text{gr. of 3})}} \tag{11.53}$$

$$Z'_{0g} = \frac{1}{2}(3r_g) + j0.8382 \log \frac{\text{GMD}_{\text{gr. 3 to gr. 2}}}{\text{GMR}_{g(\text{gr. of 2})}} \tag{11.54}$$

$$Z_{0m} = 0.286 + j0.8382 \log \frac{D_e}{\text{GMD}_{\text{gr. 3 to gr. 2}}} \tag{11.55}$$

With these values, the Z_0 of the complete circuit is given by Eq. (11.43).

11.9.6 Example: Single Three-Phase Line, Two Ground Wires

For the example of Fig. 11.8, add a second ground wire as shown in Fig. 11.9. As before, the conductors and ground wire have individual GMR = 0.0278 ft and $r_a = r_g = 0.259$ Ω/mile at 60 Hz. From Equation (11.50) and as for Section 11.9.4,

$$\text{GMR}_{a(\text{gr. of 3})} = 2.43 \text{ ft}$$

From Eq. (11.51),

$$\text{GMD}_{\text{gr. 3 to gr. 2}} = \sqrt[6]{(13.45)^4 \times (28.8)^2} = 17.34 \text{ ft}$$

From Eq. (11.52),

$$\text{GMR}_{g(\text{gr. of 2})} = \sqrt{0.0278 \times 18} = 0.707 \text{ ft}$$

From Eq. (11.53),

$$Z'_{0a} = 0.259 + j0.8382 \log \frac{17.34}{2.43}$$

$$= 0.259 + j0.715 = 0.760 \text{ at } 70.09°$$

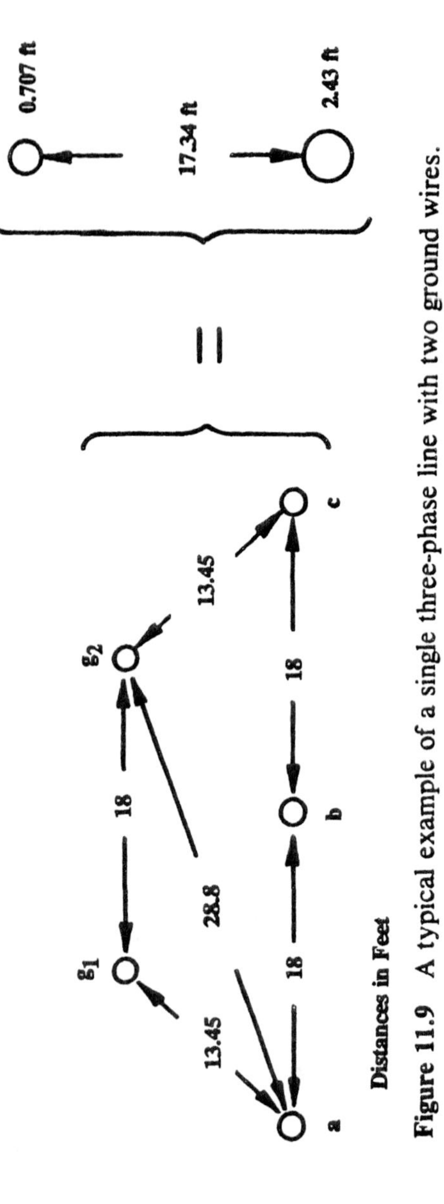

Figure 11.9 A typical example of a single three-phase line with two ground wires.

From Eq. (11.54),

$$Z'_{0g} = \frac{1}{2}(3r_g) + j0.8382 \log \frac{17.34}{0.707}$$

$$= 0.389 + j1.165 = 1.228 \text{ at } 71.53°$$

From Eq. (11.55),

$$Z_{0m} = 0.286 + j0.8382 \log \frac{2800}{17.34}$$

$$= 0.286 + j1.851 = 1.873 \text{ at } 81.22°$$

Then from Eq. (11.43),

$$Z_0 = 0.259 + j0.715 + \frac{(1.228\angle 71.53°)(1.873\angle 81.22°)}{0.675 + j3.016 = 3.091\angle 77.39°}$$

$$Z_0 = 0.447 + j1.435 = 1.5 \text{ at } 72.7° \quad \text{ohms/mile at 60 Hz}$$

$$\text{percent current in ground wire} = 100 \frac{Z_{0m}}{Z'_{0g} + Z_{0m}}$$

$$= 100 \left(\frac{1.873}{3.091}\right) = 60.6\%$$

11.9.7 Transposed Double Three-Phase Lines, No Ground Wires

When two three-phase lines are paralleled, usually on the same tower, the combined zero-sequence impedance can be determined as one value. The basic equations (11.35) and (11.36) apply with the proper values of GMR and GMD. It is convenient to reduce each three-phase line to a single equivalent, determine the GMD distance between them, and finally, combine the two to a single equivalent conductor.

Thus the self-impedance of either three-phase line is given by Eq. (11.37), and the mutual impedance between the two three-phase lines from Eq. (11.36) is

$$Z_{0m} = 0.286 + j0.8382 \log \frac{D_e}{\text{GMD}_{\text{gr. 3 to gr. 3}}} \tag{11.56}$$

where $\text{GMD}_{\text{gr. 3 to gr. 3}}$ is

$$\sqrt[9]{(D_{a1a2})(D_{a1b2})(D_{a1c2})(D_{b1a2})(D_{b1b2})(D_{b1c2})(D_{c1a2})(D_{c1b2})(D_{c1c2})}$$

Note this is the ninth root of the nine possible distances between the two three-phase lines, $a1$, $b1$, $c1$ and $a2$, $b2$, $c2$.

Now the two equivalent three-phase lines can be combined into one equivalent as

$$\text{GMR}_{\text{gr. of 6}} = \sqrt{(\text{GMR}_{\text{gr. of 3}})(\text{GMD}_{\text{gr. 3 to gr. 3}})} \tag{11.57}$$

Thus the impedance of the paralleled lines from Eq. (11.35) is

$$Z_0 = \frac{1}{2} r_a + 0.286 + j0.8382 \frac{D_e}{\text{GMR}_{\text{gr. of 6}}} \quad \text{ohms/mile at 60 Hz} \tag{11.58}$$

An alternative solution is to determine the Z_{0s} self-impedance from Eq. (11.35) or for each three-phase line from Eq. (11.37), and the Z_{0m} mutual impedance from Eq. (11.36). Then Z_0 for the parallel lines $= \frac{1}{2}(Z_{0s} + Z_{0m})$. This equation will be derived later under mutual impedance.

11.9.8 Example: Double Three-Phase Lines, No Ground Wires

Consider two three-phase lines on a single tower as shown in Fig. 11.10. The conductor GMR = 0.0278 ft, with a resistance

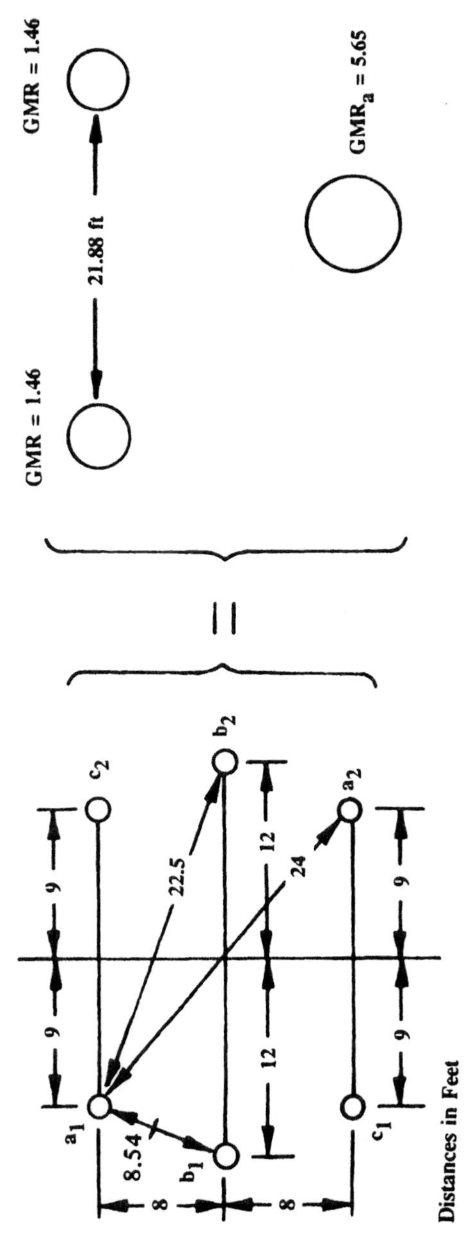

Figure 11.10 A typical example of two three-phase lines without ground wires.

of 0.259 Ω/mile at 60 Hz. As in Eq. (11.50), the GMR for each of the three-phase lines is

$$GMR_{a(\text{gr. }3)} = \sqrt[9]{(0.0278)^3 \times (8.54)^2 \times (16)^2 \times (8.54)^2}$$
$$= 1.46 \text{ ft}$$

The GMD separation between the two three-phase lines is the ninth root of the nine distances between the six wires:

$$GMD_{\text{gr. 3 to gr. 3}} = \sqrt[9]{(18)^2 \times (22.5)^4 \times (24)^3} = 21.88 \text{ ft}$$

Then from Eq. (11.57),

$$GMR_{a(\text{gr. }6)} = \sqrt{1.46 \times 21.88} = 5.65 \text{ ft}$$

and from Eq. (11.58),

$$Z_0 = \tfrac{1}{2}(0.259) + 0.286 + j0.8382 \log \frac{2800}{5.65}$$
$$= 0.416 + j2.26 = 2.30 \text{ at } 79.6° \qquad \text{ohms/mile at 60 Hz}$$

Using the alternative approach, the self-impedance of each three-phase line is

$$Z_{0s} = 0.259 + 0.286 + j0.8382 \log \frac{2800}{1.46}$$
$$= 0.545 + j2.75 \ \Omega/\text{mi at 60 Hz}$$

$$Z_{0m} = 0.286 + j0.8382 \log \frac{2800}{21.88}$$
$$= 0.286 + j1.77 \ \Omega/\text{mi at 60 Hz}$$

Then

$$Z_0 = \tfrac{1}{2}(Z_{0s} + Z_{0m}) = \tfrac{1}{2}(0.831 + j4.52) = 0.416 + j2.26$$

which is the same as above.

11.9.9 Transposed Double Three-Phase Lines, One Ground Wire

The impedance values are based on the $\mathrm{GMR}_{a(\mathrm{gr.\ of\ 6})}$, Eq. (11.57), for the double three-phase line, $\mathrm{GMR}_{g(\mathrm{gr.\ 1})}$ equal to the GMR of the ground wire and the separation between them of $\mathrm{GMD}_{\mathrm{gr.\ 6\ to\ gr.\ 1}}$. From these,

$$Z'_{0a} = \frac{1}{2} r_a + j0.8382 \log \frac{\mathrm{GMD}_{\mathrm{gr.\ 6\ to\ gr.\ 1}}}{\mathrm{GMR}_{\mathrm{gr.\ of\ 6}}} \qquad (11.59)$$

$$Z'_{0g} = 3r_g + j0.8382 \log \frac{\mathrm{GMD}_{\mathrm{gr.\ 6\ to\ gr.\ 1}}}{\mathrm{GMR}_{\mathrm{gr.\ of\ 1}}} \qquad (11.60)$$

$$Z_{0m} = 0.286 + j0.8382 \log \frac{D_e}{\mathrm{GMD}_{\mathrm{gr.\ 6\ to\ gr.\ 1}}} \qquad (11.61)$$

The combined impedance of the two lines and ground wire can be determined from Eq. (11.43) with the foregoing GMR and GMD values.

11.9.10 Transposed Double Three-Phase Lines, Two Ground Wires

The impedance values are based on $\mathrm{GMR}_{a(\mathrm{gr.\ of\ 6})}$ from Eq. (11.57), $\mathrm{GMR}_{g(\mathrm{gr.\ of\ 2})}$ from Eq. (11.52), and the separation $\mathrm{GMD}_{\mathrm{gr.\ 6\ to\ gr.\ 2}}$. The latter is the 12th root of the 12 distances between the six conductors $a1$, $b1$, $c1$ and $a2$, $b2$, $c2$ and the two ground wires $g1$, $g2$, or

$$\sqrt[12]{D_{a1g1}D_{b1g1}D_{c1g1}D_{a2g1}D_{b2g1}D_{c2g1}D_{a1g2}D_{b1g2}D_{c1g2} D_{a2g2}D_{b2g2}D_{c2g2}}$$

From these,

$$Z'_{0a} = \frac{1}{2} r_a + j0.8382 \log \frac{\text{GMD}_{\text{gr. 6 to gr. 2}}}{\text{GMR}_{\text{gr. of 6}}} \tag{11.62}$$

$$Z'_{0g} = \frac{1}{2} (3r_g) + j0.8382 \log \frac{\text{GMD}_{\text{gr. 6 to gr. 2}}}{\text{GMR}_{\text{gr. of 2}}} \tag{11.63}$$

$$Z_{0m} = 0.286 + j0.8382 \log \frac{D_e}{\text{GMD}_{\text{gr. 6 to gr. 2}}} \tag{11.64}$$

The combined impedance of the two lines and two ground wires can be determined from Eq. (11.43) with the foregoing GMR and GMD values.

11.9.11 Example: Double Three-Phase Line, Two Ground Wires

A typical configuration is illustrated in Fig. 11.11. As before, the GMR_a of the conductors and GMR_g of the ground wires are 0.0278 ft, with $r_a = r_g = 0.259\ \Omega$/mi at 60 Hz.

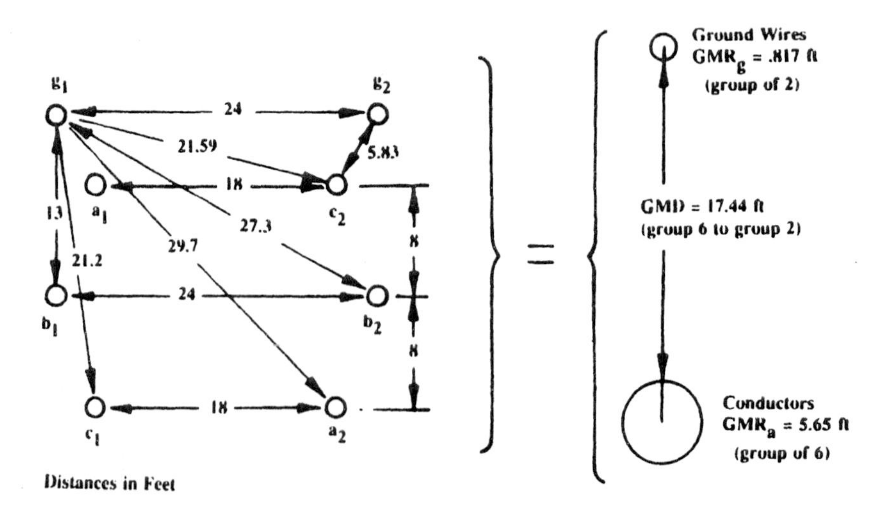

Figure 11.11 A typical example of two three-phase lines with two ground wires.

$$\text{GMR}_{a(\text{gr. of } 6)} = 5.65 \text{ ft} \quad \text{(same as in Section 11.9.8)}$$

$$\text{GMR}_{g(\text{gr. of } 2)} = \sqrt{0.0278\,24} = 0.817 \text{ ft} \quad \text{[Eq. (11.52)]}$$

$$\text{GMD}_{\text{gr. 6 to gr. 2}} = \sqrt[12]{(5.83)^2 \times (13)^2 \times (21.2)^2 \times (29.7)^2 \times (27.3)^2 \times (21.59)^2}$$

$$= 17.44 \text{ ft}$$

$$Z'_{0a} = \tfrac{1}{2}(0.259) + j0.8382 \log \frac{17.44}{5.65}$$

$$= 0.130 + j0.410\ \Omega/\text{mi at 60 Hz}$$

$$Z'_{0g} = \tfrac{1}{2}(3)(0.259) + j0.8382 \log \frac{17.44}{0.817}$$

$$= 0.389 + j1.11 = 1.18 \text{ at } 70.76°$$

$$Z_{0m} = 0.286 + j0.0.8382 \log \frac{2800}{17.44}$$

$$= 0.286 + j1.85 = 1.87 \text{ at } 81.21°$$

Then from Eq. (11.43),

$$Z_0 = 0.13 + j0.410 + \frac{(1.18\underline{/70.6°})(1.87\underline{/81.21°})}{0.675 + j2.96 = 3.04\underline{/77.15°}}$$

$$Z_0 = 0.322 + j1.11 = 1.16 \text{ at } 73.82° \qquad \text{ohms/mile at 60 Hz}$$

$$\text{percent current in ground wire} = 100\left(\frac{1.87}{3.04}\right) = 61.5\%$$

11.10 SUMMARY FOR ZERO-SEQUENCE IMPEDANCE CALCULATIONS

The various cases developed in Section 11.9 are summarized for ready reference in Table 11.3. The equations usable with the data available from conductor and spacing tables have been added. The conductors and ground wires are assumed to be transposed and the conductors in each phase the same. If there is more than one ground wire, the ground wires are assumed to be the same.

Table 11.3 Summary for Zero-Sequence Impedance Calculations

Values are in ohms/mile at 60 Hz. For kilometers, multiply mile values by 0.62137. For other frequencies, multiply the r_e, X_a, X_d, X_e, X_g values by $f/60$, where f is the desired frequency in hertz.

Sections/ Type	General equations	Eq. no.	Equations for use with tables	Eq. no.
11.9.1[1] Single circuit, no ground wire	$Z_0 = r_a + 0.286 + j0.8382 \log \frac{D_e}{\text{GMR}_{gr\ 3}}$	(11.37) (11.50)	$Z_0 = r_a + r_e + j(X_a + X_e - 2X_d)$	(11.38)
11.9.3, 11.9.4[2] Single circuit, one ground wire	$Z_{0a}^1 = r_a + j0.8382 \log \frac{\text{GMD}_{gr\ 3\ to\ gr\ 1}}{\text{GMR}_{gr\ of\ 3}}$ $Z_{0g}^1 = 3r_g + j0.8382 \log \frac{\text{GMD}_{gr\ 3\ to\ gr\ 1}}{\text{GMR}_{gr\ of\ 1}}$ $Z_{0m} = 0.286 + j0.8382 \log \frac{D_e}{\text{GMD}_{gr\ 3\ to\ gr\ 1}}$ Total Z_0 See notes	(11.45) (11.46) (11.44) (11.43)	$Z_{0a}^1 = r_a + j(X_a + 3X_{d(ag)} - 2X_d)$ $Z_{0g}^1 = 3r_g + j(3X_g + 3X_{d(ag)})$ $Z_{0m} = r_e + j(X_e - 3X_{d(ag)})$ Total Z_0 See notes	(11.47) (11.48) (11.49) (11.43)
11.9.5, 11.9.6[3] Single circuit, two ground wires	$Z_{0a}^1 = r_a + j0.8382 \log \frac{\text{GMD}_{gr\ 3\ to\ gr\ 2}}{\text{GMR}_{gr\ of\ 3}}$ $Z_{0g}^1 = \frac{3r_g}{2} + j0.8382 \log \frac{\text{GMD}_{gr\ 3\ to\ gr\ 2}}{\text{GMR}_{gr\ of\ 2}}$ $Z_{0m} = 0.286 + j0.8382 \log \frac{D_e}{\text{GMD}_{gr\ 3\ to\ gr\ 2}}$ Total Z_0 See notes	(11.53) (11.54) (11.55) (11.43)	$Z_{0a}^1 = r_a + j(X_a + 3X_{d(ag^1)} - 2X_d)$ $Z_{0g}^1 = \frac{3r_g}{2} + j\left(\frac{3}{2}X_g + 3X_{d(ag^1)} - \frac{3}{2}X_{d(g^1)}\right)$ $Z_{0m} = r_e + j(X_e - 3X_{d(ag^1)})$ Total Z_0 See notes	 (11.43)

Note 1

$$\text{GMR}_{\text{gr of 3}} = \sqrt[9]{\text{GMR}_{\text{cond}}^3 D_{ab}^2 D_{bc}^2 D_{ca}^2} = \sqrt[3]{\text{GMR}_{\text{cond}} \text{GMR}_{\text{gr of 3}}^2} \quad (11.50)$$

$$\text{where } \text{GMD}_{\text{gr of 3}} (X_d) = \sqrt[3]{D_{ab} D_{ba} D_{ca}} \quad (11.16)$$

Note 2

$$\text{GMR}_{\text{gr 1}} = \text{GMR}_{\text{gnd wire}}$$

$$\text{GMD}_{\text{gr 3 to gr 1}} ((X_{d(ag)}) = \sqrt[3]{D_{ag} D_{bg} D_{cg}}$$

$$\text{GMR}_{\text{gr 3}} = \text{same as Section 11.9.1} \quad (11.50)$$

Circuit: Z'_{0a} in series with the parallel combination of Z_{0m} and Z'_{0g}.

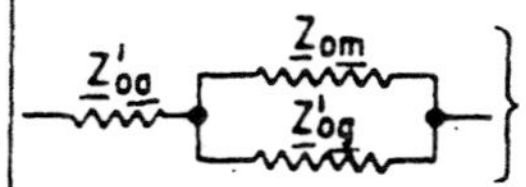

Total Z of circuit

$$= Z^1_{0a} + \frac{Z^1_{0g} Z_{0m}}{Z^1_{0g} + Z_{0m}} \quad (11.43)$$

where
- Z^1_{0a} is leakage Z of conductors $= Z_{0a} - Z_{0m}$
- Z^1_{0g} is leakage Z of ground wire(s) $= Z_{0g} - Z_{0m}$
- Z_{0m} is mutual Z between conductors and ground wire(s)
- Z_{0a} is self Z of conductors
- Z_{0g} is self Z of ground wire(s)

Note 3

$$\text{GMD}_{\text{gr 2}} (X_{d(g^1)})$$

$$\text{GMR}_{\text{gr 2}} = \sqrt{\text{GMR}_{\text{gnd wire}} D_{g^1 g^2}} \quad (11.52)$$

$$\text{GMD}_{\text{gr 3 to gr 2}} (X_{d(ag^1)}) = \sqrt[6]{D_{ag^1} D_{bg^1} D_{cg^1} D_{ag^2} D_{bg^2} D_{cg^2}} \quad (11.51)$$

$$\text{GMR}_{\text{gr 3}} = \text{same as Section 11.9.1} \quad (11.50)$$

Total Z_0 of circuit

$$= Z^1_{0a} + \frac{Z^1_{0g} Z_{0m}}{Z^1_{0g} + Z_{0m}} \quad (11.43)$$

Table 11.3 *(Continued)*

Sections/ Type	General equations	Eq. no.	Equations for use with tables
11.9.7, 11.9.8[4] Double circuit, no ground wire	$Z_0 = r_a + 0.286 + j0.8382 \log \frac{D_e}{\text{GMR}_{\text{gr } 3}}$	(11.37) (11.50)	$Z_0 = r_a + r_e + j(X_a + X_e - 2X_d)$
	$Z_{0m} = 0.286 + j0.8382 \log \frac{D_e}{\text{GMD}_{\text{gr } 3 \text{ to gr } 3}}$	(11.56)	$Z_{0m} = r_e + j(X_e - 3X_{d(s)})$
	$Z_{0(\text{parallel})} = \frac{1}{2}(Z_0 + Z_{0m}) = \frac{r_a}{2} + 0.286 + j0.8382 \log \frac{D_e}{\text{GMR}_{\text{gr of } 6}}$	(11.58)	$Z_{0(\text{parallel})} = \frac{1}{2}(Z_0 + Z_{0m}) = \frac{r_a}{2} + r_e + j\left(\frac{X_a}{2} + X_e - X_d - \frac{3}{2} X_{d(s)}\right)$
11.9.9[5] Double circuit, one ground wire	$Z_{0a}^1 = \frac{r_a}{2} + j0.8382 \log \frac{\text{GMD}_{\text{gr } 6 \text{ to gr } 1}}{\text{GMR}_{\text{gr of } 6}}$	(11.59)	$Z_{0a}^1 = \frac{r_a}{2} + j\left(\frac{X_a}{2} + 3X_{d(a^1g)} - X_d - \frac{3}{2} X_{d(s)}\right)$
	$Z_{0g}^1 = 3r_g + j0.8382 \log \frac{\text{GMD}_{\text{gr } 6 \text{ to gr } 1}}{\text{GMR}_{\text{gr of } 1}}$	(11.60)	$Z_{0g}^1 = 3r_g + j(3X_g + 3X_{d(a^1g)})$
	$Z_{0m} = 0.286 + j0.8382 \log \frac{D_e}{\text{GMD}_{\text{gr } 6 \text{ to gr } 1}}$	(11.61)	$Z_{0m} = r_e + j(X_e - 3X_{d(a^1g)})$
	Total Z_0 See notes	(11.43)	Total Z_0 See notes or (11.43)
11.9.10, 11.9.11[6] Double circuit, two ground wires	$Z_{0a}^1 = \frac{r_a}{2} + j0.8382 \log \frac{\text{GMD}_{\text{gr } 6 \text{ to gr } 2}}{\text{GMR}_{\text{gr of } 6}}$	(11.62)	$Z_{0a}^1 = \frac{r_a}{2} + j\left(\frac{X_a}{2} + 3X_{d(a^1g^1)} - X_d - \frac{3}{2} X_{d(s)}\right)$
	$Z_{0g}^1 = \frac{3r_g}{2} + j0.8382 \log \frac{\text{GMD}_{\text{gr } 6 \text{ to gr } 2}}{\text{GMR}_{\text{gr of } 2}}$	(11.63)	$Z_{0g}^1 = \frac{3r_g}{2} + j\left(\frac{3}{2} X_g + 3X_{d(a^1g^1)} - \frac{3}{2} X_{d(g^1)}\right)$
	$Z_{0m} = 0.286 + j0.8382 \log \frac{D_e}{\text{GMD}_{\text{gr } 6 \text{ to gr } 2}}$	(11.64)	$Z_{0m} = r_e + j(X_e - 3X_{d(a^1g^1)})$
	Total Z_0 See notes	(11.43)	Total Z_0 See notes or (11.43)

Note 4

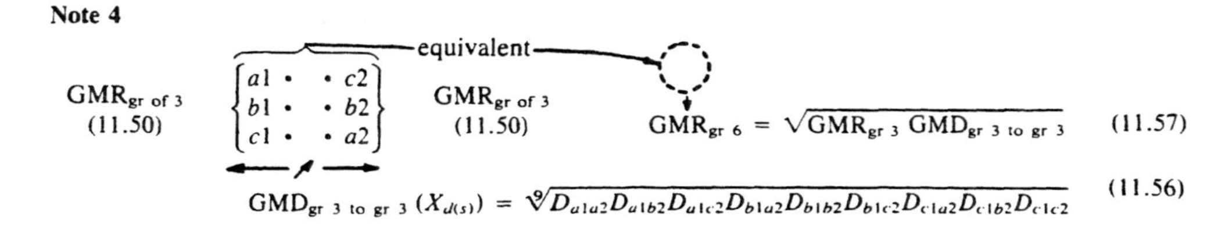

$\text{GMR}_{\text{gr of 3}}$ (11.50) $\quad \begin{Bmatrix} a1 \cdot & \cdot\ c2 \\ b1 \cdot & \cdot\ b2 \\ c1 \cdot & \cdot\ a2 \end{Bmatrix}$ $\quad \text{GMR}_{\text{gr of 3}}$ (11.50) — equivalent

$$\text{GMR}_{\text{gr 6}} = \sqrt{\text{GMR}_{\text{gr 3}}\ \text{GMD}_{\text{gr 3 to gr 3}}} \qquad (11.57)$$

$$\text{GMD}_{\text{gr 3 to gr 3}}\ (X_{d(s)}) = \sqrt[9]{D_{a1a2}D_{a1b2}D_{a1c2}D_{b1a2}D_{b1b2}D_{b1c2}D_{c1a2}D_{c1b2}D_{c1c2}} \qquad (11.56)$$

Note 5

$\cdot\ g\}$ $\quad \text{GMR}_{\text{gr of 1}} = \text{GMR}_{\text{gnd wire}}\ \{\cdot\ g$

$$\text{GMD}_{\text{gr 6 to gr 1}\ (X_{d(a1g)})} = \sqrt[6]{D_{a1g}D_{b1g}D_{c1g}D_{a2g}D_{b2g}D_{c2g}}$$

$\text{GMR}_{\text{gr of 3}}$ (11.50) $\quad \begin{Bmatrix} a1 \cdot & \cdot\ c2 \\ b1 \cdot & \cdot\ b2 \\ c1 \cdot & \cdot\ a2 \end{Bmatrix}$ $\quad \text{GMR}_{\text{gr of 3}}$ (11.50)

$\text{GMD}_{\text{gr 3 to gr 3}}$ (same as Sections 11.9.7 and 11.9.8)

$\text{GMR}_{\text{gr 6}}$ (11.57)

Total Z_0 of circuit

$$= Z^1_{0a} + \frac{Z^1_{0g}Z_{0m}}{Z^1_{0a} + Z_{0m}} \qquad (11.43)$$

Note 6

$\cdot\ g_1 \quad \cdot\ g_2\,\}$ $\quad \text{GMR}_{\text{gr of 2}}$ (11.52) $\quad \{\ \text{GMR}_{\text{gr of 2}}$ (11.52)

$$\text{GMD}_{\text{gr 6 to gr 2}}\ (X_{d(a1g1)}) = \sqrt[12]{D_{a1g1}D_{b1g1}D_{c1g1}D_{a2g1}D_{b2g1}D_{c2g1}D_{a1g2}D_{b1g2}D_{c1g2}D_{a2g2}D_{b2g2}D_{c2g2}}$$

$\text{GMR}_{\text{gr of 3}}$ (11.50) $\quad \begin{Bmatrix} a1 \cdot & \cdot\ c2 \\ b1 \cdot & \cdot\ b2 \\ c1 \cdot & \cdot\ a2 \end{Bmatrix}$ $\quad \text{GMR}_{\text{gr of 3}}$ (11.50)

$\text{GMD}_{\text{gr 3 to gr 3}}$ = same as Sections 11.9.7 and 11.9.8

$\text{GMR}_{\text{gr 6}}$ (11.57)

Total Z_0 of circuit

$$= Z^1_{0a} + \frac{Z^1_{0g}Z_{0m}}{Z^1_{0g} + Z_{0m}} \qquad (11.43)$$

The various values are summarized as follows:

r or r_a = ac resistance of the phase conductors

r_g = ac resistance of the ground wire(s)

r_e = 0.286 Ω/mile at 60 Hz

X_a or X_g = $0.2794 \log \frac{1}{\text{GMR}}$, ohms/mile at 60 Hz

X_d = 0.2794 log GMD, ohms/mile at 60 Hz

X_e = $0.8382 \log D_e$, ohms/mile at 60 Hz

$\text{GMR} = \text{GMR}_a = \text{GMR}_{\text{cond.}}$ = conductor geometric mean radius, feet

$\text{GMR}_g = \text{GMR}_{\text{gnd wire}}$ = ground wire geometric mean radius, feet

Otherwise, the GMR values are specified for various groups

GMD = geometric mean distance between groups as specified

D_e = equivalent depth of earth return

D_{ab}, D_{ag}, etc. = distances between respective conductors or ground wires, feet

12
Overhead Line Characteristics: Mutual Impedance

12.1 INTRODUCTION

Inductive coupling (mutual impedance) between lines that are paralleled for part or all their length must be considered primarily for unbalanced faults. This applies to power lines of the same or different voltages and to communication wire lines that parallel power lines.

Mutual coupling resulting from positive- and negative-sequence currents is generally negligible; nominally, the mutual impedance is on the order of 3 to 7% or less of the self-impedance. In contrast, the zero-sequence mutual impedance can be as high as 50 to 70% of the self-impedance. This is significant in fault calculations and protective relay applications. The basic fundamentals are reviewed with applications to typical power and communication lines.

12.2 MUTUAL COUPLING FUNDAMENTALS

Two paralleled circuits with self-impedances Z_{GH} and Z_{RS} and Z_m mutual impedance between them are shown in Fig. 12.1. The

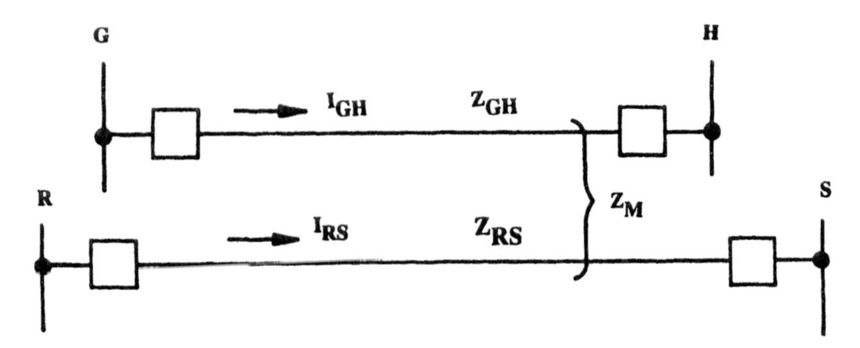

Figure 12.1 General case of paralleled lines with mutual coupling.

voltage drops across these lines are

$$V_{GH} = I_{GH}Z_{GH} + I_{RS}Z_m \tag{12.1}$$

$$V_{RS} = I_{RS}Z_{RS} + I_{GH}Z_m \tag{12.2}$$

Figure 12.2 shows the general network representation of Eqs. (12.1) and (12.2). The 1:1 ideal or perfect transformer, as discussed in Section 7.5, is used here to reflect the mutual voltage drop $(I_{GH} + I_{RS})Z_m$ in both lines.

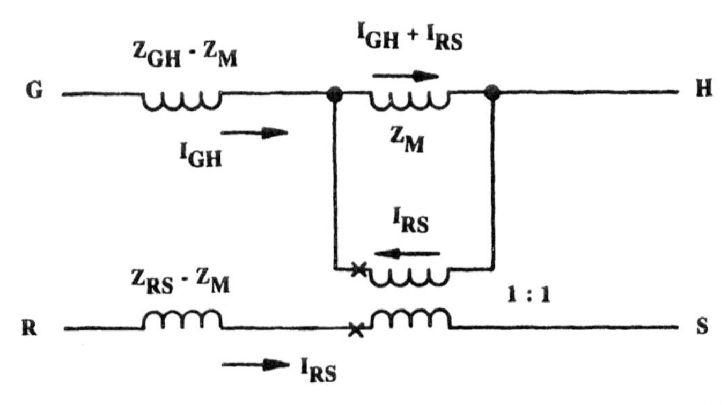

Figure 12.2 Equivalent network for the mutual coupled lines of Fig. 12.1.

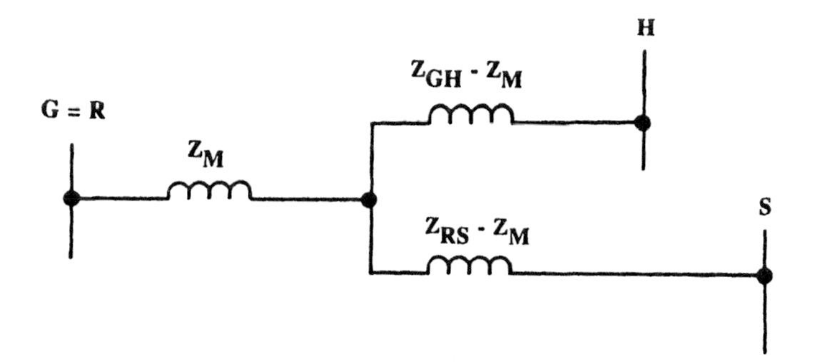

Figure 12.3 Equivalent network for the mutual coupled lines of Fig. 12.1 where there is a common bus at one end.

Figure 12.2 is general for single- or three-phase circuits, and combinations can be used where the mutual coupling is not the same for all of the line sections. If the parallel lines have a common bus at least at one terminal (bus G and R connected together as in Fig. 12.3), the equivalent circuit is as Fig. 12.3.

If both ends of the line end in common buses ($G = R$ and $H = S$) and the self-impedance $Z_{GH} = Z_{RS} = Z_L$, the total impedance of two parallel lines will reduce to

$$Z_{\text{total}} = \tfrac{1}{2}(Z_L + Z_m) \tag{12.3}$$

The application to the sequence networks and sequence quantities in three-phase circuits follows.

12.3 POSITIVE- AND NEGATIVE-SEQUENCE MUTUAL IMPEDANCE

The positive- and negative-sequence mutual impedance between paralleled three-phase circuits, even when on the same tower, is quite small and is neglected in fault calculations. With parallel lines such as shown in Fig. 12.4, the net flux available from positive- and negative-sequence current in either line to cut the adjacent circuit will be small because these currents are equal

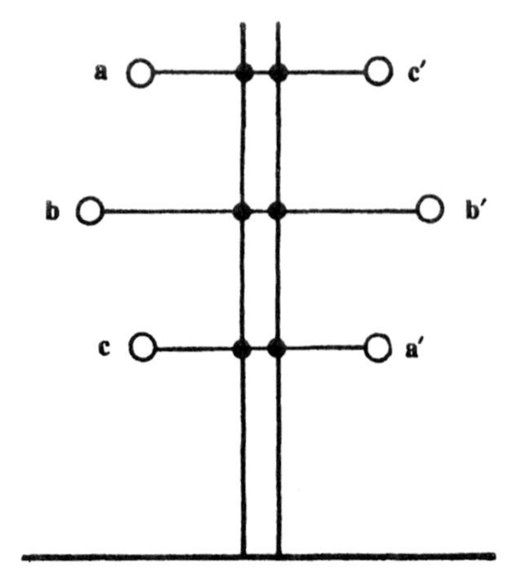

1st Transposition Section

$D_{ab} = D_{bc} = D_{a'b'} = D_{b'c'} = 8.54$ ft
$D_{ac} = D_{a'c'} = 16$ ft
$D_{ac'} = D_{ca'} = 18$ ft
$D_{ab'} = D_{cb'} = D_{ba'} = D_{bc'} = 22.5$ ft
$D_{aa'} = D_{bb'} = D_{cc'} = 24$ ft

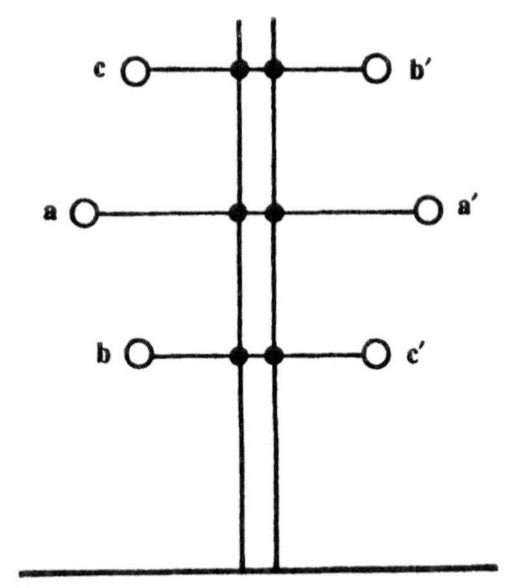

2nd Transposition Section

$D_{ab} = D_{ac} = D_{a'b'} = D_{a'c'} = 8.54$ ft
$D_{cb} = D_{c'b'} = 16$ ft
$D_{cb'} = D_{bc'} = 18$ ft
$D_{ca'} = D_{ac'} = D_{ab'} = D_{a'b} = 22.5$ ft
$D_{aa'} = D_{bb'} = D_{cc'} = 24$ ft

Figure 12.4 Transposed parallel lines showing the conductor spacings for the three transposition sections. Typical spacings are the same as in Fig. 11.10.

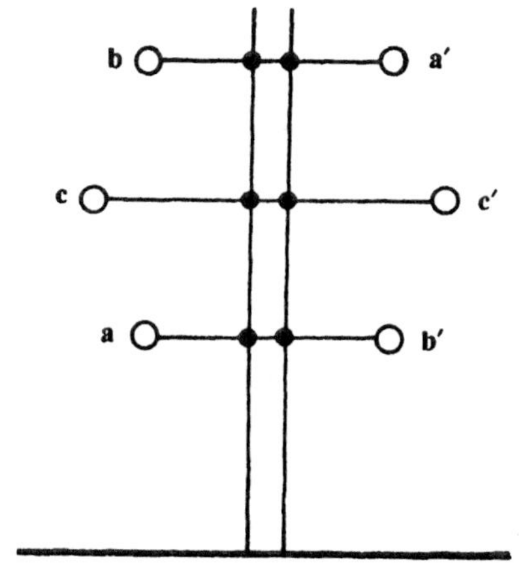

3rd Transposition Section

$D_{bc} = D_{ac} = D_{a'c'} = D_{c'b'} = 8.54$ ft
$D_{ba} = D_{a'b'} = 16$ ft
$D_{ba'} = D_{ab'} = 18$ ft
$D_{bc'} = D_{cb'} = D_{ca'} = D_{ac'} = 22.5$ ft
$D_{aa'} = D_{bb'} = D_{cc'} = 24$ ft

The conductors: ACSR 397,500 CM (30/7)

$GMR_{cond.} = 0.0278$ ft @ 60 Hz.

Figure 12.4 Continued

in magnitude and symmetrically spaced 120° in time phase. There would be zero induced voltage in the parallel circuit from the positive- and negative-sequence currents if it were physically possible to space each of the three conductors of the parallel circuit equidistant from each of the three conductors of the line itself. This would require spacings in Fig. 12.4 of

$$D_{aa'} = D_{ab'} = D_{ac'}$$

$$D_{ba'} = D_{bb'} = D_{bc'}$$

$$D_{ca'} = D_{cb'} = D_{cc'}$$

which are not possible. The nearest approach would be a triaxial cable.

With the conductors arranged symmetrically around the vertical axis as in Fig. 12.4, there will be no difference between the voltages induced in the two lines. An unsymmetrical conductor arrangement produces circulating currents between the two lines. This is documented by Blackburn in the *IEEE Transactions* (see the Bibliography).

The determination of the total positive and negative impedance of paralleled circuits is a good example of the application of the GMR and GMD techniques that were covered in Chapter 11.

Consider two typical parallel lines with three transposition sections between common buses as illustrated in Fig. 12.4. The impedance of any section as developed in Chapter 11 is

$$Z_1 = Z_2 = r_a + j0.2794 \log \frac{\text{GMD}}{\text{GMR}} \tag{11.11}$$

using 0.2794 for ohms/phase/mile at 60 Hz for convenience.

12.3.1 First Transposition Section (Fig. 12.4)

Consider conductors a–a', conductors b–b', and conductors c–c' each as one conductor instead of two, with separation of $D_{aa'} = D_{bb'} = D_{cc'} = 24$ ft. For each group,

$$\begin{aligned} \text{GMR}_{\text{gr. of 2}} &= \sqrt{(\text{GMR}_{\text{cond.}})(\text{GMD}_{\text{sep. of 2}})} \\ &= \sqrt{0.0278 \times 24} = 0.81682\,\text{ft} \qquad (12.4) \\ \text{GMD}_{aa'\text{ from }bb'} &= \sqrt[4]{D_{ab}D_{ab'}D_{a'b}D_{a'b'}} = \sqrt{D_{ab}D_{ab'}} \\ \text{GMD}_{1aa'-bb'} &= \sqrt{8.54 \times 22.5} = 13.8618\,\text{ft} \qquad (12.5) \\ \text{GMD}_{bb'\text{ from }cc'} &= \sqrt[4]{D_{bc}D_{bc'}D_{b'c}D_{b'c'}} = \sqrt{D_{bc}D_{bc'}} \\ \text{GMD}_{1bb'-cc'} &= \sqrt{8.54 \times 22.5} = 13.8618\,\text{ft} \qquad (12.6) \\ \text{GMD}_{cc'\text{ from }aa'} &= \sqrt[4]{D_{ca}D_{ca'}D_{c'a}D_{c'a'}} = \sqrt{D_{ca}D_{ca'}} \\ \text{GMD}_{1cc'-aa'} &= \sqrt{16 \times 18} = 16.9706\,\text{ft} \qquad (12.7) \end{aligned}$$

The GMD separation of the three above is

$$\begin{aligned} \text{GMD}_{\text{sep. of 3}} &= \sqrt[3]{(\text{GMD}_{1aa'-bb'})(\text{GMD}_{1bb'-cc'})(\text{GMD}_{1cc'-aa'})} \\ &= \sqrt[3]{13.8618 \times 13.8618 \times 16.9706} = 14.829\,\text{ft} \end{aligned} \tag{12.8}$$

12.3.2 Second Transposition Section (Fig. 12.4)

Similar to the first section, from Eq. (12.4),

$$\text{GMR}_{\text{gr. of 2}} = \sqrt{0.0278 \times 24} = 0.81682 \text{ ft}$$

The GMD distances between the three are the same as Eqs. (12.5) through (12.7) with possibly different numerical values:

$$\text{GMD}_{2aa' \text{ from } bb'} = \sqrt{8.54 \times 22.5} = 13.8618 \text{ ft}$$

$$\text{GMD}_{2bb' \text{ from } cc'} = \sqrt{16 \times 18} = 16.971 \text{ ft}$$

$$\text{GMD}_{2cc' \text{ from } aa'} = \sqrt{8.54 \times 22.5} = 13.8618 \text{ ft}$$

From Eq. (12.8) using the GMD_2 values above,

$$\text{GMD}_{2 \text{ sep. of 3}} = \sqrt[3]{13.8618 \times 16.971 \times 13.8618} = 14.829 \text{ ft}$$

12.3.3 Third Transposition Section (Fig. 12.4)

Again similar to above, from Eq. (12.4),

$$\text{GMR}_{\text{gr. of 2}} = \sqrt{0.0278 \times 24} = 0.81682 \text{ ft}$$

and the GMD distances are as Eqs. (12.4) through (12.6) with possible different numerical values:

$$\text{GMD}_{3aa' \text{ from } bb'} = \sqrt{16 \times 18} = 16.971 \text{ ft}$$

$$\text{GMD}_{3bb' \text{ from } cc'} = \sqrt{8.54 \times 22.5} = 13.8618 \text{ ft}$$

$$\text{GMD}_{3cc' \text{ from } aa'} = \sqrt{8.54 \times 22.5} = 13.8618 \text{ ft}$$

From Eq. (12.8);

$$\text{GMD}_{3 \text{ sep. of 3}} = \sqrt[3]{16.971 \times 13.8618 \times 13.8618} = 14.829 \text{ ft}$$

12.3.4 Total Line Reactance for the Transposed Line

With the line transposed, the reactance of the line will be equal to one-third of the sum of the values for the three sections, where the distance of each section will be one-third of the total line length. Thus, from Eq. (11.11), using GMR and GMD values of Eqs. (12.4) through (12.8),

$$X_1 = X_2 = j\frac{0.2794}{3}\left(\log\frac{\text{GMD}_{1\ \text{sep. of 3}}}{\sqrt{\text{GMR}_{\text{cond.}}D_{aa'}}}\right.$$

$$\left. + \log\frac{\text{GMD}_{2\ \text{sep. of 3}}}{\sqrt{\text{GMR}_{\text{cond.}}D_{aa'}}} + \log\frac{\text{GMD}_{3\ \text{sep. of 3}}}{\sqrt{\text{GMR}_{\text{cond.}}D_{aa'}}}\right) \quad (12.9)$$

Equation (12.9) reduces to

$$X_1 = X_2 = j0.2794\left(\frac{1}{2}\log\frac{\sqrt[3]{D_{1ab}D_{1bc}D_{1ca}}}{\text{GMR}_{\text{cond.}}}\right.$$

$$\left. - \frac{1}{12}\log\frac{D_{1aa'}^4 D_{1bb'}^2}{D_{1ab'}^2 D_{1ca'} D_{1ac'} D_{1ba'}^2}\right) \quad (12.10)$$

The first part of Eq. (12.10) is the parallel impedance of the two lines without mutual. The second part is the mutual effect, which may reduce the total reactance 3 to 5% or less. With the spacings of Fig. 12.4, Eq. (12.10) is

$$X_1 = X_2 = j0.2794\left(\frac{1}{2}\log\frac{\sqrt[3]{(8.54)^2 \times 16}}{0.0278} - \frac{1}{12}\log\frac{(24)^6}{(22.5)^4 \times (18)^2}\right)$$

$$X_1 = X_2 = j0.36019 - j0.00843 = j0.35176 \quad \text{ohms/mile at 60 Hz} \quad (12.11)$$

Alternative and more general equations for the positive- and negative-sequence mutual impedance are (Blackburn, *IEEE Transactions*, eqs. 38 and 26; see Bibliography)

$$Z_{m1} = 0.0931\left(0.866\log\frac{D_{1ab'}D_{1bc'}D_{1ca'}}{D_{1ac'}D_{1ba'}D_{1cb'}} - j\log D_Z\right) \quad (12.12)$$

$$Z_{m2} = 0.0931\left(0.866\log\frac{D_{1ac'}D_{1ba'}D_{1cb'}}{D_{1ab'}D_{1bc'}D_{1ca'}} - j\log D_Z\right) \quad (12.13)$$

where

$$D_Z = \frac{D_{1aa'}D_{1bb'}D_{1cc'}}{\sqrt{D_{1ab'}D_{1ac'}D_{1ba'}D_{1bc'}D_{1ca'}D_{1cb'}}} \tag{12.14}$$

With the lines transposed and with symmetrical spacings as in Fig. 12.4, the first part (resistance component) of Eqs. (12.12) and (12.13) is zero, so that the equations both reduce to

$$\begin{aligned} Z_{m1} = Z_{m2} &= \frac{-j0.2794}{3} \log D_Z \\ &= -j0.0931 \log \frac{24^3}{(22.5)^2 \times 18} \qquad (12.15) \\ &= -j0.01686 \qquad \text{ohms/mile at 60 Hz} \end{aligned}$$

The reactance of each line without mutual is [first part of Eq. (12.10) omitting the $\frac{1}{2}$],

$$\begin{aligned} X_{L1} = X_{L2} = X_L &= j0.2794 \log \frac{\sqrt[3]{(8.54)^2 \times 16}}{0.0278} \\ &= j0.72038 \qquad \text{ohms/mile at 60 Hz} \end{aligned}$$

Using these values in Fig. 12.3 and Eq. (12.3), we have

$$X_{\text{total}} = j\tfrac{1}{2}(0.72038 - 0.01686) \doteq j0.35176 \qquad \text{ohms/mile} \tag{12.16}$$

This checks the value determined in Eq. (12.11). The mutual reactance X_{m1} and X_{m2} are 2.34% of the self-reactance X_{L1} and X_{L2} or X_L. Detailed information on the mutual effects of all three sequence currents with both transposed and untransposed three-phase lines may be found in my article in the *IEEE Transactions*, 1963, listed in the Bibliography.

12.4 ZERO-SEQUENCE MUTUAL IMPEDANCE

The zero-sequence mutual is not negligible and should be included in all fault studies, as it can be on the order of 50 to 70% of the zero-sequence self-impedance.

The basic equations for the zero-sequence line impedances were developed in Chapter 11; the self-impedance (Z_0) as Eq. (11.35), and the mutual impedance (Z_{0m}) as Eq. (11.36). These were then applied to a number of line configurations in this chapter. These are summarized in Table 11.3. The parallel lines are as shown in Fig. 12.1, where the various quantities are zero-sequence values. Correspondingly, the equivalents of Figs. 12.2 and 12.3 apply with appropriate zero-sequence quantities.

12.5 MUTUAL IMPEDANCE BETWEEN LINES OF DIFFERENT VOLTAGES

In Fig. 12.1 assume that the top line is operating at kV_{GH} and the lower line, at kV_{RS}. The voltage drops across these two lines are

$$V_{0GH} = I_{0GH}Z_{0GH} + I_{0RS}Z_{0m} \qquad \text{volts at } kV_{GH} \tag{12.17}$$

$$V_{0RS} = I_{0RS}Z_{0RS} + I_{0GH}Z_{0m} \qquad \text{volts at } kV_{RS} \tag{12.18}$$

In these equations the currents are those flowing in the respective lines and the self-impedance in ohms per Eq. (11.35). Z_{0m} is also in ohms per Eq. (11.56), where GMD is the proper separation between the three conductors of the kV_{GH} line and the three conductors of the parallel kV_{RS} line. This is the ninth root of the nine distances for $\text{GMD}_{\text{gr. 3 to gr. 3}}$. With ground wire use Eq. (11.61) or (11.64).

From Chapter 2,

$$Z_{pu} = \frac{IZ}{V_{\text{line-to-neutral}}} \tag{12.19}$$

$$= \frac{\sqrt{3}\, IZ}{1000\ kV_{\text{line-to-line}}}$$

For the mutual part of Eq. (12.17),

$$Z_{0m}(\text{pu}) = \frac{\sqrt{3}\, I_{0RS} Z_{0m}}{1000\ \text{kV}_{GH}} \tag{12.20}$$

Since $I_{0RS} = 1000\ \text{MVA}/\sqrt{3}\ \text{kV}_{RS}$, Eq. (12.20) becomes,

$$Z_{0m}(\text{pu}) = \frac{\text{MVA}\ Z_{0m}\ (\text{ohms})}{\text{kV}_{GH}\ \text{kV}_{RS}} \tag{12.21}$$

Similarly, for the mutual part of Eq. (12.18),

$$Z_{0m}(\text{pu}) = \frac{\sqrt{3}\, I_{0GH} Z_{0m}}{1000\ \text{kV}_{RS}}$$

$$I_{0GH} = \frac{1000\ \text{MVA}}{\sqrt{3}\ \text{kV}_{GH}} \tag{12.22}$$

$$Z_{0m}(\text{pu}) = \frac{\text{MVA}\ Z_{0m}\ (\text{ohms})}{\text{kV}_{RS}\ \text{kV}_{GH}}$$

Equations (12.21) and (12.22) are identical and compatible to Eq. (2.15). Thus the per unit mutual impedance is a function of the product of the two system voltages.

12.6 POWER SYSTEM–INDUCED VOLTAGES IN WIRE COMMUNICATION LINES

The use of fiber optics in communication circuits eliminates the problem of extraneous voltages in communications circuits caused by the power circuits. Since not all wire lines have been replaced, the following outlines the problems involved in the wire circuits. Two types of extraneous voltage can result:

A. Conduction

1. *Metallic cross:* physical contact between the power line and the communication wires.

2. *Ground potential:* difference between the power station or substation ground and remote earth. This is caused by power system ground fault current flowing through the ground resistance between the station ground mat and remote earth. This is known as the rise in station ground mat; an IR-type voltage.

B. Induction

1. *Electric:* capacitive voltage effect
2. *Magnetic:* current; IX_m effect

12.6.1 Voltage Induced in a Communication Pair: General Case

A typical circuit is shown in Fig. 12.5. For the general case assume a nontransposed power line a, b, c with phase currents I_a,

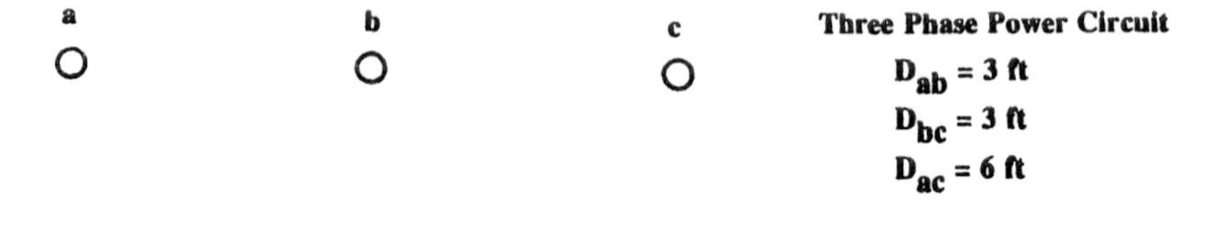

x
y

Communications Wire Pair

$D_{xy} = 1$ ft

$D_{ax} = D_{cx} = 10.42$ ft

$D_{ay} = D_{cy} = 11.42$ ft

$D_{bx} = 10$ ft

$D_{by} = 11$ ft

Figure 12.5 A typical communication pair near a three-phase power line.

I_b, I_c and a nontransposed communication pair x, y. The voltages induced are

$$V_x = I_a Z_{ax} + I_b Z_{bx} + I_c Z_{cx} \tag{12.23}$$

$$V_y = I_a Z_{ay} + I_b Z_{by} + I_c Z_{cy} \tag{12.24}$$

Applying the simplified Carson's formula (11.32) for the mutual impedance between two conductors with a common earth return, and for 60 Hz,

$$\begin{aligned} V_x = {} & I_a \left(0.0954 + j0.2794 \log \frac{D_e}{D_{ax}}\right) \\ & + I_b \left(0.0954 + j0.2794 \log \frac{D_e}{D_{bx}}\right) \\ & + I_c \left(0.0954 + j0.2794 \log \frac{D_e}{D_{cx}}\right) \quad \text{volts/mile} \end{aligned} \tag{12.25}$$

$$\begin{aligned} V_y = {} & I_a \left(0.0954 + j0.2794 \log \frac{D_e}{D_{ay}}\right) \\ & + I_b \left(0.0954 + j0.2794 \log \frac{D_e}{D_{by}}\right) \\ & + I_c \left(0.0954 + j0.2794 \log \frac{D_e}{D_{cy}}\right) \quad \text{volts/mile} \end{aligned} \tag{12.26}$$

The induced voltage between the communication pair x, y is

$$\begin{aligned} V_{xy} &= V_x - V_y \\ &= j0.2794 \left(I_a \log \frac{D_{ay}}{D_{ax}} + I_b \log \frac{D_{by}}{D_{bx}} + I_c \log \frac{D_{cy}}{D_{cx}}\right) \end{aligned} \tag{12.27}$$

Equations (12.25) through (12.27) are general equations for untransposed power and communication lines that are paralleled. They are applicable for either balanced or unbalanced currents involving any combination of positive-, negative-, and zero-sequence components.

12.6.2 Voltage Induced: Communication Pair Transposed, Power Line Untransposed

For half of the exposure length the induced voltage on wires x and y will be per Eq. (12.25), and for the other half per Eq. (12.26). Thus

$$V_x = V_y = I_a \left(0.0954 + j0.2794 \log \frac{D_e}{\sqrt{D_{ax}D_{ay}}}\right) + I_b \left(0.0954 + j0.2794 \log \frac{D_e}{\sqrt{D_{bx}D_{by}}}\right) + I_c \left(0.0954 + j0.2794 \log \frac{D_e}{\sqrt{D_{cx}D_{cy}}}\right) \text{ volts/mile} \tag{12.28}$$

$$V_{xy} = V_x - V_y = 0 \tag{12.29}$$

12.6.3 Voltage Induced: Communication Pair Untransposed, Power Line Transposed

The phase a conductor separation from wire x is D_{ax} for one-third of the line length, D_{bx} for another one-third, and D_{cx} for the last one-third. Correspondingly, b and c conductors are spaced similarly. Similar spacings exist for wire y. The induced voltages are

$$V_x = (I_a + I_b + I_c) \left(0.0954 + j0.2794 \log \frac{D_e}{\sqrt[3]{D_{ax}D_{bx}D_{cx}}}\right) \text{ volts/mile} \tag{12.30}$$

$$V_y = (I_a + I_b + I_c) \left(0.0954 + j0.2794 \log \frac{D_e}{\sqrt[3]{D_{ay}D_{by}D_{cy}}}\right) \text{ volts/mile} \tag{12.31}$$

$$V_{xy} = (I_a + I_b + I_c) \left(j0.2794 \log \frac{\sqrt[3]{D_{ay}D_{by}D_{cy}}}{\sqrt[3]{D_{ax}D_{bx}D_{cx}}}\right) \text{ volts/mile} \tag{12.32}$$

Thus the induced voltage with the power line transposed results only from the zero-sequence ($3I_0$) current. $I_a + I_b + I_c = 3I_0$.

Since the majority of power system faults involve ground, and since zero sequence is the major contributor to the induced voltage for the untransposed case, the transposition of the power circuit is not too effective in reducing the induced voltage.

12.6.4 Voltage Induced: Both Communication and Power Line Transposed

With both circuits transposed, the induced voltage is

$$V_x = V_y = (I_a + I_b + I_c) \times \left(0.0954 + j0.2794 \log \frac{D_e}{\sqrt[6]{D_{ax}D_{bx}D_{cx}D_{ay}D_{by}D_{cy}}}\right) \quad \text{volts/mile} \qquad (12.33)$$

$$V_{xy} = 0 \qquad (12.34)$$

12.6.5 Examples: Induced Voltages in a Communication Pair

A typical communication pair x, y is underneath a three-phase power line a, b, c as shown in Fig. 12.5. Assume that a ground fault occurs where $I_a = 1000$ A, $I_b = I_c = 400$ A. For simplicity the three-phase currents are assumed in phase a (worst case). D_e (mathematical depth of earth return for Carson's formulas) is 2800 ft.

With neither circuit transposed, from Eqs. (12.25) and (12.26),

$$\begin{aligned} V_x &= 1000\left(0.0954 + j0.2794 \log \frac{2800}{10.42}\right) \\ &\quad + 400\left(0.0954 + j0.2794 \log \frac{2800}{10}\right) \\ &\quad + 400\left(0.0954 + j0.2794 \log \frac{2800}{10.42}\right) \\ &= 171.72 + j1223.74 = 1235.73\angle 82^\circ \quad \text{volts/mile} \end{aligned}$$

$$
\begin{aligned}
V_y &= 1000\left(0.0954 + j0.2794 \log \frac{2800}{11.4}\right) \\
&\quad + 400\left(0.0954 + j0.2794 \log \frac{2800}{11}\right) \\
&\quad + 400\left(0.0954 + j0.2794 \log \frac{2800}{11.4}\right) \\
&= 171.72 + j1203.84 = 1216.03\underline{/81.88^\circ} \qquad \text{volts/mile}
\end{aligned}
$$

From Eq. (12.27),

$$
\begin{aligned}
V_{xy} &= j0.2794\left(1000 \log \frac{11.4}{10.42} + 400 \log \frac{11}{10}\right. \\
&\quad \left. + 400 \log \frac{11.4}{10.42}\right) = j19.9 \qquad \text{volts/mile}
\end{aligned}
$$

Similar calculations can be made for various other combinations according to Eqs. (12.28) through (12.34). The results are summarized in Table 12.1.

With uniform exposure these induced voltages to ground are illustrated in Fig. 12.6 when the communication pair is ungrounded. If one end of the pair is grounded, full V_x and V_y will

Table 12.1

Power line transposed ?	Communications pair transposed ?	Equations	Induced volts/mile of exposure		
			V_x	V_y	V_{xy}
No	No	(12.25)–(12.27)	$1235.73\underline{/82^\circ}$	$1216.03\underline{/81.88^\circ}$	$19.9\underline{/90^\circ}$
No	Yes	(12.28)–(12.29)	$1231.82\underline{/81.99^\circ}$	$1231.82\underline{/81.00^\circ}$	0
Yes	No	(12.30)–(12.32)	$1236.71\underline{/82.02^\circ}$	$1216.88\underline{/81.88^\circ}$	$20.03\underline{/90^\circ}$
Yes	Yes	(12.33)	$1226.80\underline{/81.95^\circ}$	$1226.80\underline{/81.95^\circ}$	0

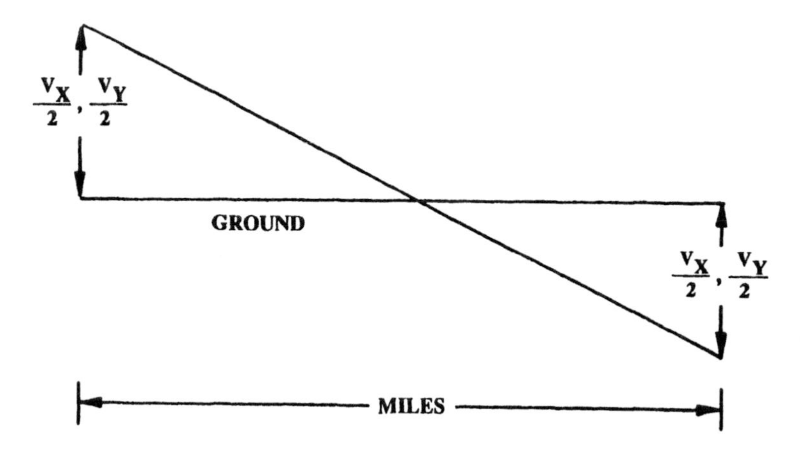

Figure 12.6 Induced voltage profile along an ungrounded communications pair.

appear at the ungrounded end. If both ends are grounded, the induced voltage causes a circulation of current as

$$I_x = \frac{V_x}{Z_x} \quad \text{and} \quad I_y = \frac{V_y}{Z_y} \tag{12.35}$$

Z_x and Z_y constitute the impedance of the communication circuit with earth return between the grounding points. This circulation reduces the voltage stress on the communication pair. Normally, ungrounded communication circuits are momentarily grounded by protector gaps, which short the wires around 300- to 600-V induction.

Since the distances D_{ax}, D_{bx}, D_{cx} and D_{ay}, D_{by}, D_{cy} are not equal, there will be small induced voltages for the positive- and negative-sequence currents flowing in the power system. However, the prime induction voltage results from the zero-sequence ($3I_0$) current. This is because the communication circuit is directly in the flux path between the three power conductors and the ground return. Ground wires, which are closer to the power circuit than the communication circuit and reduce the earth return current, act as a shield to reduce the induced voltage in the communication circuits.

12.6.6 Voltage Induced: Effect of a Communication Circuit Sheath

A metallic sheath grounded at frequent intervals and surrounding the communication circuit aids in reducing the induced voltage. A typical example with the equivalent circuit is shown in Fig. 12.7, R_{pc} and X_{pc} are the resistance and reactance of the power

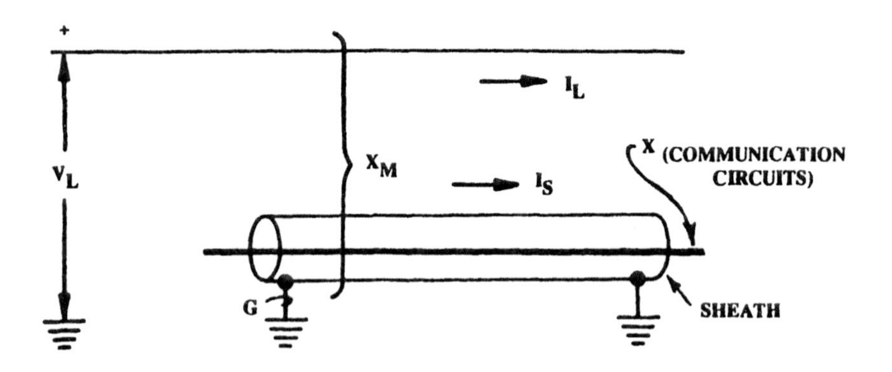

From this, an equivalent circuit can be derived:

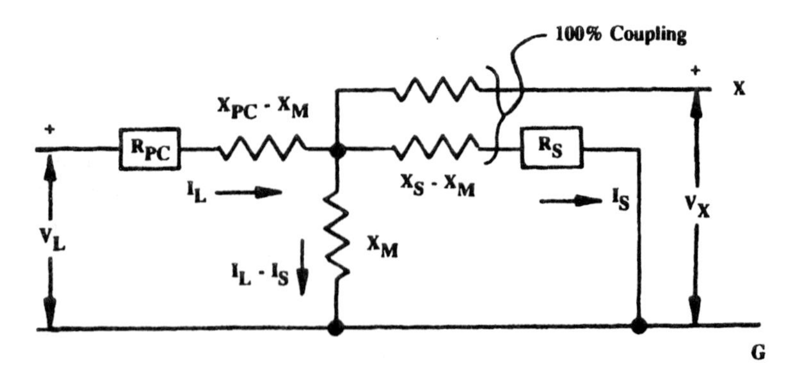

Figure 12.7 Induced voltages with a sheath around the communication circuit.

circuit per Eq. (11.31). With the power circuit transposed, these are the zero-sequence values, Eq. (11.35). The mutual impedance is found using Eq. (11.32) or with transposed power line, Eq. (11.36). R_s and X_s are the resistance and reactance of the sheath. The coupling between the communication circuit and it sheath is represented by the perfect or ideal transformer. If the sheath surrounds the wires concentrically, the coupling is 100%.

The logic of this diagram is apparent by assuming no sheath. Then $I_S = 0$ and $V_L = I_L(R_{pc} + jX_{pc})$, the drop along the line. The voltage induced in the communication circuit is $V_X = I_L X_M$. The capital subscripts indicate total values, in contrast to lower case subscripts, which indicate values per mile.

The drop for any current in the sheath is $I_S(R_S + jX_S)$, which is zero with the sheath grounded. From the equivalent diagram (Fig. 12.7) the voltage induced in the communication circuits is V_{XG}.

$$V_{XG} = (I_L - I_S)X_M - I_S(X_S - X_M) \tag{12.36}$$

$$(I_L - I_S)X_M - I_S(X_S - X_M) - I_S R_S = 0 \tag{12.37}$$

from which

$$V_{XG} = I_S R_S \tag{12.38}$$

Thus the induced voltage is the resistance drop along the sheath. This can vary considerably, depending on the sheath.

12.7 SUMMARY

Mutual coupling between lines paralleled in the available right-of-ways even for relative short distances can cause protection problems. Most of the problems result from zero-sequence mutual. The principles of determining the mutual impedances for the sequence networks is covered as an extension of Chapter 11. The positive- and negative-sequence mutual impedance, while small and generally negligible, is also documented. The determination of induced voltages in wire communication circuits that parallel power circuits is outlined.

13
Overhead Line Characteristics: Capacitive Reactance

13.1 INTRODUCTION

The capacitance is usually negligible for most low-, medium-, and high-voltage power lines that are operated grounded. For EHV lines, generally around 100 miles or more, the capacitance can be important. These lines should be represented by generalized networks involving *ABCD* constants, and so on. A brief review of line capacitance follows.

13.2 CAPACITANCE OF OVERHEAD CONDUCTORS

Consider a single isolated conductor with a charge of q. The voltage between any two points p_1 and p_2 outside the conductor is defined as the work done in moving a unit charge of 1 coulomb from point p_1 to point p_2 through the electric field produced by the charge q. This is expressed as

$$V_{p1p2} = q \times 18 \times 10^9 \ln \frac{y}{x} \qquad \text{volts} \tag{13.1}$$

where x is the distance of p_1 from the conductor and y is the distance of p_2 from the conductor.

With this basic equation and the superposition theorem, the capacitances of various paralleled conductors can be determined. The distribution of the charge on the conductor is assumed to be uniform. This is practical because the spacing between the line conductors in an overhead line is large relative to the conductor radius. Hence there is negligible distortion of the charge distribution from the associated conductors.

A single-phase circuit is shown in Fig. 13.1. First assume only a charge q_1 on conductor 1 with no charge on conductor 2. Then

$$V_{21} = q_1 \times 18 \times 10^9 \ln \frac{D_{12}}{r_1} \quad \text{volts} \tag{13.2}$$

This represents the work done in moving a unit charge from conductor 1 over a distance of D_{12} meters to the surface of conductor 2 through the electric field produced by q_1.

Now consider only a charge q_2 on conductor 2 and no charge on conductor 1. The voltage is

$$V_{21} = q_2 \times 18 \times 10^9 \ln \frac{r_2}{D_{12}} \quad \text{volts} \tag{13.3}$$

With both charges present together, by superposition,

$$V_{21} = 18 \times 10^9 (q_1 \ln \frac{D_{12}}{r_1} + q_2 \ln \frac{r_2}{D_{12}} \tag{13.4}$$

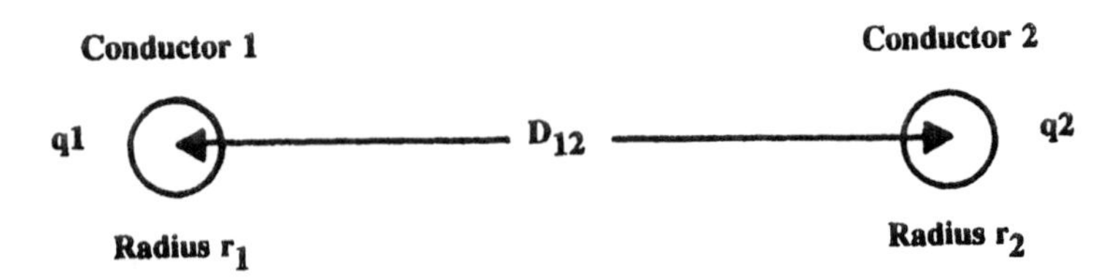

Figure 13.1 A general single-phase circuit.

If the charges on the two conductors are equal and their sum is zero, $q_2 = -q_1$. Substituting in Eq. (13.4),

$$V_{21} = 36 \times 10^9 \ln \frac{D_{12}}{\sqrt{r_1 r_2}} \quad \text{volts} \tag{13.5}$$

The capacitance between the two conductors is the ratio of the charge to the voltage:

$$\frac{q_1}{V_{21}} = C_{21} = C_{12} = \frac{1}{36 \times 10^9 \ln \dfrac{D_{12}}{\sqrt{r_1 r_2}}} \quad \text{farads/meter} \tag{13.6}$$

The capacitance to neutral (C_n) in the single-phase circuit is twice this value since the voltage to neutral is $\frac{1}{2}V_{21}$.

$$C_n = \frac{1}{18 \times 10^9 \ln \dfrac{D_{12}}{\sqrt{r_1 r_2}}} \quad \text{farads/meter} \tag{13.7}$$

Converting from $\ln_{(e)}$ to $\log_{(10)}$, and so on, Eq. (13.7) becomes

$$C_n = \frac{0.03883}{\log D_{12}/\sqrt{r_1 r_2}} \quad \text{microfarads/mile to neutral} \tag{13.8}$$

In the equations for capacitance, r is the actual radius of the conductor, not the GMR, as required for inductance. For capacitance the charge is distributed over the *surface* of the conductor, in contrast to inductance, where the current is distributed *throughout* the conductor and involves skin effect, and other factors. Where $r_1 = r_2 = r$, then $\sqrt{r_1 r_2} = r$.

Since $X'_c = -j1/2\pi fC$, $jb = -j1/X'_c$, and substituting GMR_r for $\sqrt{r_1 r_2}$, Eq. (13.8) can be written in more general terms as

$$X'_c = -j\frac{4.09876 \times 10^6}{f} \log \frac{\text{GMD}}{\text{GMR}_r} \quad \text{ohms/mile} \tag{13.9}$$

or as

$$b = -j2.43976f \times 10^{-7}/\log \frac{\text{GMD}}{\text{GMR}_r} \quad \text{siemens/mile} \quad (13.10)$$

13.3 POSITIVE- AND NEGATIVE-SEQUENCE CAPACITANCE

The basic equation (13.9) or (13.10) can be applied to a transposed three-phase line with three conductors 1, 2, 3 of radius $r = \text{GMR}_r$ separated by D_{12}, D_{23}, D_{31}. Then, as in Table 11.1, $\text{GMD} = \text{GMD}_{\text{gr. of 3}} = \sqrt[3]{D_{12}D_{23}D_{31}}$.

In terms of the X_a and X_d constants, as in Section 11.4, with X'_a as the capacitance reactance of the conductor to a radius of 1 ft, and X'_d as the capacitance reactance from 1 ft to the return conductor or conductors,

$$X'_a = -j\left(\frac{4.09876}{f}\right)\log\frac{1}{\text{GMR}_r} \quad \text{megohms/mile} \quad (13.11)$$

$$X'_d = -j\left(\frac{4.09876}{f}\right)\log \text{GMD} \quad \text{megohms/mile} \quad (13.12)$$

Then

$$X'_c = X'_1 = X'_2 = X'_a + X'_d \quad \text{megohms/mile} \quad (13.13)$$

This is similar to Eq. (11.12), used for inductance calculations. The conductor tables when available provide values of X'_a and X'_d.

For a line of n miles, then

$$X'_1 = X'_2 = \frac{X'_c}{n} \quad \text{megohms} \quad (13.14)$$

where X'_c is either Eq. (13.9) or (13.13).

Equation (13.9) can be rewritten in more general terms as

$$X'_c = -j\frac{4.09876}{fn}\log\frac{\text{GMD}}{\text{GMR}_r} \quad \text{megohms} \quad (13.15)$$

similarly, for Eq. (13.10),

$$b = j\frac{0.243976fn \times 10^{-6}}{\log(\text{GMD/GMR}_r)} \qquad \text{siemens} \tag{13.16}$$

where n is in miles. For kilometers, $n = 1.60934$:

$$X_c' = -j\frac{0.06381}{n}\log\frac{\text{GMD}}{\text{GMR}_r} \qquad \text{megohms at 60 Hz} \tag{13.17}$$

$$b = j\frac{14.63857n \times 10^{-6}}{\log(\text{GMD/GMR}_r)} \qquad \text{siemens at 60 Hz} \tag{13.18}$$

where b is the capacitive susceptance. These equations are useful to provide a good approximation of the line charging current I_c as

$$I_c = \frac{V_{\text{line-to-neutral}}}{X_1'} = bV_{LN} \qquad \text{amperes} \tag{13.19}$$

Note that X_1' must be in ohms and b in siemens. This simplified equation ignores the effect of the other conductors and ground.

Another approximation is

$$\text{charging kVA}_{3\phi} = \frac{2.05\ \text{kV}^2}{Z_0} \tag{13.20}$$

where Z_0 is the characteristic impedance of the overhead three-phase line. The average value of this for lines above 69 kV is around 386 Ω (varying from 360 to 430 Ω).

13.4 EXAMPLE: THREE-PHASE CIRCUIT CAPACITIVE REACTANCE

A three-phase 115-kV 60-Hz line has flat horizontal spacings of 10 ft and conductors with a radius of 0.0475 ft. For this line,

$$\text{GMD}_{\text{sep. of 3}} = \sqrt[3]{10 \times 10 \times 20}$$

$$= 12.6 \text{ ft} \quad \text{and} \quad \text{GMR}_r = 0.0475 \text{ ft}$$

The capacitive reactance from Eq. (13.17) or capacitive susceptance from Eq. (13.18):

$$X_c' = -j0.06831 \log \frac{12.6}{0.0475} = -j0.1656 \qquad \text{megohms/mile}$$

or

$$b = j14.63857 \times \frac{10^{-6}}{\log 12.6/0.0475} = 6.04 \times 10^{-6} \text{ S/mile}$$

The charging current, from Eq. (13.19), is

$$I_c = 115{,}000 \times \frac{10^{-6}}{\sqrt{3} \times 0.1656} = j0.401 \qquad \text{amperes/mile}$$

$$= 115{,}000 \times 6.04 \times \frac{10^{-6}}{\sqrt{3}} = j0.401 \qquad \text{amperes/mile} \tag{13.21}$$

From Eq. (13.20), a rough approximation:

$$\text{kVA}_{3\phi} = 2.05 \times \frac{115^2}{380} = 71.345$$

$$I_c = \frac{71.345}{115\sqrt{3}} = 0.358 \text{ A/mi}$$

Wagner and Evans (see the Bibliography) in Figure 100 on page 185 provide complex formulas for calculating the capacitive susceptance between the phases, including the effect of the ground, which has been neglected so far. Applying these formulas, the susceptances and phase charging currents were calculated for the 115-kV line of this example. This gave charging currents for the three phases of the line in amperes/mile at 60 Hz as $I_a = 0.395$, $I_b = 0.429$, and $I_c = 0.395$. These average to 0.406 A/mi, very close to the 0.401 A/mi determined above, neglecting the effect of the ground or earth plane.

Thus the presence of the earth has some effect on the capacitance values, but generally it is small and may be neglected. This also applies to the effect of ground wires. The effects are reduced if the phase conductors are transposed.

13.5 EXAMPLE: DOUBLE-THREE-PHASE-CIRCUIT CAPACITIVE REACTANCE

Determine the capacitive reactance of the two three-phase 115-kV 60-Hz lines shown in Fig. 12.4. The radius of all the conductors is 0.03358 ft = GMR_r.

Combining both three-phase lines in a single equivalent line per Eq. (12.4), with $\text{GMR}_{aa'} = \sqrt{\text{GMR}_{r(a)} D_{aa'}}$, $\text{GMR}_{bb'} = \sqrt{\text{GMR}_{r(b)} D_{bb'}}$ and $\text{GMR}_{cc'} = \sqrt{\text{GMR}_{r(c)} D_{cc'}}$

$$\text{then GMR}_{\text{gr. }3} = \sqrt[3]{\text{GMR}_{aa'}\text{GMR}_{bb'}\text{GMR}_{cc'}}$$

$$\text{or GMR}_{\text{gr. }3} = \sqrt[6]{\text{GMR}_{r(a)} D_{aa'} \text{GMR}_{r(b)} D_{bb'} \text{GMR}_{r(c)} D_{cc'}}$$

$$= \sqrt[6]{(0.03358)^3 \times (24)^3} = 0.89773 \text{ ft} \qquad (13.22)$$

The separation between the three aa', bb', cc' equivalent conducters is, per Eq. (12.8):

$$\text{GMD}_{\text{sep. of }3} = \sqrt[3]{(\text{GMD}_{aa'-bb'})(\text{GMD}_{bb'-cc'})(\text{GMD}_{cc'-aa'})}$$

$$= 14.829 \text{ ft} \qquad (13.23)$$

From Eq. (13.17),

$$X'_c = -j0.06381 \log \frac{14.829}{0.89773} = 0.0832 \text{ M}\Omega/\text{mi}$$

or (13.24)

$$b = j12.019 \times 10^{-6} \text{ s/mile}$$

$$I_c = 115{,}000 \times 12.019 \times \frac{10^{-6}}{\sqrt{3}} = 0.798 \text{ A/mi}$$

This is charging current to neutral for both lines in parallel. One-half of this equals 0.399 A/mi.

For each line alone, either a, b, c or a', b', c':

$$\text{GMR}_r = 0.03358 \text{ ft} \quad \text{and} \quad \text{GMD} = \sqrt[3]{(8.54)^2 \times 16} = 10.528 \text{ ft}$$

and from Eq. (13.18),

$$b = \frac{j14.63857 \times 10^{-6}}{\log 10.528/0.03358} = 5.864 \times 10^{-6} \text{ s/mile}$$

$$I_c = 115{,}000 \times 5.864 \times \frac{10^{-6}}{\sqrt{3}} = 0.389 \text{ A/mi}$$

Comparing $\frac{1}{2}(0.789) = 0.399$ and 0.389, the parallel line has very little effect on the capacitance of each line alone.

13.6 ZERO-SEQUENCE CAPACITANCE

The capacitance to ground and the zero-sequence capacitance reactance or susceptance can be calculated by assuming the earth to be an equipotential plane between the charge on the conductors and the opposite charge on the images of the conductors. These images are assumed to be the same distance below the earth or ground plane as the actual conductors are above the plane. Under these ideal conditions the capacitive or subceptive reactances between the conductors and the earth are given by Eq. (13.15) or (13.16) with proper interpretation of the GMR_r and GMD quantities. For these cases the GMR_r is the equivalent GMR_r of the group of phase conductors, the same as used for zero-sequence inductive reactance except that the individual conductor radius is used instead of the equivalent conductor GMR. The GMD is the equivalent separation between the conductors and their images.

The zero-sequence susceptances are one-third of (capacitances three times) the previous values since $3I_0$ actually flow in the system, but the zero-sequence networks are I_0 networks.

Thus for zero sequence, Eq. (13.15) becomes

$$X'_{0c} = -j\,\frac{12.2963}{fn}\log\frac{\text{GMD}}{\text{GMR}_r} \qquad \text{megohms} \tag{13.25}$$

and Eq. (13.16) becomes

$$b_0 = j\,\frac{0.08133fn \times 10^{-6}}{\log(\text{GMD}/\text{GMR}_r)} \qquad \text{siemens} \tag{13.26}$$

13.7 ZERO-SEQUENCE CAPACITANCE: TRANSPOSED THREE-PHASE LINE

A typical three-phase transposed line without ground wires is shown in Fig. 13.2 with their image conductors. The GMD and GMR_r for this configuration from Eqs. (11.8) and (11.10) are

$$\begin{aligned}\text{GMR}_{r(\text{gr. of }3)} &= \sqrt[9]{\text{GMR}_r{}^3 D_{ab}^2 D_{bc}^2 D_{ca}^2} \\ &= \sqrt[3]{\text{GMR}_r \text{GMD}_{\text{sep. of }3}^2}\end{aligned} \tag{13.27}$$

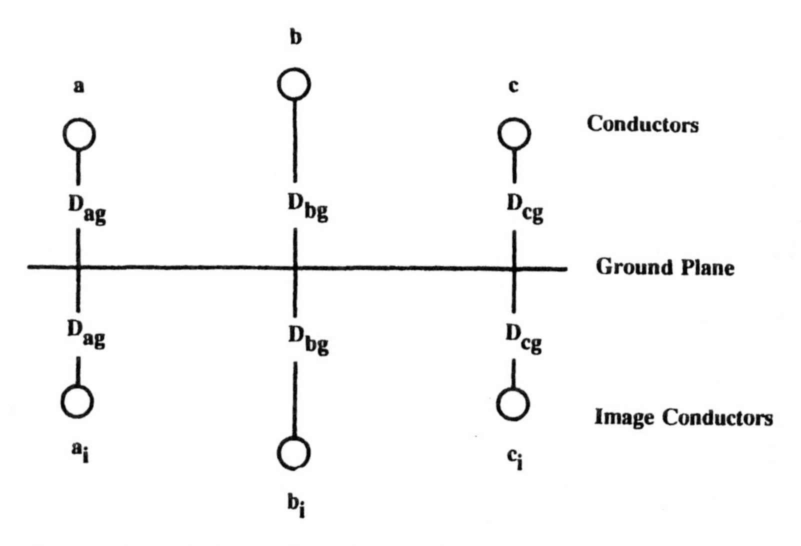

Figure 13.2 A three-phase line and its ground image for zero-sequence capacitance.

where

$$\mathrm{GMD}_{\text{sep. of 3}} = \sqrt[3]{D_{ab}D_{bc}D_{ca}} \quad \text{(Table 11.1)} \tag{13.28}$$

$$\mathrm{GMD}_{\text{gr. 3 to gr. 3}} = \sqrt[9]{D_{aa_i}D_{ab_i}D_{ac_i}D_{ba_i}D_{bb_i}D_{bc_i}D_{ca_i}D_{cb_i}D_{cc_i}}$$

13.8 EXAMPLE: ZERO-SEQUENCE CAPACITANCE, TRANSPOSED THREE-PHASE LINE

Using the example of Section 13.4 with the three conductors 40 ft above the earth, the zero-sequence susceptance is

$$\mathrm{GMR}_{r(\text{gr. of 3})} = \sqrt[3]{0.0475 \times (12.6)^2} = 1.96\,\text{ft}$$

$$\mathrm{GMD}_{\text{gr. 3 to gr. 3}} = \sqrt[9]{(80)^3 \times (80.62)^4 \times (82.46)^2} = 80.817\,\text{ft}$$

From Eq. (13.26),

$$\begin{aligned} b_0 &= \frac{4.88 \times 10^{-6}}{\log 80.817/1.96} = j3.02 \times 10^{-6}\ \text{s/mi at 60 Hz} \\ I_{0c} &= 115{,}000 \times 3.02 \times \frac{10^{-6}}{\sqrt{3}} = 0.200\ \text{A/mi at 60 Hz} \end{aligned} \tag{13.29}$$

13.9 SUMMARY

Capacitance is a shunt effect, in contrast to inductance, which is a series effect. Thus currents (charging currents) will flow in either symmetrical or unsymmetrical systems whenever the line or system is energized by voltage. Their magnitude is a function of the configuration of the circuit, the earth, ground wires, and adjacent circuits. In a three-phase line most of the capacitance results from the associated phase(s), as shown above.

Fault calculations seldom include the capacitive reactance values; they are very high relative to the inductive reactance values. One exception is for ungrounded or high-impedance (resistance) grounded systems. Here the capacitive reactance determines or has a major influence on a ground fault current, and the positive sequence values from Eq. (13.15) or (13.16) are used. As has been pointed out, the capacitance between the phase conductors is the major capacitance with that to the earth plane, other circuits, and ground wires relatively small. Thus when one phase is grounded in an ungrounded power system, fault current into the ground returns through the phases. Alternatively, the capacitance reactance value can be determined from the charging current when this is known. More details are provided in Blackburn, *Protective Relaying* (see the Bibliography), in the system grounding principles chapter.

For rough approximations useful for estimating when specific data are not available, for overhead three-phase 60-Hz lines: typical shunt capacitance reactance $X'_1 = X'_2 = 0.20$ MΩ/mi for single-conductor lines, and 0.14 MΩ/mi for bundled conductors.

14
Cable Characteristics

14.1 INTRODUCTION

The general methods of determining the constants of overhead lines applies to cables. However, the close proximity of the sheath introduces additional losses that must be considered. It is customary with cables to express the dimensions in inches rather than in feet, and this convention will be followed in these notes. The types of cable generally encountered for power transmission are shown Fig. 14.1.

(a) Single-conductor solid type (up to 69 kV)
(b) Belted three-conductor (up to 15 kV)
(c) Type H or shielded three-conductor (15 to 35 kV)
(d) Three conductors in steel pipe

Either round or sector-type conductors are used in the three-conductor types. The higher-voltage cables, either single- or three-conductor, are usually oil- or gas-filled, with the channels for the fluid built into the cable. The oil or gas is maintained at

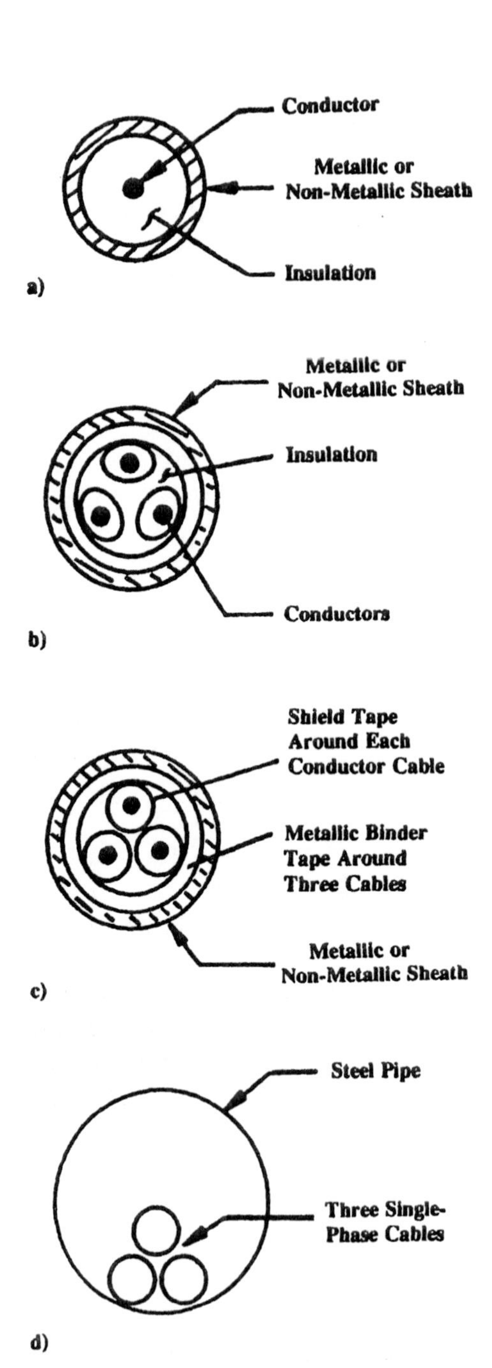

Figure 14.1 Power transmission cable.

a low pressure of 10 to 15 psi or at a high pressure of about 200, depending on the application.

Single-conductor, low-pressure oil-filled cables are used between 69 and 230 kV and three-conductor cables between 23 and 69 kV. Single- conductor high-pressure oil-filled pipe-type cables are used between 69 and 345 kV. Both single- and three-conductor low-pressure gas-filled cables are used up to and including 46 kV. Single-conductor high-pressure, gas-filled cables, either pipe-type or self-contained in an aluminum sheath, are used at 139 kV.

The following discussion is based on paper-insulated lead-covered cables, but the results can be applied to other insulations, such as rubber, varnish cambric, and so on.

14.2 POSITIVE- AND NEGATIVE-SEQUENCE CONSTANTS

As indicated previously, the three-phase circuits should be symmetrical in order to properly apply the method of symmetrical components. With the same size and type of conductor in each phase of the line, symmetry between phases can be obtained in two ways:

1. Transpositions of the three conductors
2. Equilateral triangle spacings of the three conductors

For power cables the second method is most practical. Three-phase cables are built with equal spacing between conductors. Where neither method is convenient, the error in assuming transpositions or equilateral spacing is small. The equivalent equilateral spacing is obtained by using the GMD between the conductors. For three conductors,

$$\mathrm{GMD}_{\mathrm{sep.\ of\ 3}} = \sqrt[3]{D_{12}D_{23}D_{31}} \tag{14.1}$$

where

$$D_{12} = D_{23} = D_{31} = S \qquad \mathrm{GMD}_{\mathrm{sep.\ of\ 3}}\ \sqrt[3]{S^3} = S \tag{14.2}$$

14.2.1 Single-Conductor Cables

Consider the section of cable shown in Fig. 14.2. The drop along the conductor is

$$V = IZ_c - jI_sX_m \tag{14.3}$$

and the drop along the sheath in zero with the sheath grounded.

$$0 = (r_s + jX_m)I_s - jIX_m \tag{14.4}$$

The sheath reactance is X_m on the basis that the conductor is concentric within the sheath, either actually or approximately.

From Eq. (14.4),

$$I_s = \frac{+jIX_m}{r_s + jX_m}$$

Substitution in Eq. (14.3) gives

$$\begin{aligned} V &= IZ_c - jX_m \frac{+jIX_m}{r_s + jX_m} = \left(Z_c + \frac{X_m^{\ 2}}{r_s + jX_m}\right) I \\ &= \left(r_c + jX_c + \frac{X_m^{\ 2}}{r_s + jX_m} \cdot \frac{r_s - jX_m}{r_s - jX_m}\right) I \\ &= \left[\left(r_c + \frac{r_s X_m^{\ 2}}{r_s^{\ 2} + X_m^{\ 2}}\right) + j\left(X_c - \frac{X_m^{\ 3}}{r_s^{\ 2} + X_m^{\ 2}}\right)\right] I \end{aligned} \tag{14.5}$$

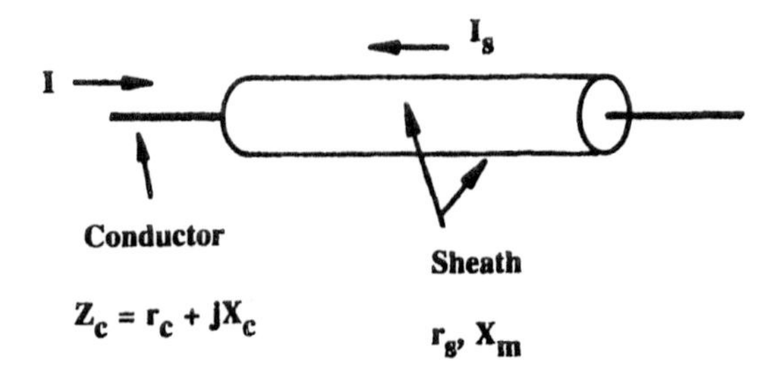

Figure 14.2 Single-conductor cable.

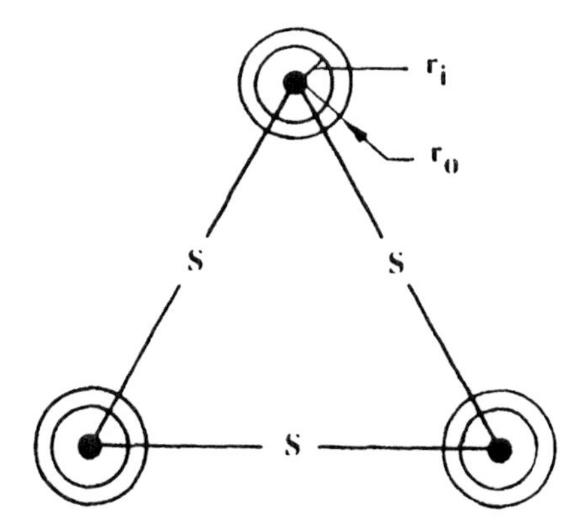

Figure 14.3 Cable configuration.

The second part of the resistance expression of Eq. (14.5) represents the sheath loss. This results from the conductor currents inducing voltages in the sheath. In turn, these voltages create sheath currents. These sheath losses increase the conductor resistance in the manner shown above.

The second term of the reactance expression of Eq. (14.5) is a correction for the presence of sheath currents and is subtracted because the flow of these sheath currents is opposite to that in the main conductors. Thus it tends to limit the flux to the region between the conductor and the sheath.

Resistance

The ac resistance of a group of three conductors (Fig. 14.3) from Eq. (14.5) is

$$r_a = r_c + \frac{r_s X_m^2}{r_s^2 + X_m^2}$$

where

r_c = ac resistance of the conductor, including skin effect
X_m = mutual reactance between the conductor and sheath
r_s = ac resistance of the sheath

For the circuit shown,

$$X_m = 0.004657f \log \frac{2S}{r_o + r_i} \quad \text{ohms/phase/mile} \tag{14.6}$$

$$r_s = \frac{0.200}{(r_o + r_i)(r_o - r_i)} \quad \text{ohms/phase/mile for lead} \tag{14.7}$$

where

S = spacing between conductors, inches
r_o = outer radius of the lead sheath, inches
r_i = inner radius of the lead sheath, inches

$(r_o + r_i)/2$ in the reactance equation is the mean radius of the sheath and is an approximate GMR of the sheath. As indicated previously, $GMD_{cond.\ sep.}$ may be used in place of S to give an equivalent where there are unequal spacings between the phase cables.

When the sheath is bonded together solidly, the sheath resistance may be 60 to 90% of the conductor resistance such that it is not negligible. As a result of the high sheath loss, several means are used to limit the less, such as:

1. Insulating the sheath may produce high sheath voltages which are a hazard to life and increase electrolysis
2. Series bonding impedances across the insulated cable joints to reduce the circulating current
3. Bonding transformers

Reactance of Single-Conductor Cables

The reactance of single-conductor cables from Eq. (14.5) is

$$X = 0.004657f \log_{10} \frac{S}{GMR_{cond.}} - \frac{X_m^3}{r_s^2 + X_m^2} \quad \text{ohms/phase/mile} \tag{14.8}$$

The conductor GMR is defined and used in the same manner as discussed previously for overhead transmission lines. The conductor with skin effect and perhaps noncircular and nonuniform cross section can be replaced as far as the formulas are concerned by a thin-walled circular tube with radius equal to GMR. GMR values are given in many tables of conductor characteristics.

The reactance equation can be rewritten as

$$X = 0.004657f \log \frac{S}{12} + 0.004657f \log \frac{12}{\mathrm{GMR_{cond.}}} - \frac{X_m{}^3}{r_s{}^2 + X_m{}^2}$$

and as for open wire, X_a and X_d are defined as

$$X_a = 0.004657f \log \frac{12}{\mathrm{GMR_{cond.}}} \tag{14.9}$$

the reactance of the conductor up to 12 in.

$$X_d = 0.004657f \log \frac{S}{12} \tag{14.10}$$

the reactance from 12 in. to the equivalent return conductors. Also, the sheath reactance, X_m, can be written

$$X_m = X_s + X_d \tag{14.11}$$

where X_d is as defined above and

$$X_s = 0.004657f \log \frac{2 \times 12}{r_o + r_i} \quad \text{ohms/phase/mile} \tag{14.12}$$

As before, S can be replaced with a GMD equivalent spacing between the several conductors (three conductors in the case above). Hence the total positive- and negative-sequence impedance of single-conductor cables, including sheath current with solidly bonded sheaths, is

$$Z_1 = Z_2 = r_c + \frac{r_s X_m{}^2}{r_s{}^2 + X_m{}^2} + j\left(X_a + X_d - \frac{X_m{}^3}{r_s{}^2 + X_m{}^2}\right)$$

ohms/phase/mile (14.13)

The r_c, r_s, X_a, X_s, and X_d values are often conveniently obtained from tables. It should be noted that X_d is in terms of inches and that for spacings of less than 12 in., the formulas apply, with X_d being negative. If the sheath is not continuous and the sheath currents can be neglected, then

$$Z_1 = Z_2 = r_c + j(X_a + X_d) \qquad \text{ohms/phase/mile} \qquad (14.14)$$

Example

Determine the 60-Hz impedance of 3-1 MCM single-conductor cables that have been drawn into fiber conduits in the same horizontal plane 4.125 in. between adjacent conductors. The conductor insulation is $\frac{30}{64}$ in. (46 kV) and the lead sheath is $\frac{1}{8}$ in. thick. The conductor radius is 0.576 in. The conductor radius of 0.576 in. plus the insulation thickness of 0.469 makes $r_i = 1.045$ in. r_i plus the sheath thickness makes $r_o = 1.170$ in.

$$S = \text{GMD} = \sqrt[3]{4.125 \times 4.125 \times 8.25} = 5.20 \text{ in.} \qquad (14.1)$$

$$X_m = 0.2794 \log \frac{2 \times 5.2}{1.045 + 1.17} = 0.188\ \Omega/\text{phase/mile} \qquad (14.6)$$

$$r_s = \frac{0.200}{(1.17 + 1.045)(1.17 - 1.045)} = 0.722\ \Omega/\text{phase/mile} \qquad (14.7)$$

r_c = 0.070 Ω/phase/mile and $GMR_{cond.}$ = 0.576 × 0.769 = 0.445 in.

$$\frac{r_s X_m{}^2}{r_s{}^2 + X_m{}^2} = \frac{0.722 \times (0.188)^2}{(0.722)^2 + (0.188)^2} = 0.046\ \Omega\text{/phase/mile}$$

$$\frac{X_m{}^3}{r_s{}^2 + X_m{}^2} = \frac{(0.188)^3}{0.557} = 0.0119\ \Omega\text{/phase/mile}$$

$$\begin{aligned} Z_1 = Z_2 &= 0.070 + 0.046 \\ &\quad + j\left(0.2794 \log \frac{5.2}{0.445} - 0.0119\right) \\ &= 0.116 + j0.2864 = 0.31\underline{/68^\circ}\ \Omega\text{/phase/mile} \end{aligned} \tag{14.13}$$

Neglecting sheath currents yields

$$Z_1 = Z_2 = 0.070 + j0.298 = 0.305\underline{/76.8^\circ}\ \Omega\text{/phase/mile} \tag{14.14}$$

Alternative Method Using Cable Characteristic Tables

From reference tables X_a = 0.400, X_s = 0.290, r_c = 0.070, and r_s = 0.752 for the 46-kV 1-MCM single-conductor cable. It will be noted that the cable listed in the particular table used had a slightly different insulation and sheath thickness, hence r_s is larger. Also from tables, X_d equals

$$\begin{aligned} X_d \text{ for } 4.125 \text{ in.} &= -0.130 \\ X_d \text{ for } 4.125 \text{ in.} &= -0.130 \\ X_d \text{ for } 8.25 \text{ in.} &= \underline{-0.045} \\ X_d &= \tfrac{1}{3}(-0.305) \\ &= -0.102 \end{aligned}$$

$$X_m = 0.290 - 0.102 = 0.188\ \Omega\text{/phase/mile} \tag{14.11}$$

$$Z_1 = Z_2 = 0.070 + \frac{(0.188)^2\,0.752}{(0.188)^2 + (0.752)^2}$$

$$+ j\left(0.400 - 0.102 - \frac{(0.188)^3}{(0.188)^2 + (0.752)^2}\right) \quad (14.13)$$

$$= 0.070 + 0.034 + j(0.298 - 0.0086)$$

$$= 0.104 + j0.2894 = 0.309\underline{/70.3^\circ}$$

14.3 THREE-CONDUCTOR CABLES

14.3.1 Resistance

The sheath loss for three conductor cables is usually negligible except for the very large sizes, where the loss is on the order of 3 to 5%. Where necessary to calculate the resistance with sheath loss, the following formula is used:

$$r_a = r_c + \frac{44{,}160 S_1^{\,2}}{r_s(r_o + r_i)^2} \times 10^{-6}\ \Omega/\text{phase/mile} \quad (14.15)$$

where S_1 is the effective center of each conductor and the cable center.

For round conductors,

$$S_1 = \frac{1}{\sqrt{3}}(d + 2t) \quad (14.16)$$

where d is the conductor diameter and t is the conductor insulation thickness. For sector-shaped conductors use Eq. (14.16) but with d = 82 to 86% of the diameter of a round conductor having the same cross-sectional area. For type H cables, the additional losses in the shielding tapes of the individual conductors and the binding tape are negligible. Consequently, the resistance to positive and negative sequence can be calculated as though the shields were not present.

14.3.2 Reactance of Three-Conductor Cables

The reactance is

$$X = 0.004657f \log \frac{S}{\text{GMR}_{\text{cond.}}} \tag{14.17}$$

where S = GMD, the geometric mean distance between the three conductors. For sector-shaped conductors the reactance is from 5 to 10% less than for round conductors of the same area and insulation thickness. For type H cables, the reactance can be calculated as though the shields were not present because the effect on reactance of the circulating currents in the shielding tapes is negligible.

Example

Find the 60-Hz impedance of a 750-MCM three-conductor belted cable having $\frac{10}{64}$ (0.1563)-in. conductor insulation and $\frac{8.5}{64}$ (0.133)-in. lead sheath (15 kV). The overall diameter of the cable is 2.833 in. and the conductors are sector shaped with a sector depth of 0.78 in., a resistance of 0.091 Ω/mile, and a GMR of 0.366 in. The diameter of an equivalent round conductor with same cross-sectional area is 0.998 in.

$$S_1 = \frac{1}{\sqrt{3}}(0.998 \times 0.84 + 2 \times 0.156) = 0.664 \text{ in.} \tag{14.16}$$

Since the overall diameter is 2.833, then r_o = 1.417 in. and r_i = 1.284 in.

$$r_s = \frac{0.200}{(1.417 + 1.284)(1.417 - 1.284)} = 0.557\ \Omega/\text{phase/mile} \tag{14.7}$$

The equivalent distance S between the sectors is difficult to obtain because of the nonsymmetry of the sectors. As a rough approximation, this distance can be taken as the sector depth plus twice the conductor insulation, or for the example, $S = 0.78 + 2 \times 0.156 = 1.09$ in.

$$Z_1 = Z_2 = 0.091 + \frac{44{,}160 \times (0.664)^2}{0.557(1.417 + 1.284)} \times 10^{-6}$$

$$+ j0.2794 \log \frac{1.09}{0.366} \quad \text{[Eqs. (14.15) and (14.17)]}$$

$$= 0.091 + 0.0048 + j0.132$$

$$= 0.096 + j0.132 = 0.164\angle 54^\circ \ \Omega\text{/phase/mile}$$

14.3.3 Three Conductors in Steel Pipe

The positive- and negative-sequence impedance of three conductors in a steel pipe is very difficult to calculate. Estimating curves for the resistance and the reactance are given in the Westinghouse *Electrical Transmission and Distribution Reference Book*, figures 13 and 14 of Chapter 4 (see the Bibliography).

14.4 ZERO-SEQUENCE CONSTANTS OF CABLES

Zero-sequence currents flowing in the three-phase conductors must have a return path. Three cases outline the problem:

1. All return current in the sheath, none in the ground.
2. Return current in sheath and ground in parallel.
3. All return current in the ground, none in the sheath.

The actual zero-sequence impedance will approach one of these conditions, depending on how the sheath is bonded and grounded, the nature of the earth return path, and the size and length of the cable involved. If the cable is solidly bonded and not grounded, or if the resistance of the ground is relatively high, so that no appreciable current flows in the ground, case 1 applies. Where the sheath is broken by insulating sleeves, or the sheath has a high resistance, the third case gives the best approximation.

Carson's simplified formulas, modified for zero sequence, are also applicable to cable circuits. The impedance of a group of three paralleled conductors with earth return (no sheath return) is

$$Z_{0c} = r_c + 0.00477f + j0.01397f \log \frac{D_e}{\text{GMR}_c} \quad \text{ohms/phase/mile} \quad (14.18)$$

$$= r_c + r_e + j(X_a + X_e - 2X_d) \quad \text{ohms/phase/mile} \quad (14.19)$$

The impedance of the sheath with earth return (no conductor groups) is

$$Z_{0s} = N_s r_s + 0.00477f + j0.01397f \log \frac{D_e}{\text{GMR}_s} \quad \text{ohms/phase/mile} \quad (14.20)$$

$$= 3_{rs} + r_e + j(3X_s + X_e) \quad \text{for a three-phase cable}$$

$$= r_s + r_e + j(X_s + X_e - 2X_d) \quad (14.21)$$

$$\text{for three single-phase cables} \quad (14.22)$$

The mutual impedance between conductors and sheath with common earth return is

$$Z_{0m} = 0.00477f + j0.01397f \log \frac{D_e}{\text{GMD}_{\text{sheath to cond.}}} \quad \text{ohms/phase/mile} \quad (14.23)$$

$$= r_e + j(3X_s + X_e) \quad \text{for a three-phase cable} \quad (14.24)$$

$$= r_e + j(X_e + X_s - 2X_d) \quad \text{for three single-phase cables} \quad (14.25)$$

These formulas are applicable to either three single-phase cables or one three-phase cable. These are shown schematically in Fig. 14.4. Either circuit can be represented by the same equivalent circuit as that in Fig. 14.5. This will be recognized as the same

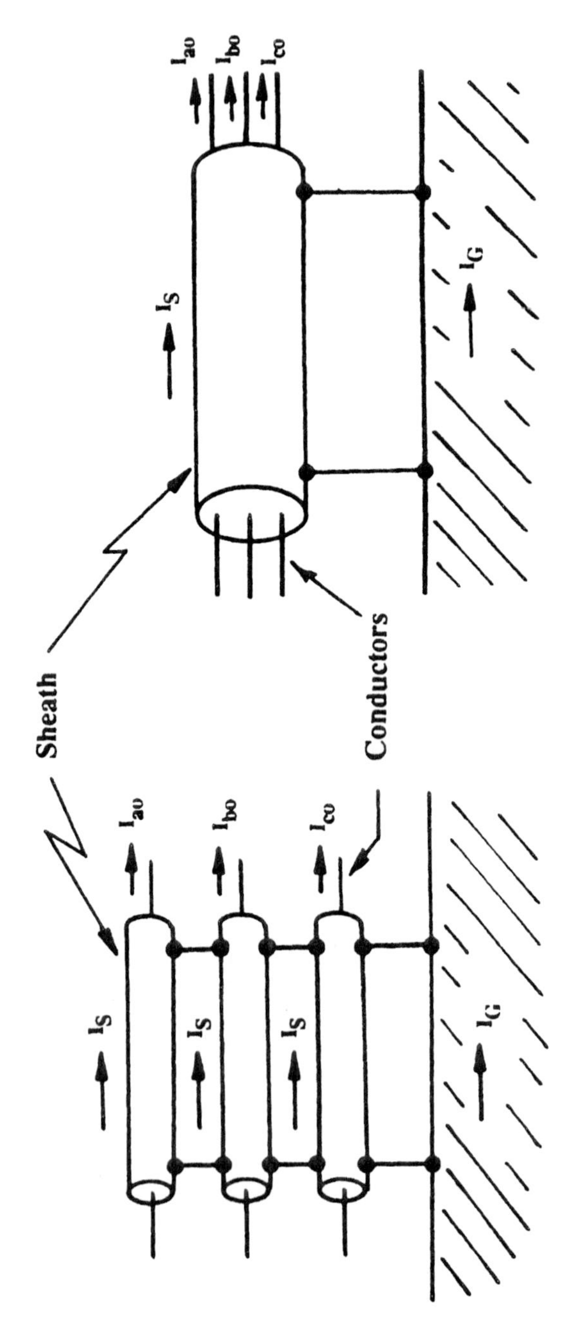

Figure 14.4 Mutual impedance in cables.

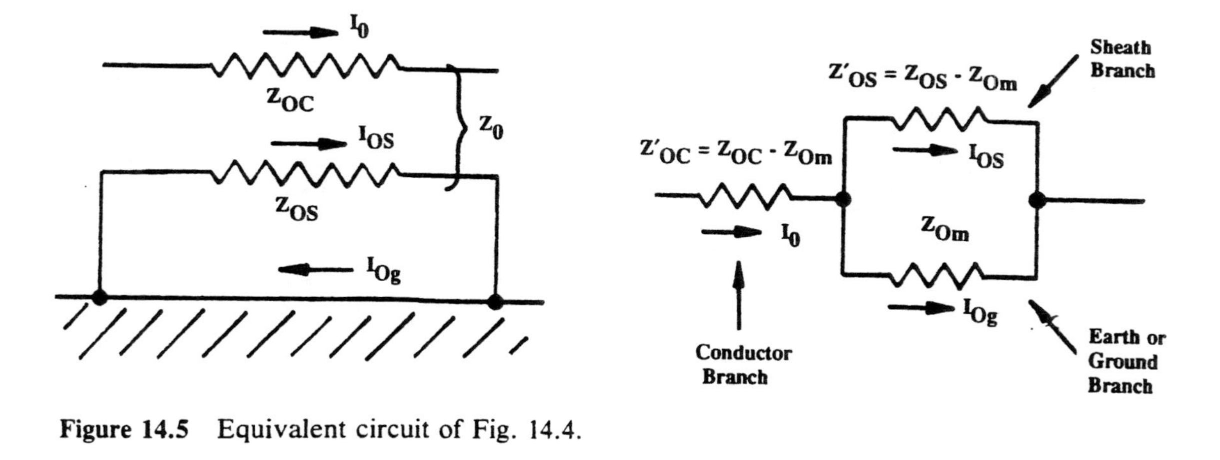

Figure 14.5 Equivalent circuit of Fig. 14.4.

type of equivalent as given for the overhead lines with ground wires.

The values in the preceding formulas may or may not be the same for each type of cable. The terminology and comparisons of the values for both types follow:

r_c = ac resistance of one conductor, ohms/mile

r_e = ac resistance of earth return,
= $0.00477f$ ohms/mile (14.26)

D_e = distance to equivalent earth return path (see Chapter 11)

X_e = reactance of earth return (14.27)
= $0.01397f \log D_e$ ohms/mile (D_e in feet)

GMR_c = geometric mean radius of the three conductors as a group, inches

$$= \sqrt[3]{\text{GMR}_{\text{cond.}} \times \text{GMD}^2_{\text{sep. of 3}}} \quad (14.28)$$

$\text{GMD}_{\text{sep. of 3}}$ = geometric mean distance between conductor centers, inches
= $S = d + 2t$ for round conductors in three-conductor cables (14.29)
= $\sqrt[3]{S_{ab}S_{bc}S_{ca}}$ for three single-conductor cables, a, b, c (14.1)

X_a = reactance of an individual phase conductor at 12-in. spacing, ohms/mile

$$= 0.004657f \log \frac{12}{\text{GMR}_{\text{cond.}}} \quad (14.9)$$

$$X_d = 0.004657f \log \frac{\text{GMD}_c}{12} \quad \text{ohms/mile} \quad (14.10)$$

N_s = number of sheaths = 1 for a three-phase cable
= 3 for three single-phase cables

r_s = ac resistance of the sheath, ohms/mile

$$= \frac{0.20}{(r_o + r_i)(r_o - r_i)} \quad \text{for lead sheaths} \quad (14.7)$$

r_o = outside radius of sheath, inches

r_i = inside radius of sheath, inches

X_s = reactance of the sheath $= 0.004657f \log \dfrac{24}{r_o + r_i}$ ohms/mile (14.12)

GMR_s = geometric mean radius of sheath return

$$= \frac{r_o + r_i}{2} \quad \text{inches for three-phase cables}$$

$$= \sqrt[3]{\frac{(r_o + r_i)}{2}\,(\text{GMD}^2 \text{ sep. of } 3)} \tag{14.30}$$

inches for three single-phase cables

$\text{GMD}_{\text{sheath to cond.}}$ = geometric mean distance between sheaths and conductors (14.31)

$$= \frac{r_o + r_i}{2} \quad \text{inches for three-phase cables}$$

$$= \sqrt[3]{\left(\frac{r_o + r_i}{2}\right)(\text{GMD}^2_{\text{sep. of } 3})} \tag{14.32}$$

inches for three single-phase cables

The equivalent circuit above reduces as follows:
Return current in sheath and ground in parallel:

$$Z_0 = Z_{0c} - Z_{0m} + \frac{(Z_{0s} - Z_{0m})Z_{0m}}{Z_{0s}} \tag{14.33}$$

$$= Z_{0c} - \frac{Z_{0m}^2}{Z_{0s}} \quad \text{ohms/phase/mile}$$

Return current in sheath only:

$$Z_0 = Z_{0c} - Z_{0m} + Z_{0s} - Z_{0m} \tag{14.34}$$

$$= Z_{0c} + Z_{0s} - 2Z_{0m} \quad \text{ohms/phase/mile}$$

Return current in ground only:

$$Z_0 = Z_{0c} - Z_{0m} + Z_{0m} = Z_{0c} \qquad \text{ohms/phase/mile} \qquad (14.35)$$

For submarine cables at a considerable depth below the water, $D_e = 280$ ft as an average value, and assume practically all of the return flows through the water. The zero sequence of type H cables can be calculated as though the shielding tapes were not present with very little error. For larger cables, the absolute value of Z_0 is virtually unchanged by the nature of the ground connection, and thus it is customary to calculate Z_0 based on all return in the sheath.

Example

Find the Z_0 of a three-conductor belted cable of No. 2 AWG (seven strands) size. The conductors are round with a diameter of 0.292 in., conductor insulation $\frac{10}{64}$ in. thick, belt insulation $\frac{5}{64}$ in., lead sheath $\frac{7}{64}$ in. thick, and overall diameter is 1.732 in. Assume that the resistance of one conductor = 0.987 Ω/mile at 60 Hz and that $D_e = 2800$ ft. (See Fig. 14.6.) The GMR of a seven-strand conductor is $0.292/2 \times 0.726 = 0.106$ in.

Distance between conductor centers S

$$= 0.292 + 2\left(\frac{10}{64}\right) = 0.605 \text{ in.} \qquad (14.29)$$

$$\text{GMR}_c = \sqrt[3]{0.106 \times (0.605)^2} = 0.338 \text{ in.} \qquad (14.28)$$

$$r_o = \frac{1.732}{2} = 0.866 \text{ in.}$$

$$r_i = 0.866 - \tfrac{7}{64} = 0.757 \text{ in.}$$

$$\text{GMR}_s = \text{GMD}_{\text{sheath to cond.}} = \frac{0.866 + 0.757}{2} = 0.811 \text{ in.} \qquad (14.31)$$

$$Z_{0c} = 0.987 + 0.286 + j0.838 \log \frac{2800 \times 12}{0.388} \qquad (14.18)$$

$$= 1.273 + j4.19 = 4.38\underline{/73.1^\circ}\ \Omega/\text{mi}$$

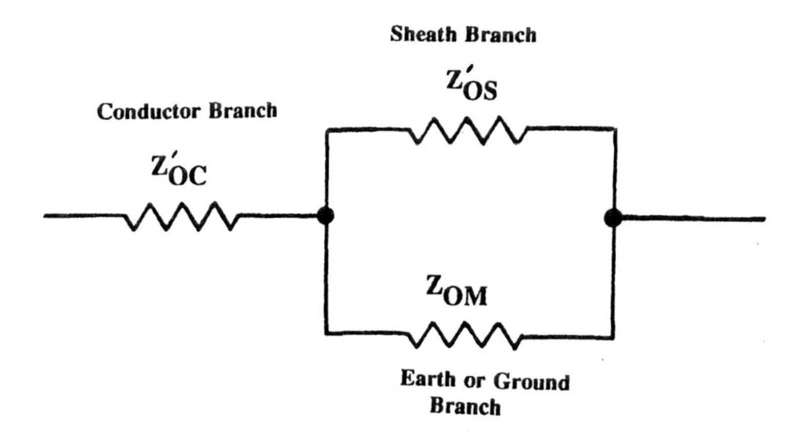

$$Z_{OC} = \frac{r_C \text{ of one cond.}}{\text{No. of Circuits}} + j\,0.01397\, f \,\text{Log}\, \frac{GMD_{sheath - conductor}}{GMR_{conductor}}$$

$$Z_{OS} = \frac{0.60}{(r_o + r_1)(r_o - r_1)(\text{No. of Sheaths})}$$

$$Z_{Om} = 0.00477\, f + j\,0.01597\, f\, \text{Log} \frac{D_e}{GMD_{sheath - conductor}}$$

Figure 14.6 Equivalent circuit for parallel three-phase cables.

$$Z_{Os} = \frac{3 \times 0.20}{(0.866 + 0.757)(0.866 - 0.757)} + 0.286$$

$$+ j0.838 \log \frac{2800 \times 12}{0.811} \qquad \text{[Eqs. (14.7) and (14.8)}$$

$$= 3.39 + 0.286 + j3.87$$

$$= 3.676 + j3.87 = 5.32\angle 46.5^\circ$$

$$Z_{0m} = 0.286 + j0.838 \log \frac{2800 \times 12}{0.811} \tag{14.23}$$

$$= 0.286 + j3.87 = 3.88\angle 85.8°$$

Return current in sheath and ground in parallel:

$$Z_0 = 4.38\angle 73.1° - \frac{(3.88)^2\angle 171.6°}{5.32\angle 46.5°}$$

$$= 4.38\angle 73.1° - 2.84\angle 125.1° \tag{14.33}$$

$$= 1.27 + j4.19 + 1.64 - j2.32$$

$$= 2.91 + j1.87 = 3.46\angle 32.7° \ \Omega/\text{mi}$$

Return current in sheath only:

$$Z_0 = 1.27 + j4.19 + 3.68 + j3.87 - 0.572 - j7.74$$

$$= 4.38 + j0.32 = 4.4\angle 4.2° \tag{14.34}$$

Return current in ground only:

$$Z_0 = 4.38\angle 73.1° \ \Omega/\text{mi} \tag{14.35}$$

14.4.1 Paralleled Groups of Three-Phase Cables

The equivalent circuit can be used to determine the zero-sequence impedance of paralleled conductors. For this purpose it can be written in a slightly different form. Z'_{0c} is derived from Eqs. (14.18) and (14.23); Z'_{0s}, from Eqs. (14.20) and (14.23) referring to Figs. 14.4 and 14.5. For three-phase cable $N_s = 3$ and $GMR_s = GMD_{\text{sheath to cond.}}$ per Eqs. (14.30) and (14.32): hence is derived Eq. (14.36).

$$Z'_{0s} = \frac{0.60}{(r_o + r_i)(r_o - r_i)(\text{no. of sheaths})} \tag{14.36}$$

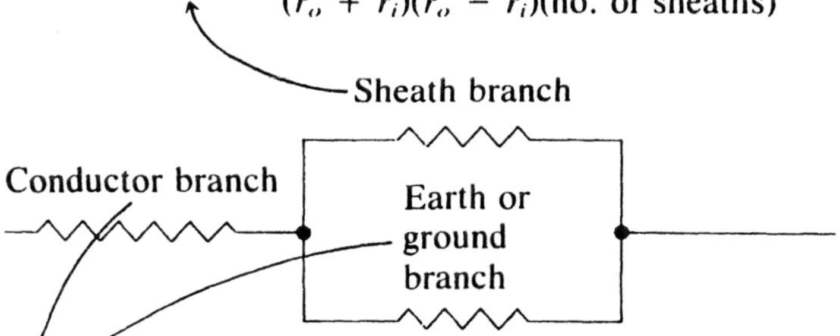

Sheath branch

Conductor branch

Earth or ground branch

$$Z_{0m} = 0.00477f + j0.01597f \log \frac{D_e}{\text{GMD}_{\text{sheath to cond.}}} \tag{14.37}$$

$$Z'_{0c} = \frac{r_{c \text{ of one cond.}}}{\text{no. of circuits}} + j0.01397f \log \frac{\text{GMD}_{\text{sheath to cond.}}}{\text{GMR}_{\text{cond.}}} \tag{14.38}$$

Example

Determine the total zero-sequence impedance of a group of four cables arranged in a duct bank with 5-in. horizontal separation (Fig. 14.7). The cables are the same as described in the preceding example, where

$$\text{GMR}_{\text{each cable cond.}} = 0.338 \text{ in.} \tag{14.28}$$

$$\text{GMD}_{\text{sheath to cond.}} = 0.811 \text{ in.} \tag{14.31}$$

$$\text{GMR}_{\text{four circuits}} = \sqrt[16]{(0.338)^4 \times 5^6 \times 10^4 \times 15^2}$$

$$= 3.479 \text{ in.}$$

$$\text{GMD}_{\text{sep. 4 cond. to sheaths}}$$

$$= \sqrt[16]{(0.811)^4 \times 5^6 \times 10^4 \times 15^2}$$

$$= 4.33 \text{ in.}$$

$$\text{Conductor branch } Z'_{0c} = \frac{0.987}{4} + j0.838 \log \frac{4.33}{3.479}$$

$$= 0.247 + j0.080 \ \Omega/\text{mi} \tag{14.38}$$

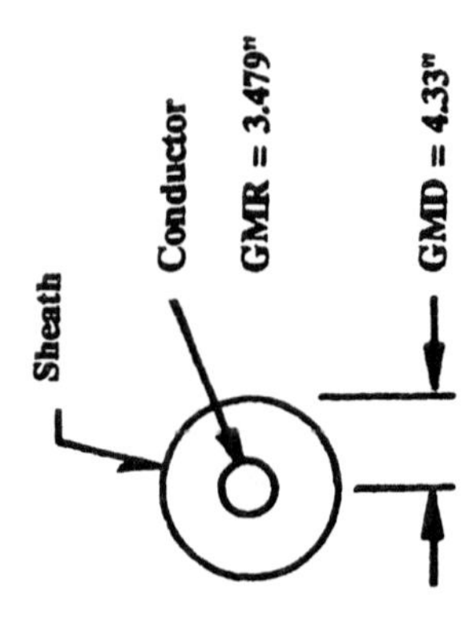

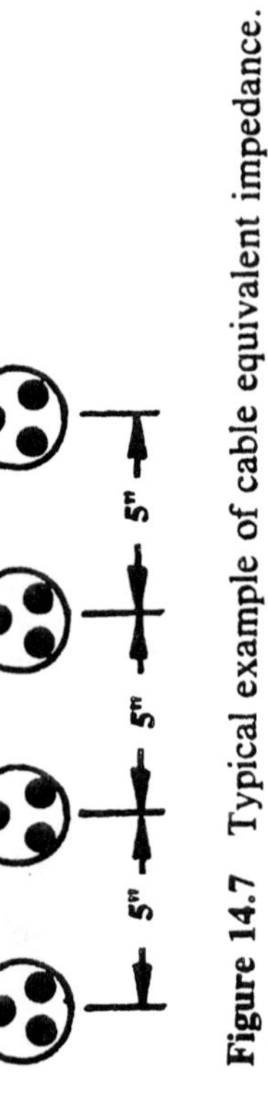

Figure 14.7 Typical example of cable equivalent impedance.

$$\text{Sheath branch } Z'_{0s} = \frac{0.60}{1.623 \times 0.109 \times 4} \tag{14.36}$$

$$= 0.848\ \Omega/\text{mi}$$

$$\text{Ground branch } Z_{0m} = 0.286 + j0.838 \log \frac{2800 \times 12}{4.33}$$

$$= 0.286 + j3.26\ \Omega/\text{mi} \tag{14.37}$$

$$\text{Total } Z_0 = \frac{(Z'_{0s})(Z_{0m})}{Z'_{0s} + Z_{0m}} + Z'_{0c}$$

$$= \frac{0.848(0.286 + j3.26)}{0.848 + 0.286 + j3.26} + 0.247 + j0.08$$

$$= 1.022 + j0.275 = 1.06\underline{/15^\circ}\ \Omega/\text{mi}$$

Z_0 all return in sheath.

$$\text{Total } Z_0 = Z'_{0c} + Z_{0m} = 0.247 + j0.080 + 0.848$$

$$= 1.095 + j0.08$$

$$= 1.1\underline{/0^\circ}\ \Omega/\text{mi}$$

hence most returns in the sheath with the four cables in parallel.

14.4.2 Zero Sequence of Cables in Steel Pipes

No method is available for calculating this complicated arrangement. Some tests have been made and typical values shown in Table 2 of Chapter 4 of the Westinghouse *Transmission and Distribution Reference Book* (see the Bibliography).

Problems

This series of problems provides the opportunity to apply the concepts and information developed in the various chapters to enhance the learning process. They are practical problems, many coming from actual cases. A good "RH" factor is intended: Reasonable minimum labor and High educational value.

CHAPTER 2

2.1 Two transformers are connected in series. One transformer is rated 15,000 kVA, 66 kV:132 kV, 10% X, and the other 10 MVA, 13.8 kV:69 kV, 8% X. Determine the reactance of each transformer and the total reactance (sum of the two in series) in percent on a 30-MVA, 138-kV base.

2.2 The impedance of a transmission line is $9.7 + j25.86$ ohms. What is the impedance of the line in percent on a 100-MVA, 115-kV base? Express your answer in both rectangular and polar forms.

2.3 In problem 2.2, what is the per unit impedance of the line on a 50-MVA, 115-kV base?

2.4 Three 5-MVA single-phase transformers, each rated 8:1.39 kV, have a leakage impedance of 6%. These can be connected in a number of different ways to supply three identical 5-ohm resistive loads. Various transformer and load connections are outlined in the Table, P2.4. Complete the table columns. Use a three-phase base of 15 MVA.

2.5 Three generators connect to a common 13.8-kV bus at a power station. Their constants are

Generator 1	20 MVA	13.2 kV	$X''_d = 24\%$
Generator 2	50 MVA	13.8 kV	$X''_d = 20\%$
Generator 3	80 MVA	13.8 kV	$X''_d = 12\%$

Determine the equivalent reactance (parallel of the three) of the generators to the bus in per unit at 100 MVA, 13.8 kV.

2.6 A three-phase generator feeds three large synchronous motors over a 16-km, 115-kV transmission line, through a transformer bank, as shown in Fig. P2.6. Draw an equivalent single-line reactance diagram with all reactances indicated in per unit on a 100-MVA, 13.8-kV or 115-kV base.

Table P2.4

Case	Transformer connection		Load connection	Line-to-line base kV		Load R in	Total Z as viewed from the high side	
no.	Primary	Secondary	to secondary	HV	LV	per unit	Per Unit	Ohms
1	Wye	Wye	Wye					
2	Wye	Wye	Delta					
3	Wye	Delta	Wye					
4	Wye	Delta	Delta					
5	Delta	Wye	Wye					
6	Delta	Wye	Delta					
7	Delta	Delta	Wye					
8	Delta	Delta	Delta					

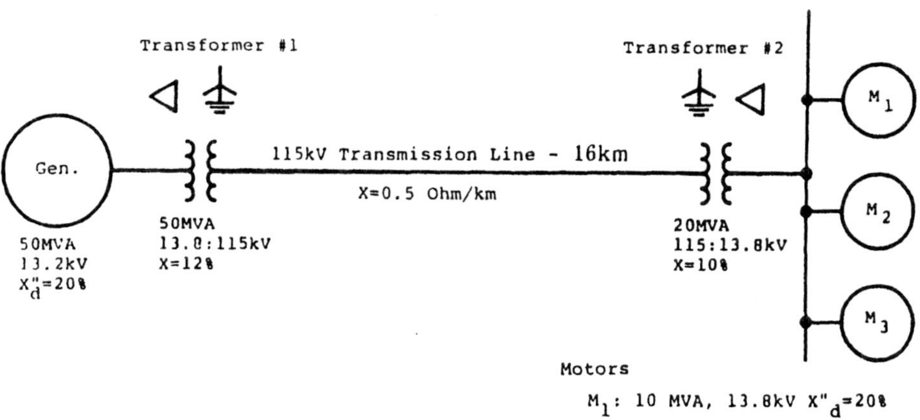

Figure P2.6

2.7 In the system shown in Figure P2.6, determine the equivalent source reactance to the right-hand motor bus (sum of generator, transformer 1 and 2, and line).

2.8 In the system of Problem 2.6 (Fig. P2.6) it is desired to maintain the voltage at the motor bus of $1.\underline{/0^\circ}$ per unit. The three motors are operating at full rating and 90% PF.

a. Determine the voltage required at the generator terminals assuming that voltage is not regulating taps or similar equipment in this system.

b. What is the voltage required behind the subtransient reactance?

CHAPTER 3

3.1 Four boxes represent an ac generator, reactor, resistor, and capacitor and are connected to a source bus XY as shown. From the circuit and phasor diagrams in Fig. P3.1, identify each box.

3.2 Two transformer banks are connected to a common bus as shown in Fig. P3.2. What are the phase relations between the voltages V_{AN} and $V_{A'N'}$; V_{BN} and $V_{B'N'}$, V_{CN} and $V_{C'N'}$?

3.3 Reconnect transformer bank number 2 of Problem 3.2 with the left windings in wye instead of delta, and the right windings in delta instead of wye so that V_{AN} and $V_{A'N'}$ are in phase, V_{BN} and $V_{B'N'}$ are in phase, and V_{CN} and $V_{C'N'}$ are in phase.

3.4 Draw the three-phase connections for three transformers connected in wye with phase rotation a, b, c and the delta with phase rotation A, C, B. The wye side leads the delta side by 30°.

3.5 Draw the three-phase connections for three transformers connected in wye with phase rotation a, c, b and the delta with phase rotation A, B, C. The delta side leads the wye side by 30°.

3.6 Plot the primary and secondary currents and voltages phasors required to supply a 13.8-kV, 30-MW load at unity power factor through a 50-MVA, 115-kV delta: 13.8-kV wye transformer bank with $X = 10\%$.

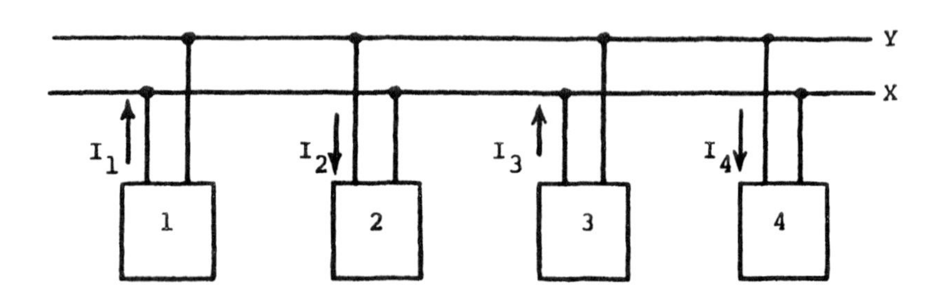

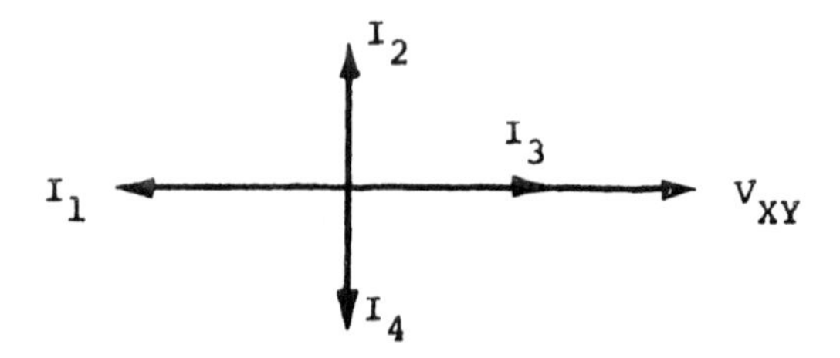

Figure P3.1

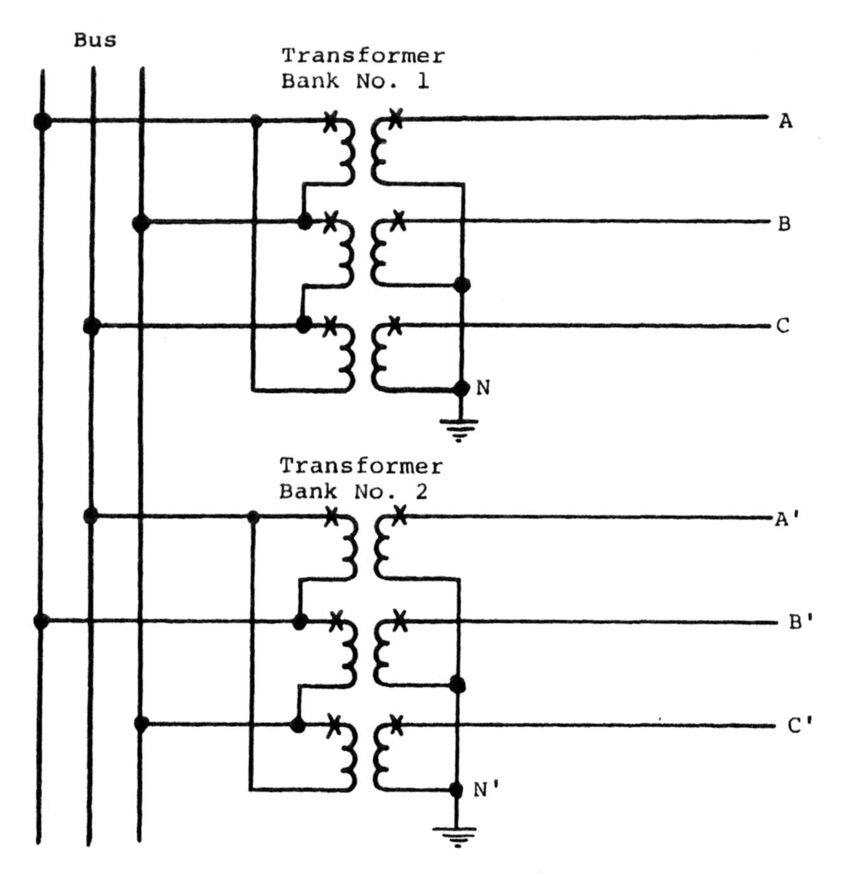

Figure P3.2

CHAPTER 4

4.1 Plot the following phasors where a = $1\angle 120°$:

No.	Phasor	No.	Phasor	No.	Phasor	No.	Phasor
1	$1\angle 1°$	6	$-a$	11	$1/a$	16	$j(a - a^2)(a^2 - a)$
2	a	7	$-a^2$	12	$1/a^2$	17	$1 + a$
3	a^2	8	$a + a^2$	13	$j\sqrt{3}$	18	$1 - a$
4	a^3	9	$a - a^2$	14	$-j\sqrt{3}$	19	$(1 - a)/(1 - a^2)$
5	a^4	10	$a^2 - a$	15	ja	20	$(a^2 - 1)\sqrt{3}$

4.2 Calculate the three-phase quantities when

a. $I_1 = I_2 = I_0 = 5.5$ pu.
b. $I_1 = -I_2 = 8.0$ pu.
c. $I_1 = 9.0$, $I_2 = -4.0$, $I_0 = -5.0$ pu.
d. $I_1 = 12.0$, $I_2 = I_0 = 0$ pu.

4.3 Determine the positive, negative, and zero sequence components of the following three-phase quantities:

a. $V_a = 6.0\angle 90°$, $V_b = V_c = 0$ pu.
b. $I_a = 9.0\angle 0°$, $I_b = 4.0\angle -60°$, $I_c = 5.4\angle 150°$ pu.
c. $V_a = V_c = 0$, $V_b = 8.0\angle -30°$ pu.
d. $V_a = 6.0\angle 90°$, $V_b = 9.0\angle -30°$, $V_c = 3.0\angle 210°$ pu.

4.4 In a three-phase system with a, b, c phase sequence the voltage and current transformers measure balanced voltages ($V_{an} = 69\angle 0°$ V) and balanced currents ($I_a = 4.5\angle 120°$ A). To test the relay connections the phase-*a* current transformer is shorted to the CT neutral and the lead to the relay switchboard is opened ($I_{an} = 0$ to relays). Then the phase-*b* voltage lead is opened and shorted to neutral (ground) on the relay side ($V_{bn} = 0$). What are the I_1, I_2, $3I_0$, V_1, V_2, and $3V_0$ quantities that are available at the relays?

4.5 For the system shown in Fig. P4.5 set up the positive, negative, and zero sequence networks in per unit on a 30-MVA base. Assuming no load, reduce the three networks to equivalent single circuits, as shown in Fig. 4.13 for a fault at bus *H*. (This problem is continued in Problems 6.2, 6.3, and 7.4.)

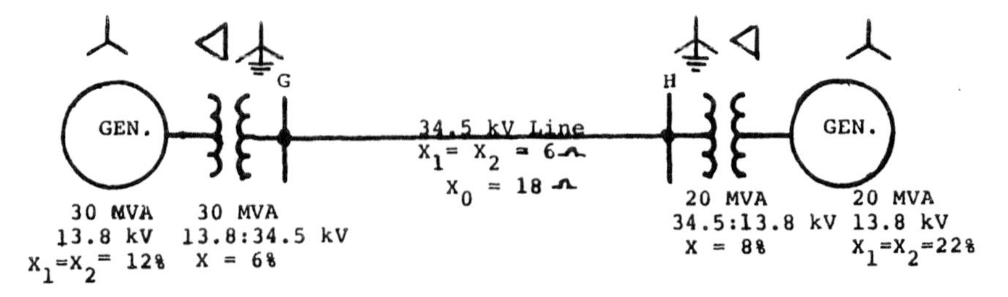

Figure P4.5

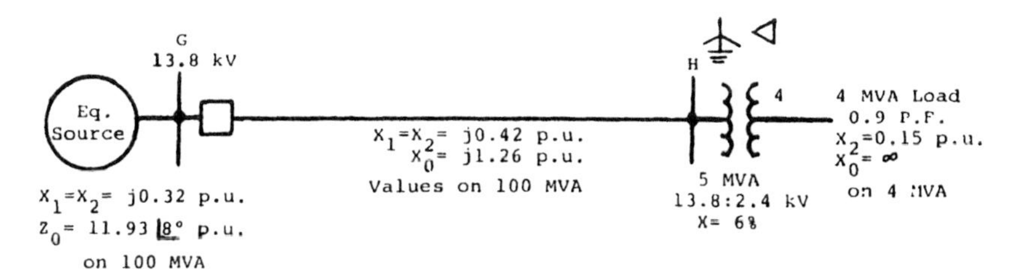

Figure P4.6

4.6 In the distribution system of Fig. P4.6, determine the Thevenin equivalent voltage and reactance at the 13.8-kV bus H. The load is served at 1.0 pu voltage. (This problem is continued in Problems 6.5, 6.6, 7.1, 7.2, and 7.3.)

Additional problems involving setting up and reducing the sequence networks are in the problems for chapter 6.

CHAPTER 5

Problems relating to this chapter are given in the problems for Chapter 6.

CHAPTER 6

6.1 For the system shown in Fig. P6.1, convert the constants to per unit values at 100-MVA base. Then,

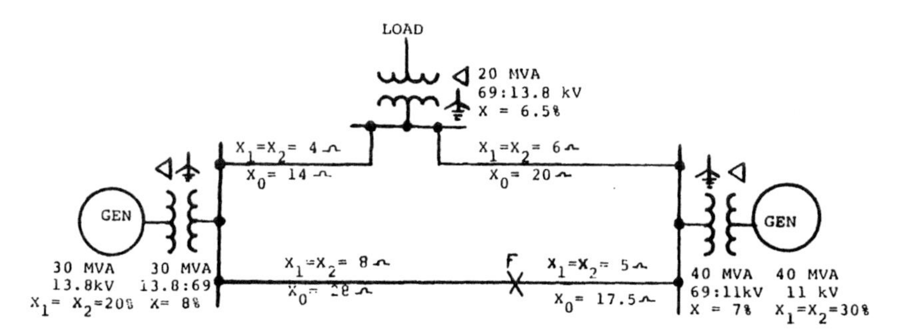

Figure P6.1

a. Draw the positive-, negative-, and zero-sequence networks with appropriate constants.
b. Assuming no load, reduce the three networks to single-circuit equivalents as in Fig. 4.13 for a fault at point *F*. Also, determine the current distribution factors.
c. For a phase-*a*-to-ground fault at *F*, calculate the total fault current in the fault.
d. For the fault of part c, calculate the currents in all three phases and the $3I_0$ flowing in the faulted line from bus *G* to the fault.
e. Calculate voltages V_2 and $3V_0$ at bus *G*, bus *H*, and the load bus for the fault at *F*.

6.2 For Problem 4.5, Fig. P4.5, calculate the fault currents at bus *H* for the following faults:
a. I_{aF} for a phase fault
b. I_{aF}, I_{bF}, I_{cF} for a *bc* phase fault
c. I_{aF}, I_{bF}, I_{cF} for a *bc* phase-to-ground fault
d. I_{aF}, I_{bF}, I_{cF} for an *a*-to-ground fault.

6.3 For Problem 6.2, Fig. P4.5,
a. Calculate the currents flowing through bus *G* for the four types of faults at bus *H*.
b. Calculate V_{an}, V_{bn}, and V_{cn} at bus *G* for the four types of faults at bus *H*.
c. Draw a phasor diagram showing the voltages and currents for each type of fault at bus *H*.

6.4 On multigrounded systems (a section is shown in Fig. P6.4), the variation in negative-sequence energy for operating ground di-

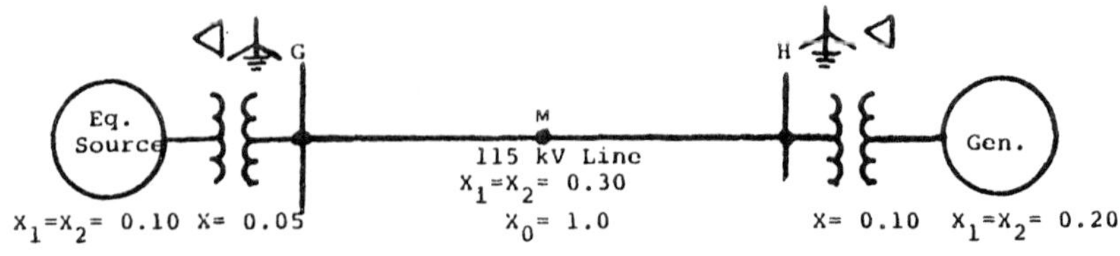

Figure P6.4

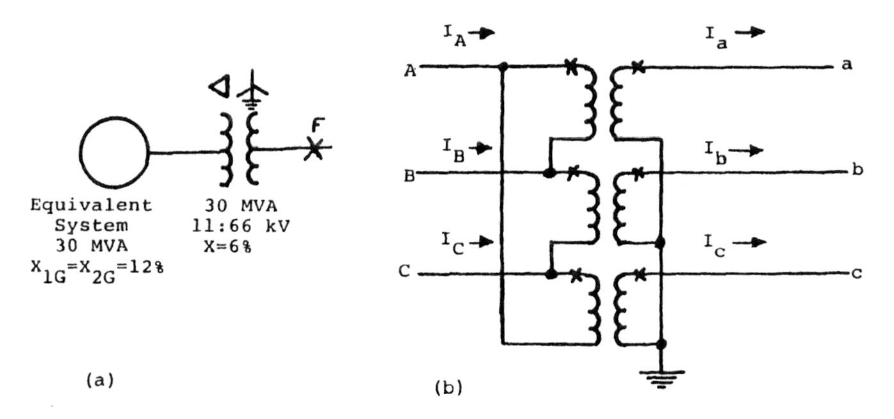

Figure P6.7

rectional relays generally is less than the variations for zero sequence. To demonstrate this,

a. Calculate line-to-ground faults at buses G and H, and at M, the midpoint of the line.
b. Determine the values of I_2, V_2, $3I_0$, and $3V_0$ at bus G for the three faults of part a.
c. Compare the products of V_2I_2 and $9V_0I_0$ for the three fault locations.

6.5 For Problem 4.6, Fig. P4.6,

a. Calculate a phase-a-to-ground fault at bus H with load using the Thevenin equivalents.
b. What are the phase currents through bus G to the fault.
c. Repeat parts a and b with load neglected.
d. Compare the two sets of values.

6.6 Solve parts a and b of Problem 6.5, Fig. P4.6, by the alternate method of using the voltage at the source necessary to supply the 4-MVA load at 1.0 per unit voltage.

6.7 For the delta-grounded-wye transformer bank of Fig. P6.7;

a. Calculate the phase fault currents on the 66-kV side of the transformer for a phase-a-to-ground fault at F. Assume the source voltage is $V_{AN} = j13{,}800$ V.

b. Calculate the 66-kV phase-to-neutral voltages for this fault. Draw a phasor diagram for these voltages and I_{aF}.

c. The transformer is connected per ANSI Standards, the high-side voltage leading the low-side voltage by 30°. Calculate I_A, I_B, and I_C at the 11-kV terminals for the fault at F.

d. Determine V_A, V_B, and V_C at the 11-kV side remembering that $V_{A1} = 6350\underline{/60^\circ}$ V $- I_{A1}(jX_{1G})$. Draw a phasor diagram of these 11-kV currents and voltages.

e. Compare the phasor quantities on the two sides of the transformer.

6.8 Repeat Problem 6.7 but for a phase-to-phase (bc) fault at the 66-kV terminals.

6.9 Before generator 4 was added to the system of Fig. P6.9, the

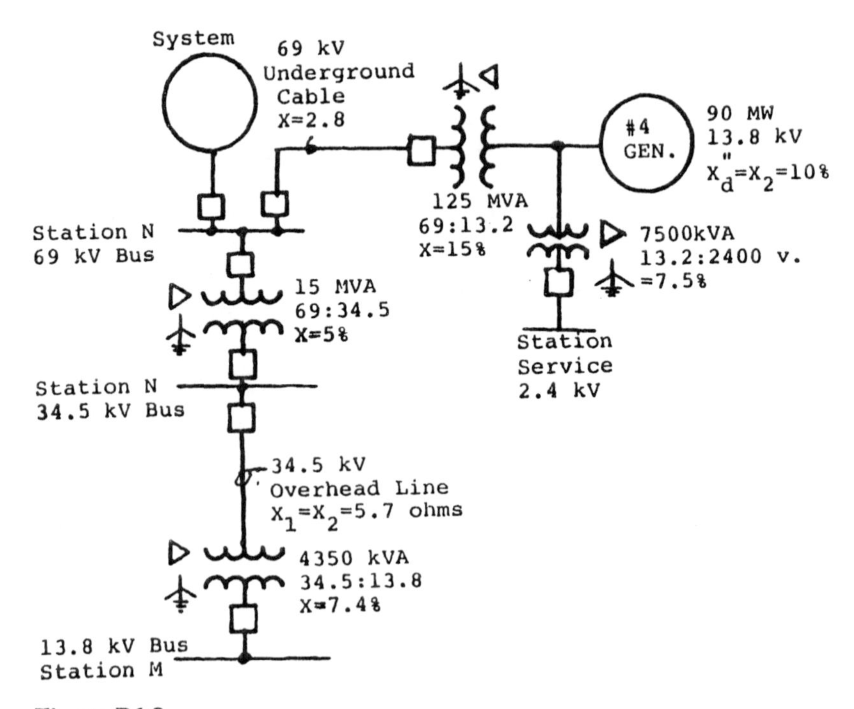

Figure P6.9

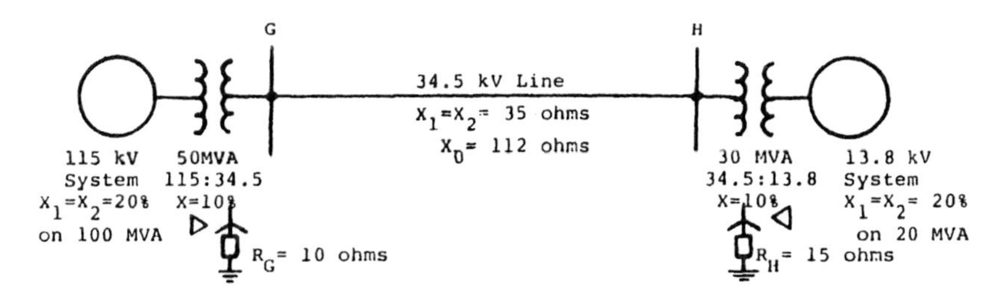

Figure P6.10

MVA_{SC} (three-phase short-circuit MVA) at station *N* 69-kV bus was 1000 MVA. Calculate

a. The new MVA_{SC} at station *N* 69-kV bus with generator 4.
b. The MVA_{SC} at station *M* 13.8-kV bus.
c. The single-phase-to-ground fault short-circuit MVA ($MVA_{\phi GSC}$) at station *M* 13.8-kV bus.

6.10 In the system of Fig. P6.10 the no-load voltages of the two sources are 120 kV and 14.4 kV. For this system calculate

a. The current for a three-phase fault at bus *H*.
b. The current for a phase-to-ground fault at bus *H*.
c. The $3I_0$ currents in the two transformer neutrals.

6.11 Three 21,875-kVA generators with $X''_d = 13.9\%$ are connected to individual buses from which various loads are supplied. These buses are connected to another bus through 0.25-ohm reactors as shown in Fig. P6.11. The generators are all ungrounded. In this system calculate

a. A three-phase fault at the terminals of one of the generators.
b. A phase-*a*-to-ground fault at the terminals of one of the generators.
c. A phase-to-ground fault at the right-hand bus.

6.12 The unit generator of Fig. P6.12 has the following capacitance-to-ground values in microfarads per phase:

Generator windings	0.240
Generator surge capacitors	0.250
Generator-to-transformer leads	0.004
Power transformer low-voltage windings	0.030

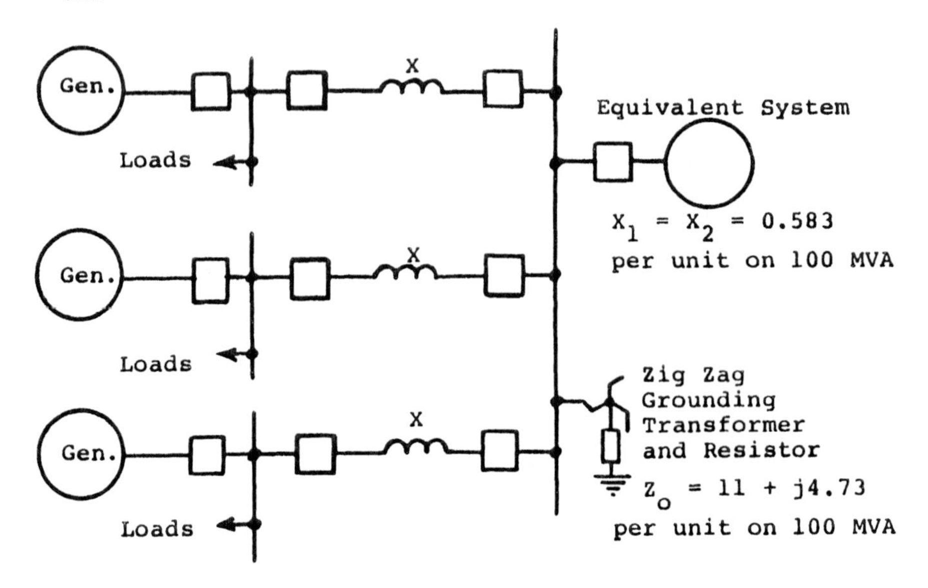

Figure P6.11

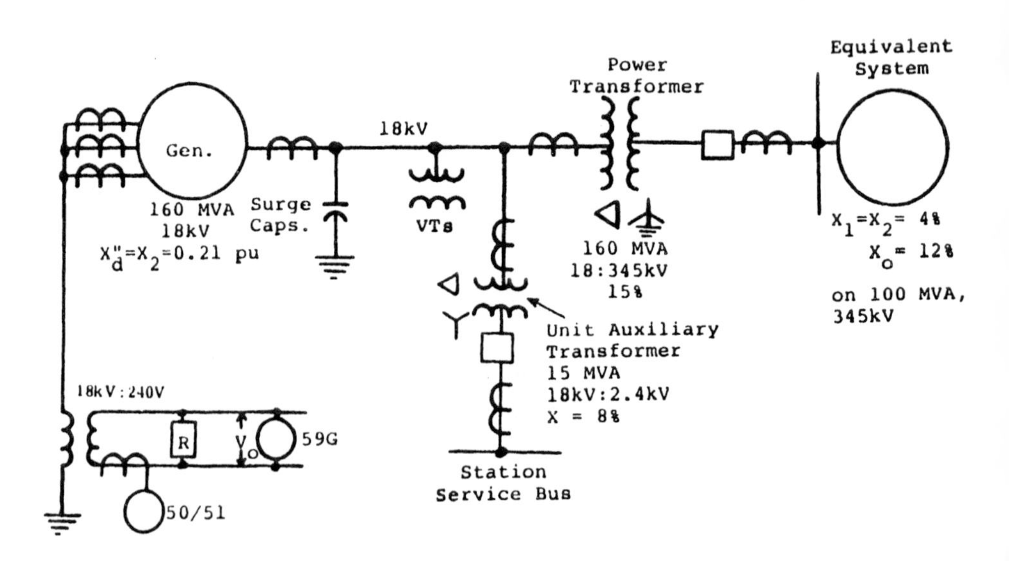

Figure P6.12

Station service transformer high-voltage windings	0.004
Voltage transformer windings	0.0005

Size the ground resistor R to be equal in magnitude to the total shunt capacitance to limit the potential transient overvoltages to not more than 2.5 times normal crest value to ground. In this system calculate the

a. Maximum three-phase fault at the generator terminals.
b. Current for a solid phase-a-to-ground fault at the terminals of the generator.
c. The current in the generator neutral for fault c.
d. The current in the resistor and the $3V_0$ for fault c.

6.13 Two 3600-rpm cross-compound generators, each rated 185 MVA, 18 kV, are paralleled and connected to a 375-MVA unit transformer, as shown in Fig. P6.13. Generator 1 is ungrounded and generator 2 grounded through a 14.4 kV:240 V distribution transformer with a 1.02-ohm resistor across the secondary. The

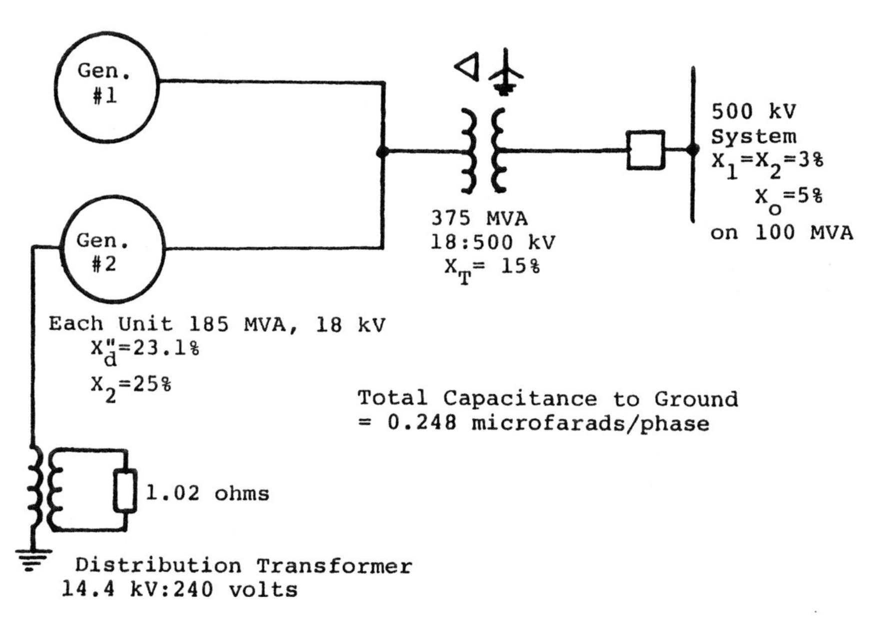

Figure P6.13

total capacitance-to-ground on the 18-kV side is 0.248 microfarad per phase. In the system, calculate

a. The current for a generator terminal three-phase fault.
b. The current for a solid ground fault on the 18-kV system.
c. The current in the grounding resistor and $3V_0$ across the resistor.

6.14 a. In order to limit ground faults, a reactor is to be connected in the grounded neutral of the 13.8-kV winding of the transformer (Fig. 6.14). Calculate the value of the reactor in ohms required to limit the solid single-phase-to-ground current on the 13.8-kV side to 4000 A.

b. What percentage reduction would this represent if the wye winding were solidly grounded instead of through the reactor?

c. Repeat part a above, but use a resistor instead of a reactor. Determine the resistor value in ohms.

6.15 Phase a of a three-phase 4.16-kV ungrounded system is solidly grounded. For this fault, calculate the magnitude of the positive-, negative-, and zero-sequence voltages at the fault. Explain your answers with reference to the sequence networks and interconnections used to calculate line-to-ground faults on three-phase systems.

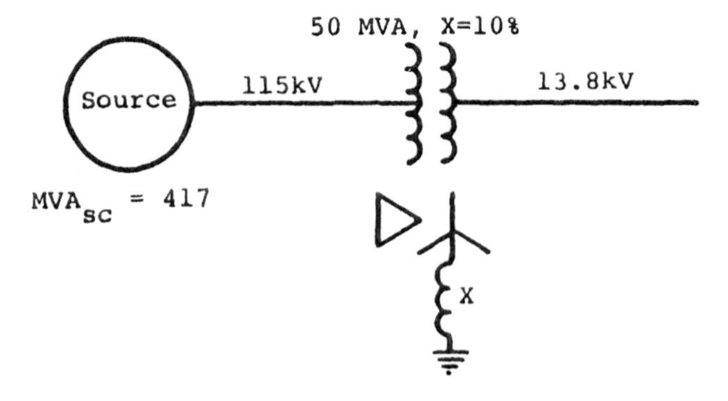

Figure P6.14

6.16 An ungrounded 4.16-kV system has a capacitance to neutral of 0.4 microfarads per phase. In this system, calculate the following:

a. The normal charging current in amperes per phase.
b. The fault current for a phase-a-to-ground fault.
c. It has been decided to ground this system with a zig-zag transformer. X_1 from the source to the 4.16-kV bus is 10% on 5000 kVA. If $X_1 = 2.4\%$ on the zig-zag bank rating, what is the kVA of the zig-zag transformer?
d. To limit the overvoltage on the unfaulted phases to a maximum of 250% for possible restriking of ground faults, the zig-zag transformer reactance is 6.67% and the ground resistor $0.292 + j0.124$ per unit both on the zig-zag transformer rating determined in part c.
e. The solid phase-a-to-ground fault current in the 4.16-kV system with the zig-zag transformer and resistor.
f. Specifications for purchasing the resistor and zig-zag transformer.

Problems involving three-winding and autotransformers are included in the problems for Chapter 9.

CHAPTER 7

7.1 Continuing Problem 4.6, Fig. P4.6 (and related Problems 6.5 or 6.6), calculate the currents flowing in the line GH when an open phase a circuit occurs in the line just at station H.

7.2 Calculate the currents flowing if phase a of line GH in Problem 7.1, Fig. P4.6, opens and falls to ground on the bus H side of the open.

7.3 Calculate the currents flowing if phase a of line GH in Problem 7.1, Fig. P4.6, opens and falls to ground on the line side at bus H.

7.4 In Fig. P4.5 assume that the generator at G is operating at $13.8\ \text{kV}\angle 0°$ and the generator at H is operating at $13.8\ \text{kV}\angle -30°$.

a. Calculate the current flowing from G to H under normal balanced conditions.

b. Calculate the three-phase and neutral currents flowing in line *GH* and the transformers when phase *a* of the line is open at bus *H*.
c. Repeat part b with phases *b* and *c* of the line open at bus *H*.
d. Compare these values with the line-to-ground fault values calculated in Problems 6.2 and 6.3.

CHAPTER 8

8.1 The per unit currents for a phase-*a*-to-ground fault are shown in Fig. P8.1. Assume that the system is reactive with all resistances neglected, and that the generator(s) are operating at $j1.0$ per unit voltage.

Draw the positive-, negative-, and zero-sequence diagrams and describe the system that must exist to produce the current flow as shown in Fig. P8.1.

CHAPTER 9

9.1 For the system shown in Fig. P9.1:

a. Determine the source and equivalent star reactances of the transformer on a 30-MVA base.
b. Set up the positive-, negative-, and zero-sequence networks. There are no fault sources in the 13.8- and 6.9-kV systems.

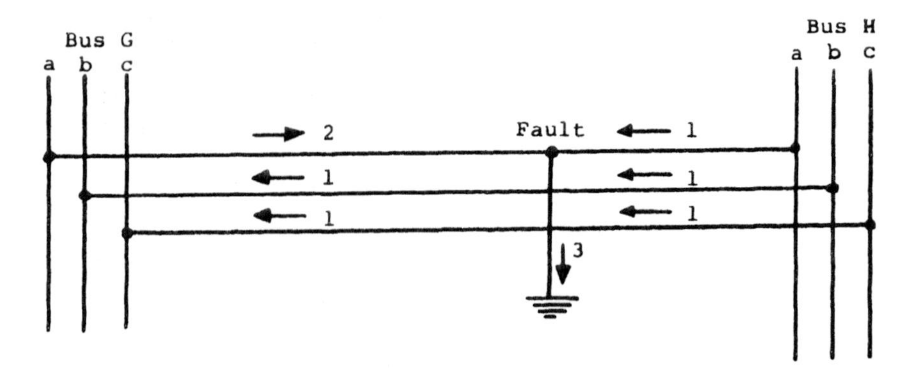

Figure P8.1

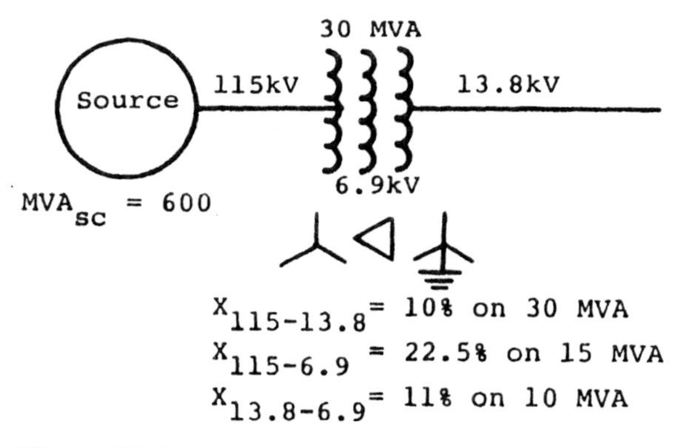

Figure P9.1

Reduce these networks to single-sequence reactances for faults on the 13.8-kV side.

c. Calculate a three-phase fault at the 13.8-kV terminals of the transformer.
d. Calculate a single-phase-to-ground fault at the 13.8-kV transformer terminals.
e. For the fault of part d, determine the phase-to-neutral voltages at the fault.
f. For the fault of part d, determine the phase currents and the phase-to-neutral voltages on the 115-kV side.
g. For the fault of part d, determine the current flowing in the delta winding of the transformer in per unit and amperes.
h. Make an ampere-turn check for the fault currents flowing in the 115 kV, 13.8 kV, and 6.9 kV windings of the transformer.

9.2 The three-winding transformer shown in Fig. P9.2 is connected per ANSI Standards. Calculate the currents flowing in amperes in the 230-kV, 69-kV and 13.8-kV circuits for a phase-*a*-to-ground fault at *F*. Show all the values in a three-phase diagram.

9.3 The system shown in Fig. P9.3 has generation connected only to the 115-kV bus at station G. Calculate

a. The total fault current in amperes for a phase *a*-to-ground fault on the 115-kV bus at station *H*.
b. For the fault of part a, the fault currents in amperes in all three phases at station *G*.

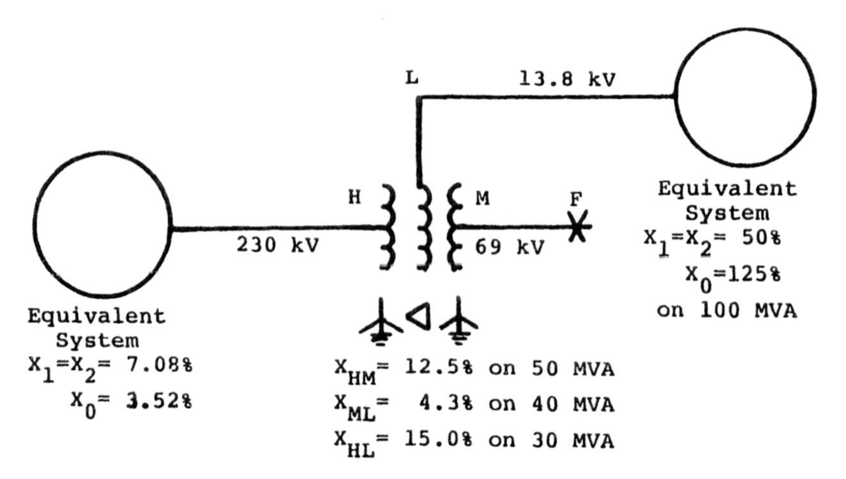

Figure P9.2

c. For the fault of part a, the fault currents in amperes in line *PM*.

d. For the fault of part a, the fault currents in amperes flowing in all three transformer bank neutrals and in the tertiary windings. Show both the magnitude and direction of these neutral currents.

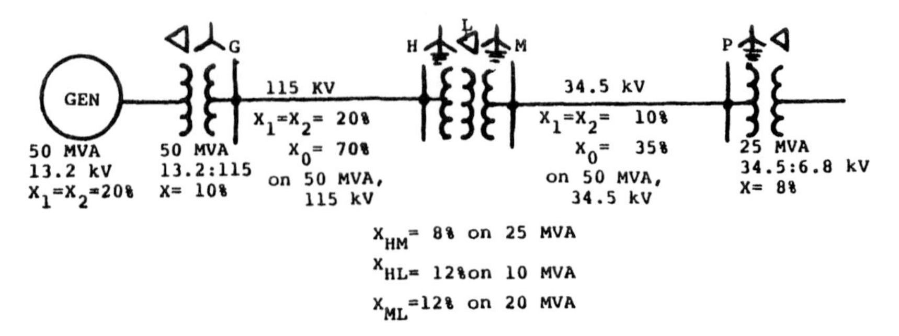

Figure P9.3

9.4 In the system of Fig. P9.4 the bank ratings and reactances are

Bank A: one 6000-kVA unit, 2.4 kV:13.2 kV, 5.0%
Bank B: two 6000-kVA units, 2.4 kV:34.5 kV, 5.6%
Bank C: one 12-MVA unit, 2.4 kV, 5.0%
Bank D: one autotransformer, 34.5 kV:69 kV, 4.6% on 12 MVA
69 kV:13.2 kV, 5.82% on 12 MVA
34.5 kV:13.3 kV, 7.87% on 12 MVA

For a phase-*a*-to-ground fault at *F*, calculate

a. The total fault current at *F*.
b. The magnitudes and direction of the line and neutral currents flowing in the 34.5-kV and 69-kV systems.
c. The currents flowing inside the 13.2-kV delta of the autotransformer and in the 13.2-kV circuits between banks *A* and *D*.

9.5 The 69-kV and 138-kV systems are connected through an autotransformer as shown in Fig. P9.5. Calculate

a. The total fault current for a phase-*a*-to-ground fault at *F*.
b. The 69-kV and 138-kV currents flowing to the fault at *F*.

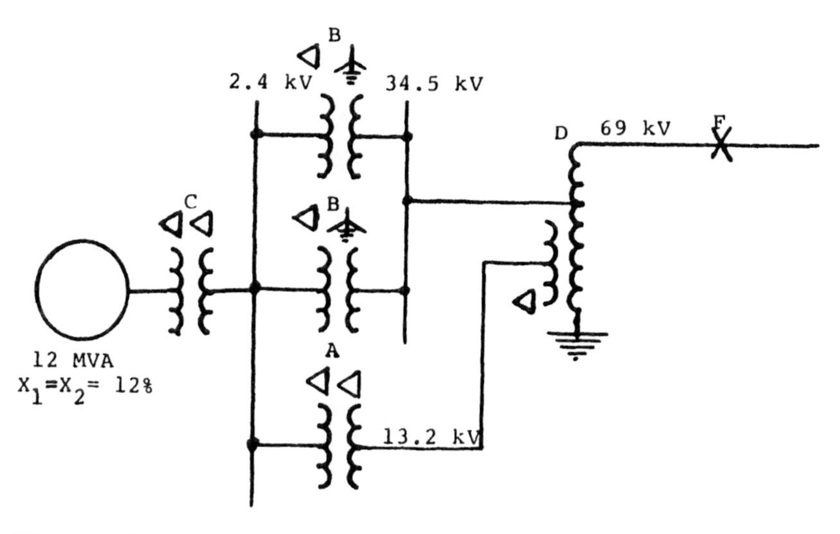

Figure P9.4

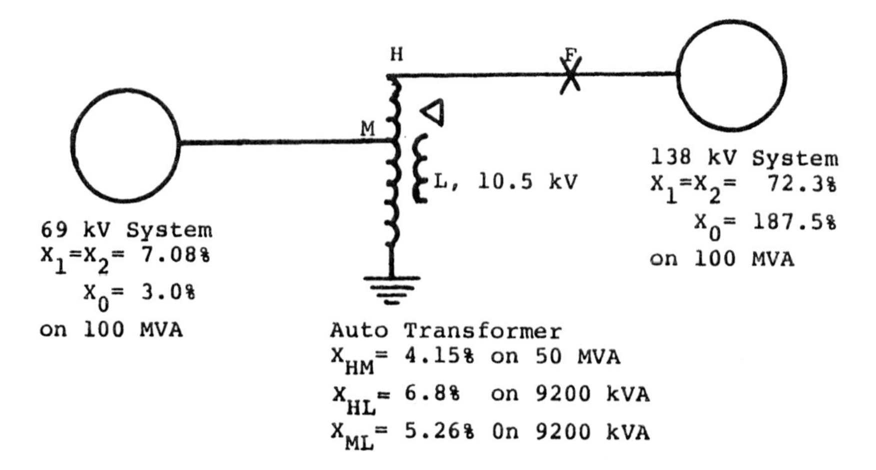

Figure P9.5

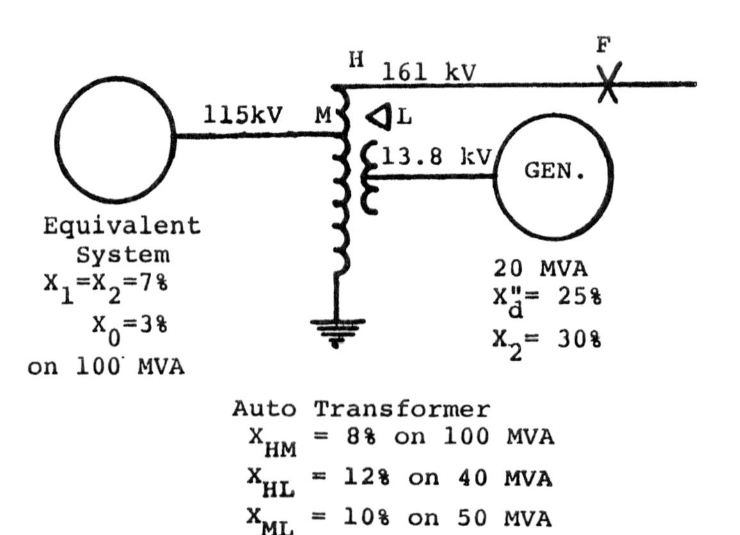

Figure P9.7

c. The magnitude and direction of the neutral ($3I_0$) currents.
d. The current flowing in the 10.5-kV tertiary.
e. The V_2 and $3V_0$ voltages at the 69-kV and 138-kV terminals during the fault.

9.6 Repeat Problem 9.5 for a phase-*a*-to-ground fault at the 69-kV terminals.

9.7 In the system of Fig. P9.7, a generator is connected to the delta winding of the autotransformer. For a phase-*a*-to-ground fault (*F*), calculate
a. The total fault current at *F*.
b. The direction and magnitude of the current in the autotransformer neutral.
c. The $3I_0$ current flowing in the 115-kV system.
d. The currents flowing inside and outside the 13.8-kV delta.
e. The V_2 and $3V_0$ voltages at the 69-kV and 115-kV terminals.

CHAPTER 10

10.1 The per unit generator constants in Fig. P10.1 are

$$X''_d = X_2 = j0.145$$
$$X'_d = j0.240$$
$$X_d = j1.10 \text{ (unsaturated)}$$

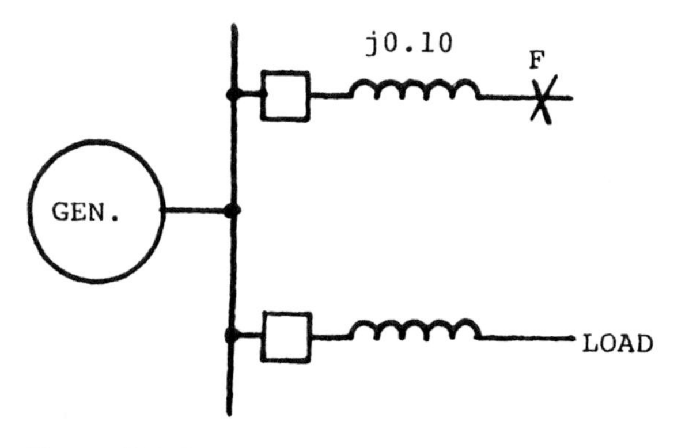

Figure P10.1

Assume that the generator is unloaded and that the terminal voltage is 1.0 per unit when a three-phase fault occurs at F. Determine the subtransient, transient, and steady-state fault currents in per unit.

10.2 Repeat Problem 10.1 but with the generator operating at full load at 80% power factor to the load shown in Fig. P10.1. The steady-state synchronous voltage for this load is 2.25 per unit.

10.3 Calculate the maximum dc component of the generator current and the total initial generator current for Problem 10.1.

10.4 Calculate the maximum dc component of the generator current and the total initial generator current for Problem 10.2.

10.5 The constants of a 160-MVA, 18-kV generator are

$$X''_d = 21\%, \qquad T''_d = 0.035 \text{ sec}$$

$$X'_d = 30\%, \qquad T'_d = 0.730 \text{ sec}$$

A three-phase fault occurs at the terminals with the unit operating at no load. Determine the symmetrical fault current

a. At the instant of the fault.
b. After 0.08 sec.
c. After 0.20 sec.

10.6 The instantaneous symmetrical three-phase fault current at the terminals of an unloaded 50-MVA, 13.2-kV generator is 20,000 A.

a. What is the per unit X''_d of the generator?
b. The sustained short circuit currents for terminal faults are 2000 A for 3ϕ, 3149 A for $\phi\phi$, and 5240 amperes for ϕ-ground. What are the X_d, X_2 and X_0 per unit values for the generator?

10.7 A three-phase 25-MVA, 13.8-kV, 60-Hz generator is operating at full load 80% power factor. The generator constants are

$$X''_d = X_2 = 15\%$$

$$X'_d = 25\%$$

$$X_d = 130\%$$

$$X_0 = 5\%$$

Determine the neutral reactance necessary to limit the single-phase-to-ground current to the three-phase fault value.

Problems 6.11, 6.12, and 6.13 also involve faults on generators.

CHAPTER 11

11.1 Typical conductors, characteristics, and spacings for 60-Hz transmission lines are:

Voltage (kV)	Code	Conductor size	Equivalent spacing (ft)
12.5	Iris	2 AA	3.5
23	Iris	2 AA	4.5
34.5	Poppy	1/0 AA	7.0
69	Aster	2/0 AA	10.5
115	Oxlip	4/0 AA	14.0
138	Goldentuft	450 kcmil, AA/19	17.0
161	Hawk	477 kcmil, ACSR 26/7	19.5
230	Mallard	795 kcmil, ACSR 30/19	27.0
345	Pheasant	1272 kcmil, ACSR 54/19	30.0
500	Plover	1431 kcmil, ACSR 54/19	35.0

Calculate the impedance per phase per mile at 60 Hz, 50°C, for each of the lines. Compare the values of the inductive reactance and the line angles.

11.2 The 60-Hz inductive reactance at 1-foot spacing of a solid conductor is 0.602 ohm per mile. Calculate the reactance for a spacing of six feet. What is the cross-sectional area of this conductor in circular mils (circular mils = diameter^2)?

11.3 An equilaterally spaced three-phase line has solid wire conductors of 0.204-in. diameter. The spacing is 10 ft. Find the reactance per mile.

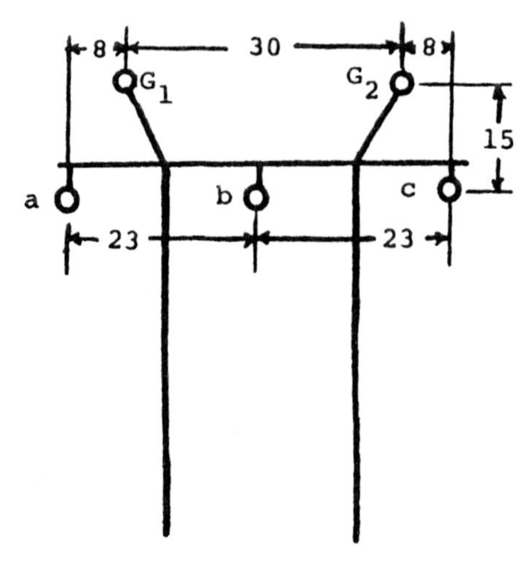

Figure P11.5

11.4 Suppose the line of Problem 11.3 is rebuilt with the horizontal spacing the distance between the center and the outer conductors of D. Determine the value of D for the transposed line to have the same reactance as Problem 11.3.

11.5 Calculate the 60-Hz positive- and negative-sequence impedance per mile of the transmission line with spacings as shown in Fig. P11.5. Assume that the line is transposed. The phase conductors are code Hen, 477 kcmil, 30/7.

11.6 Calculate the 60-Hz zero-sequence impedance per mile of the transmission line of Problem 11.5, shown in Fig. P11.5. Assume the earth resistivity is 100 meter-ohms:

a. With only one ground wire on the towers of code Oriole, 336.4 kcmil 30/7 ACSR.
b. With two ground wires as in part a.
c. Compare the zero sequence values of parts a and b with the positive sequence of Problem 11.5.

CHAPTER 12

12.1 A 50-mile, three-phase, 60-Hz, double-circuit 115-kV line has the configuration shown in Fig. P12.1. The conductors are code Ostrich, 300 kcmil 26/7 ACSR. Assume that the two lines are transposed.

a. Calculate the positive-sequence impedance of each line independently.
b. Calculate the positive-sequence impedance of the two lines in parallel.
c. What is the error when the impedances of the two lines are calculated independently and one-half of the total is used as the parallel line impedance?
d. Calculate the zero sequence of the parallel lines. Assume 100 meter-ohms resistivity.

12.2 As shown in Fig. P12.2, 345-kV and 115-kV lines, 50 miles long, are in the same right-of-way. The 345-kV line has code Bluebird, 2156 kcmil (84/19) ACSR conductors. The 115-kV line has code

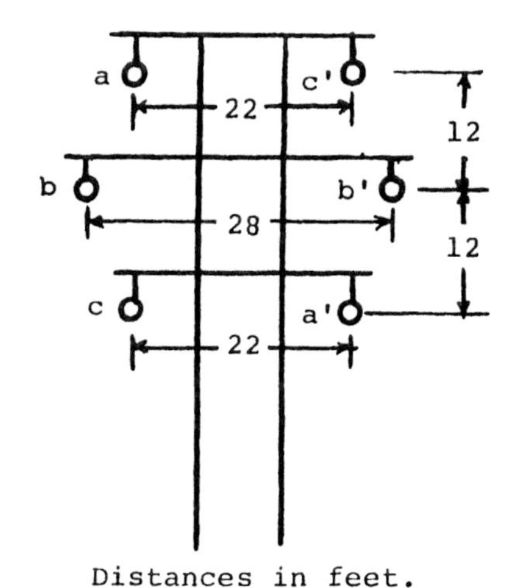

Figure P12.1

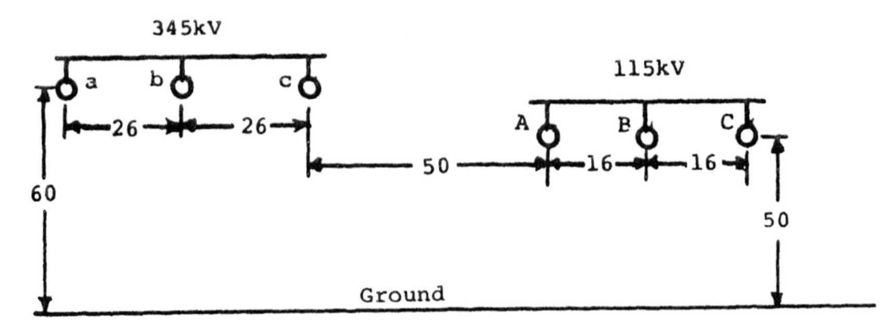

Figure P12.2

Narcissus, 1272 kcmil AA conductors. Assume both lines are transposed.

a. Calculate the total zero-sequence self-impedance for the 115-kV and 345-kV lines in ohms and percent.
b. Calculate the total zero-sequence mutual impedance between the 115-kV and 345-kV lines in ohms and percent.
c. Set up the zero-sequence network for these lines as in Fig. 12.2 for both ohms and percent.

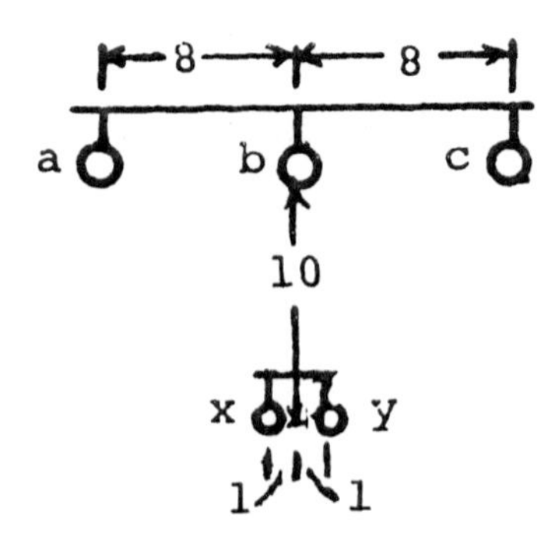

Figure P12.3

d. A ground fault occurs in the 115-kV system resulting in a $3I_0$ current in the 115-kV line of 3012 A. Determine the $3V_0$ voltage induced in the 345-kV system.

e. Each end of the 345-kV line has a 100 MVA, 12% transformer bank solidly grounded on the 345-kV side. Determine the total ohms in the zero-sequence current path.

f. From the values determined in parts d and e calculate the induced current that will flow in the 345-kV system from the 115-kV fault. What percent is this of the 100-MVA, 345-kV maximum load current?

12.3 The horizontal, flat, three-phase line in Fig. P12.3 has a pilot pair supported on a horizontal cross-arm 10 ft below the power line. The pilot pair (X, Y) are 14 AWG and the power line 4 AWG, code Rose.

a. If the load current is balanced in the three-phase line with $I_a = 340\angle{-15°}$ A, find the voltage induced per mile in the pilot pair. What is the phase relation of the induced voltage with respect to the power line current?

b. Calculate the induced voltage per mile for a phase-a-to-ground fault where $I_a = 1411\angle{-90°}$ A, $I_b = 213\angle{45°}$ A, and $I_c = 723\angle{90°}$ A.

CHAPTER 13

13.1 For the lines in Problem 11.1,

a. Calculate the capacitive reactance per phase per mile at 60 Hz for each of the transmission lines.

b. Calculate the charging current per mile for these lines.

Appendix: Overhead Line Conductor Characteristics

Aluminum conductors have replaced copper conductors because they offer lighter weight, lower cost, and reduced corona effects. The four general types in use are designated by the following:

AAC: all aluminum conductor
AAAC: all aluminum-alloy conductor
ACSR: aluminum conductor, steel-reinforced
ACAR: aluminum conductor, alloy-reinforced

The AAAC aluminum-alloy conductors have a higher tensile strength than the AAC all aluminum conductors. The ACSR conductors have a central core of steel strands surrounded by layers of aluminum strands. The ACAR conductors have a central core of higher-strength aluminum surrounded by layers of aluminum strands.

The following Tables A1 to A9 provide the code identification and size of the cables used in overhead electrical lines, along with the conductor GMR, their resistance r_a, inductive reactance X_a, and capacitive reactance X'_a.

Wire sizes are designated in circular mils for sizes larger than 4/0 AWG. 1 cir. mil is the area of a circle of 1 mil (0.001 in.) diameter. And 1 cir. mil = $\pi/4$ = 0.7854 mils^2. Conductor size conversion is

$$\begin{aligned} \text{Area in cir. mil} &= 1.2632\ 10^6 \text{ (area, in.}^2\text{)} \\ &= 10^6\ D^2 \text{ (diameter, in.)} \\ &= 1973.5 \text{ (area, mm}^2\text{)} \end{aligned}$$

With the American Wire Gage (AWG), the following approximate relations are

1. An increase of three gage numbers (such as 10 to 7) doubles the area and weight and halves the dc resistance.
2. An increase of six gage numbers (such as 10 to 4) doubles the diameter.
3. An increase of ten gages (such as 10 to 1/0) multiples the area and weight by 10 and divides the resistance by 10.

Table A.1 All-Aluminum Concentric-Lay Class AA and A Stranded Bare Conductors

1350-H19 ASTM B 231

Bold face code words indicate sizes most often used

		Resistance					Phase-to-Neutral 60 Hz Reactance at One ft Spacing	
		dc	ac – 60 hz					
Code Word	Size AWG or kcmil	20°C Ohms/ Mile	25°C Ohms/ Mile	50°C Ohms/ Mile	75°C Ohms/ Mile	GMR ft	Inductive Ohms/Mile X_a	Capacitive Megohm-Miles X'_a
Peachbell	6	3.481	3.551	3.903	4.255	0.00555	0.630	0.145
Rose	4	2.188	2.232	2.453	2.674	0.00700	0.602	0.138
Iris	2	1.374	1.402	1.541	1.680	0.00883	0.574	0.131
Pansy	1	1.091	1.114	1.224	1.334	0.00991	0.560	0.127
Poppy	1/0	0.8646	0.882	0.970	1.057	0.0111	0.546	0.124
Aster	2/0	0.6856	0.700	0.769	0.838	0.0125	0.532	0.120
Phlox	3/0	0.5441	0.556	0.611	0.665	0.0140	0.518	0.117
Oxlip	4/0	0.4311	0.441	0.484	0.528	0.0158	0.504	0.114
Daisy	266.8	0.3418	0.350	0.384	0.419	0.0177	0.489	0.110
Laurel	266.8	0.3421	0.350	0.384	0.419	0.0187	0.483	0.110
Tulip	336.4	0.2711	0.278	0.305	0.332	0.0210	0.469	0.106
Canna	397.5	0.2294	0.235	0.258	0.282	0.0228	0.459	0.104
Cosmos	477.0	0.1914	0.197	0.216	0.235	0.0250	0.448	0.101
Syringa	477.0	0.1915	0.197	0.216	0.235	0.0254	0.446	0.101
Dahlia	556.5	0.1641	0.169	0.186	0.202	0.0270	0.438	0.0989
Mistletoe	556.5	0.1641	0.169	0.186	0.202	0.0275	0.436	0.0988
Orchid	636.0	0.1435	0.149	0.163	0.177	0.0294	0.428	0.0968
Violet	715.5	0.1275	0.132	0.145	0.158	0.0312	0.421	0.0951
Nasturtium	715.5	0.1276	0.133	0.145	0.158	0.0314	0.420	0.0950
Arbutus	795.0	0.1148	0.120	0.131	0.142	0.0328	0.415	0.0935
Lilac	795.0	0.1147	0.120	0.131	0.142	0.0331	0.414	0.0935
Anemone	874.5	0.1043	0.109	0.120	0.130	0.0344	0.409	0.0921
Crocus	874.5	0.1043	0.109	0.120	0.130	0.0347	0.408	0.0920
Magnolia	954.0	0.09563	0.101	0.110	0.120	0.0360	0.403	0.0908
Goldenrod	954.0	0.09560	0.101	0.110	0.119	0.0362	0.403	0.0908
Bluebell	1033.5	0.08823	0.0933	0.102	0.111	0.0374	0.399	0.0896
Larkspur	1033.5	0.08826	0.0933	0.102	0.111	0.0377	0.398	0.0896
Marigold	1113.0	0.08197	0.0872	0.0951	0.103	0.0391	0.393	0.0885
Hawthorn	1192.5	0.07655	0.0819	0.0893	0.0968	0.0405	0.389	0.08[illegible]5
Narcissus	1272.0	0.07175	0.0772	0.0841	0.0911	0.0418	0.385	0.0865
Columbine	1351.5	0.06748	0.0731	0.0795	0.0861	0.0431	0.381	0.0856
Carnation	1431.0	0.06375	0.0695	0.0756	0.0817	0.0444	0.378	0.0847
Gladiolus	1510.5	0.06039	0.0663	0.0720	0.0778	0.0456	0.375	0.0839
Coreopsis	1590.0	0.06736	0.0634	0.0688	0.0743	0.0468	0.372	0.0832

1 Direct current (dc) resistance is based on 16.946 ohm-cmil/ft at 20°C for 1350 aluminum nominal area of conductor with standard stranding increments ASTM B 231

2 Alternating current (ac) resistance is based on dc resistance corrected for temperature, using 0.00404 as temperature coefficient of resistivity per degrees C at 20°C and for skin effect

Source: Aluminum Association, *Aluminum Electrical Conductor Handbook*, 3rd ed., 1989, Table 4.6.

Table A.2 All-Aluminum Concentric-Lay Class AA and A Bare Stranded Conductors 1350-H19 ASTM B 231

Bold face code words indicate sizes most often used

Code Word	Size kcmil	Stranding Class	Total Number of Wires	Wire Diameter in.	Conductor Diameter in.	dc Ohms per Mile 20°C	Resistance ac—60 Hz 25°C Ohms/Mile	50°C Ohms/Mile	75°C Ohms/Mile	Phase-to-Neutral 60 Hz Reactance at One Foot Spacing GMR ft	Inductive Ohms per Mile X_a	Shunt Capacitive Megohm-Miles X'_a
Sneezewort	250	AA	7	0.1890	0.567	0.3650	0.373	0.410	0.447	0.0171	0.493	0.111
Valerian	250	A	19	0.1147	0.574	0.3651	0.373	0.410	0.447	0.0181	0.487	0.111
Peony	300	A	19	0.1257	0.629	0.3040	0.311	0.342	0.372	0.0198	0.476	0.108
Daffodil	350	A	19	0.1357	0.679	0.2609	0.267	0.294	0.320	0.0214	0.466	0.106
Goldentuft	450	AA	19	0.1539	0.770	0.2028	0.208	0.229	0.249	0.0243	0.451	0.102
Zinnia	500	AA	19	0.1622	0.811	0.1826	0.188	0.206	0.225	0.0256	0.445	0.101
Hyacinth	500	A	37	0.1162	0.813	0.1827	0.188	0.206	0.225	0.0260	0.443	0.100
Meadowsweet	600	AA,A	37	0.1273	0.891	0.1522	0.157	0.172	0.188	0.0285	0.432	0.0977
Verbena	700	AA	37	0.1375	0.963	0.1305	0.135	0.148	0.161	0.0308	0.422	0.0954
Flag	700	A	61	0.1071	0.964	0.1304	0.135	0.148	0.161	0.0310	0.421	0.0954
Petunia	750	AA	37	0.1424	0.997	0.1216	0.127	0.139	0.151	0.0319	0.418	0.0944
Cattail	750	A	61	0.1109	0.998	0.1217	0.127	0.139	0.151	0.0321	0.417	0.0943
Cockscomb	900	AA	37	0.1560	1.092	0.1014	0.106	0.116	0.126	0.0349	0.407	0.0917
Snapdragon	900	A	61	0.1215	1.094	0.1014	0.106	0.116	0.126	0.0352	0.406	0.0916
Hawkweed	1,000	AA	37	0.1644	1.151	0.09126	0.0963	0.105	0.114	0.0368	0.401	0.0901
Camellia	1,000	A	61	0.1280	1.152	0.09132	0.0964	0.105	0.114	0.0371	0.400	0.0901
Jessamine	1,750	AA	61	0.1694	1.525	0.05214	0.0585	0.0634	0.0683	0.0490	0.366	0.0818
Cowslip	2,000	A	91	0.1482	1.630	0.04566	0.0525	0.0567	0.0609	0.0526	0.357	0.0798
Lupine	2,500	A	91	0.1657	1.823	0.03689	0.0446	0.0479	0.0512	0.0588	0.344	0.0765
Trillium	3,000	A	127	0.1537	1.998	0.03072	0.0392	0.0418	0.0445	0.0646	0.332	0.0737
Bluebonnet	3,500	A	127	0.1660	2.158	0.02659	0.0357	0.0379	0.0402	0.0697	0.323	0.0715

1. Data shown are subject to normal manufacturing tolerances.
2. Class AA stranding is usually specified for bare conductors used on overhead lines. Class A stranding is usually specified for conductors to be covered with weather-resistant (weatherproof) materials and for bare conductors where greater flexibility is required than afforded by Class AA. The direction of lay of the outside layer of wires with Class AA and Class A will be right-hand unless otherwise specified.
3. Direct current (dc) resistance is based on 16.946 ohm-cmil/ft at 20°C for nominal area of the conductor with standard stranding increments ASTM B 231.
4. Alternating current (ac) resistance is based on dc resistance corrected for temperature, using 0.00404 as temperature coefficient of resistivity per degrees C at 20 C and for skin effect.

Source: Aluminum Association, *Aluminum Electrical Conductor Handbook*, 3rd ed., 1989, Table 4.7.

Table A.3 All-Aluminum Shaped-Wire Concentric-Lay Compact Conductors AAC/TW

ASTM B 778 in Fixed Diameter Increments

Code Word	Size kcmil	No. of Wires	No. of Layers	Conductor Diameter in.	Weight per 1000 ft lb	Rated Strength lb	Resistance				GMR ft	Phase-to-Neutral 60 Hz, Resistance at One ft Spacing	
							dc	ac—60 Hz					
							20°C Ohms/Mile	25°C Ohms/Mile	50°C Ohms/Mile	75°C Ohms/Mile		Inductive Ohms/Mile X_a	Capacitive Megohm-Miles X_a'
Logan/TW	322.5	17	2	0.60	302.1	5,960	0.2789	0.2855	0.3140	0.3425	0.0188	0.482	0.1094
Wheeler/TW	449.4	17	2	0.70	421.0	8,030	0.2001	0.2055	0.2259	0.2463	0.0220	0.463	0.1048
Robson/TW	595.8	17	2	0.80	558.2	10,700	0.1510	0.1557	0.1710	0.1864	0.0252	0.447	0.1009
McKinley/TW	761.5	17	2	0.90	713.3	13,400	0.1181	0.1225	0.1344	0.1464	0.0284	0.432	0.0974
Rainier/TW	918.8	31	3	1.00	864.3	16,100	0.0983	0.1030	0.1129	0.1227	0.0319	0.418	0.0912
Helens/TW	1123.1	31	3	1.10	1056	19,700	0.0804	0.0853	0.0932	0.1012	0.0352	0.406	0.0915
Baker/TW	1346.8	31	3	1.20	1267	23,600	0.0670	0.0722	0.0788	0.0854	0.0385	0.395	0.0889
Hood/TW	1583.2	34	3	1.30	1489	27,200	0.0570	0.0625	0.0680	0.0736	0.0419	0.385	0.0865
Whitney/TW	1812.7	49	4	1.40	1713	31,100	0.0501	0.0563	0.0610	0.0658	0.0454	0.375	0.0843
Powell/TW	2093.6	49	4	1.50	1978	35,900	0.0433	0.0499	0.0540	0.0581	0.0489	0.366	0.0822
Jefferson/TW	2388.1	52	4	1.60	2256	40,100	0.0380	0.0450	0.0485	0.0520	0.0524	0.357	0.0803
Shasta/TW	2667.2	71	5	1.70	2528	45,200	0.0341	0.0418	0.0449	0.0480	0.0561	0.349	0.0785
Adams/TW	3006.2	71	5	1.80	2848	51,000	0.0303	0.0384	0.0411	0.0438	0.0578	0.342	0.0768

1 Data shown are subject to normal manufacturing tolerances

2 Direct current (dc) resistance is based on 16 727 ohm-cmil/ft at 20° 62% IACS conductivity.

3 Alternating current (ac) resistance is based on dc resistance corrected for temperature and skin effect.

4 Properties of the industrial wires are those of the equivalent round wires of ASTM B 230

Source: Aluminum Association, *Aluminum Electrical Conductor Handbook*, 3rd ed., 1989, Table 4.10.

Table A.4 All-Aluminum Shaped-Wire Concentric-Lay Compact Conductors AAC/TW

ASTM B 778
Area Equal to Standard AAC Sizes

							Resistance					Phase-to-Neutral 60 Hz, Resistance at One ft Spacing	
								ac—60 Hz					
Code Word	Size kcmil	No. of Wires	No. of Layers	Conductor Diameter in.	Weight per 1000 ft lb	Rated Strength lb	dc 20°C Ohms/Mile	25°C Ohms/Mile	50°C Ohms/Mile	75°C Ohms/Mile	GMR ft	Inductive Ohms/Mile X_a	Capacitive Megohm-Miles X_a
Tulip TW	336.4	17	2	0.612	315.2	6,220	0.2673	0.2737	0.3010	0.3284	0.0192	0.480	0.1088
Canna TW	397.5	17	2	0.661	372.4	7,230	0.2262	0.2319	0.2551	0.2781	0.0208	0.470	0.1066
Cosmos TW	477.0	17	2	0.720	446.9	8,530	0.1885	0.1936	0.2128	0.2321	0.0226	0.460	0.1040
Zinnia TW	500.0	17	2	0.736	468.4	8,940	0.1798	0.1848	0.2031	0.2215	0.0232	0.457	0.1033
Mistletoe TW	556.5	17	2	0.775	521.3	9,950	0.1616	0.1664	0.1829	0.1993	0.0244	0.451	0.1018
Meadowsweet TW	600.0	17	2	0.803	562.1	10,700	0.1498	0.1545	0.1697	0.1850	0.0253	0.446	0.1008
Orchid TW	636.0	17	2	0.825	595.8	11,400	0.1414	0.1460	0.1603	0.1747	0.0260	0.443	0.0999
Verbena TW	700.0	17	2	0.864	655.7	12,500	0.1285	0.1329	0.1459	0.1590	0.0272	0.437	0.0986
Nasturtium TW	750.0	17	2	0.893	702.6	13,400	0.1199	0.1243	0.1364	0.1486	0.0281	0.433	0.0976
Arbutus TW	795.0	17	2	0.919	744.7	13,900	0.1131	0.1175	0.1289	0.1404	0.0290	0.430	0.0968
Cockscomb TW	900.0	17	3	0.990	846.6	15,800	0.1004	0.1051	0.1152	0.1253	0.0316	0.419	0.0946
Magnolia TW	954.0	31	3	1.018	897.4	16,700	0.0946	0.0994	0.1089	0.1184	0.0325	0.416	0.0938
Hawkweed TW	1000.0	31	3	1.041	940.6	17,500	0.0903	0.0951	0.1041	0.1131	0.0333	0.413	0.0931
Bluebell TW	1033.5	31	3	1.057	972.2	18,100	0.0874	0.0922	0.1009	0.1096	0.0338	0.411	0.0927
Marigold TW	1113.0	31	3	1.095	1047	19,500	0.0811	0.0860	0.0941	0.1022	0.0350	0.407	0.0916
Hawthorn TW	1192.5	31	3	1.132	1122	20,900	0.0757	0.0807	0.0882	0.0957	0.0362	0.403	0.0906
Narcissus TW	1272.0	31	3	1.168	1196	22,300	0.0710	0.0760	0.0830	0.0901	0.0374	0.399	0.0896
Columbine TW	1351.5	31	3	1.202	1271	23,700	0.0668	0.0720	0.0785	0.0851	0.0386	0.395	0.0888
Carnation TW	1431.0	31	3	1.236	1346	24,600	0.0631	0.0684	0.0745	0.0807	0.0397	0.391	0.0880
Coreopsis TW	1590.0	49	4	1.315	1503	27,300	0.0570	0.0629	0.0684	0.0739	0.0425	0.383	0.0861
Jessamine TW	1750.0	49	4	1.377	1654	30,000	0.0518	0.0579	0.0629	0.0679	0.0446	0.377	0.0848
Cowslip TW	2000.0	49	4	1.468	1890	34,500	0.0453	0.0518	0.0561	0.0604	0.0478	0.369	0.0829
Lupine TW	2500.0	71	5	1.648	2369	42,400	0.0364	0.0439	0.0472	0.0505	0.0543	0.353	0.0794
Trillium TW	3000.0	71	5	1.799	2843	50,900	0.0303	0.0385	0.0412	0.0439	0.0597	0.342	0.0768

1. Data shown are subject to normal manufacturing tolerances.
2. Direct current (dc) resistance is based on 16.727 ohm-cmil/ft at 20° 62% IACS conductivity.
3. Alternating current (ac) resistance is based on dc resistance corrected for temperature and skin effect.
4. Properties of the industrial wires are those of the equivalent round wires of ASTM B 230.

Source: Aluminum Association, *Aluminum Electrical Conductor Handbook*, 3rd ed., 1989, Table 4.11.

Table A.5 Bare Aluminum Conductors, Steel-Reinforced (ACSR) Electrical Properties of Single-Layer Sizes

					Resistances				(Approximate) Inductive Reactance X, 1 ft Equivalent Spacing 60 Hz			Capacitive Reactance X, 1 ft Equivalent Spacing—60 Hz
		Stranding			dc	(Approximate) ac—60 Hz						
Code Word	Conductor Size kcmil	Al.	Steel	Assumed 75°C Current Amps	20°C Ohms/Mile	25°C Ohms/Mile	50°C Ohms/Mile	75°C Ohms/Mile	25°C Ohms/Mile	50°C Ohms/Mile	75°C Ohms/Mile	Megohm-Miles
Turkey	6	6	1	110	3.3893	3.460	3.960	4.308	0.634	0.734	0.760	0.1423
Swan	4	6	1	145	2.1291	2.175	2.531	2.755	0.608	0.694	0.723	0.1354
Swanate	4	7	1	145	2.1060	2.150	2.446	2.727	0.598	0.654	0.688	0.1345
Sparrow	2	6	1	195	1.3381	1.368	1.626	1.774	0.580	0.652	0.674	0.1285
Sparate	2	7	1	195	1.3230	1.353	1.566	1.741	0.574	0.621	0.637	0.1276
Robin	1	6	1	220	1.0617	1.087	1.306	1.427	0.564	0.629	0.646	0.1250
Raven	1/0	6	1	255	0.8410	0.862	1.041	1.141	0.549	0.601	0.614	0.1216
Quail	2/0	6	1	295	0.6679	0.687	0.853	0.929	0.537	0.590	0.599	0.1182
Pigeon	3/0	6	1	340	0.5297	0.546	0.638	0.763	0.524	0.572	0.578	0.1147
Penguin	4/0	6	1	390	0.4199	0.434	0.563	0.611	0.509	0.553	0.556	0.113
Grouse	80	8	1	200	1.0901	1.114	1.247	1.380	0.553	0.596	0.607	0.1240
Petrel	101.8	12	7	250	0.8360	0.858	1.094	1.264	0.538	0.613	0.677	0.1173
Minorca	110.8	12	7	265	0.7678	0.787	1.020	1.179	0.537	0.621	0.670	0.1160
Leghorn	134.6	12	7	300	0.6323	0.651	0.865	1.000	0.527	0.606	0.648	0.1131
Guinea	159	12	7	330	0.5353	0.552	0.753	0.873	0.517	0.590	0.628	0.1107
Dotterel	176.9	12	7	350	0.4812	0.499	0.687	0.799	0.512	0.582	0.617	0.1091
Dorking	190.8	12	7	370	0.4460	0.462	0.649	0.752	0.505	0.577	0.607	0.1079
Cochin	211.3	12	7	390	0.4027	0.418	0.594	0.692	0.499	0.567	0.596	0.1064
Brahma	203.2	16	19	380	0.4035	0.417	0.575	0.712	0.493	0.553	0.592	0.1043

1 Direct current (dc) resistance is based on 16.946 ohm-cmil/ft (61.2% IACS) at 20°C for the nominal aluminum area of the conductors and 129.64 ohm-circular mil ft (8.0% IACS) for the nominal steel area, with standard increments for stranding ASTM B 232

2 Alternating current (ac) resistance is based on dc resistance corrected for temperature using 0.00404 as temperature coefficient of resistivity per degree C for aluminum 1350 and 0.0029 per degree C for steel core, and for effect of core magnetization using method of Lewis and Tuttle Power Apparatus and Systems, Feb 1959, pp 1189-1214 Currents assumed for magnetization calculations in percent of assumed 75 C current 25 C—10%, 50 C—75%

3 Inductive reactance includes magnetization effect of steel core calculated using method of Lewis and Tuttle, Power Apparatus and Systems Feb 1959 pp 1189-1214 Currents assumed in calculating magnetization effect in percent of assumed 75 C current 25°C—10%, 50 C—75%

Source: Aluminum Association, *Aluminum Electrical Conductor Handbook*, 3rd ed., 1989, Table 4.15.

Table A.6 Bare Aluminum Conductors, Steel-Reinforced (ACSR) Electrical Properties of Multilayer Sizes

				Resistance					Phase-to-Neutral, 60 Hz Reactance at One ft Spacing	
				dc	ac–60 Hz					
Code Word	Size kcmil	Stranding Al./St.	Number of Aluminum Layers	20°C Ohms/Mile	25°C Ohms/Mile	50°C Ohms/Mile	75°C Ohms/Mile	GMR ft	Inductive Ohms/Mile X_a	Capacitive Megohm-Miles X_a
Waxwing	266.8	18/ 1	2	0.3398	0.347	0.382	0.416	0.0197	0.477	0.109
Partridge	266.8	26/ 7	2	0.3364	0.344	0.377	0.411	0.0217	0.465	0.107
Ostrich	300.	26/ 7	2	0.2993	0.306	0.336	0.366	0.0230	0.458	0.106
Merlin	336.4	18/ 1	2	0.2693	0.276	0.303	0.330	0.0221	0.463	0.106
Linnet	336.4	26/ 7	2	0.2671	0.273	0.300	0.327	0.0244	0.451	0.104
Oriole	336.4	30/ 7	2	0.2650	0.271	0.297	0.324	0.0255	0.445	0.103
Chickadee	397.5	18/ 1	2	0.2279	0.234	0.257	0.279	0.0240	0.452	0.103
Ibis	397.5	26/ 7	2	0.2260	0.231	0.254	0.277	0.0265	0.441	0.102
Lark	397.5	30/ 7	2	0.2243	0.229	0.252	0.274	0.0277	0.435	0.101
Pelican	477	18/ 1	2	0.1899	0.195	0.214	0.233	0.0263	0.441	0.100
Flicker	477	24/ 7	2	0.1889	0.194	0.213	0.232	0.0283	0.432	0.0992
Hawk	477	26/ 7	2	0.1883	0.193	0.212	0.231	0.0290	0.430	0.0988
Hen	477	30/ 7	2	0.1869	0.191	0.210	0.229	0.0304	0.424	0.0980
Osprey	556.5	18/ 1	2	0.1629	0.168	0.184	0.200	0.0284	0.432	0.0981
Parakeet	556.5	24/ 7	2	0.1620	0.166	0.183	0.199	0.0306	0.423	0.0969
Dove	556.5	26/ 7	2	0.1613	0.166	0.182	0.198	0.0313	0.420	0.0965
Eagle	556.5	30/ 7	2	0.1602	0.164	0.180	0.196	0.0328	0.415	0.0957
Peacock	605	24/ 7	2	0.1490	0.153	0.168	0.183	0.0319	0.418	0.0957
Squab	605	26/ 7	2	0.1485	0.153	0.167	0.182	0.0327	0.415	0.0953
Teal	605	30/19	2	0.1475	0.151	0.166	0.181	0.0342	0.410	0.0944
Kingbird	636	18/ 1	2	0.1420	0.147	0.162	0.175	0.0301	0.425	0.0951
Rook	636	24/ 7	2	0.1417	0.146	0.160	0.174	0.0327	0.415	0.0950
Grosbeak	636	26/ 7	2	0.1411	0.145	0.159	0.173	0.0335	0.412	0.0946
Swift	636	36/ 1	3	0.1410	0.148	0.162	0.176	0.0300	0.426	0.0964
Egret	636	30/19	2	0.1403	0.144	0.158	0.172	0.0351	0.406	0.0937
Flamingo	666.6	24/ 7	2	0.1352	0.139	0.153	0.166	0.0335	0.412	0.0943
Crow	715.5	54/ 7	3	0.1248	0.128	0.141	0.153	0.0372	0.399	0.0920
Starling	715.5	26/ 7	2	0.1254	0.129	0.142	0.154	0.0355	0.405	0.0928
Redwing	715.5	30/19	2	0.1248	0.128	0.141	0.153	0.0372	0.399	0.0920
Coot	795	36/ 1	3	0.1146	0.119	0.130	0.142	0.0335	0.412	0.0932
Cuckoo	795	24/ 7	2	0.1135	0.118	0.128	0.140	0.0361	0.403	0.0917
Drake	795	26/ 7	2	0.1129	0.117	0.128	0.139	0.0375	0.399	0.0912
Mallard	795	30/19	2	0.1122	0.116	0.127	0.138	0.0392	0.393	0.0904
Tern	795	45/ 7	3	0.1143	0.119	0.130	0.141	0.0352	0.406	0.0925
Condor	795	54/ 7	3	0.1135	0.117	0.129	0.140	0.0368	0.401	0.0917
Crane	874.5	54/ 7	3	0.1030	0.107	0.117	0.127	0.0387	0.395	0.0902
Ruddy	900	45/ 7	3	0.1008	0.106	0.115	0.125	0.0374	0.399	0.0907
Canary	900	54/ 7	3	0.1002	0.104	0.114	0.124	0.0392	0.393	0.0898

Table A.6 *(Continued)*

Code Word	Size kcmil	Stranding Al./St.	Number of Aluminum Layers	Resistance: dc 20°C Ohms/Mile[1]	Resistance: ac-60 Hz[2] 25°C Ohms/Mile	50°C Ohms/Mile	75°C Ohms/Mile	GMR ft	Phase-to-Neutral, 60 Hz Reactance at One ft Spacing: Inductive Ohms/Mile X_a	Capacitive Megohm-Miles X_a'
Corncrake	954	20/7	2	0.0950	0.099	0.109	0.118	0.0378	0.396	0.0898
Rail	954	45/7	3	0.09526	0.0994	0.109	0.118	0.0385	0.395	0.0897
Towhee	954	48/7	3	0.0950	0.099	0.108	0.118	0.0391	0.393	0.0896
Redbird	954	24/7	2	0.0945	0.098	0.108	0.117	0.0396	0.392	0.0890
Cardinal	954	54/7	3	0.09452	0.0983	0.108	0.117	0.0404	0.389	0.0890
Ortolan	1033.5	45/7	3	0.08798	0.0922	0.101	0.110	0.0401	0.390	0.0886
Curlew	1033.5	54/7	3	0.08728	0.0910	0.0996	0.108	0.0420	0.385	0.0878
Bluejay	1113	45/7	3	0.08161	0.0859	0.0939	0.102	0.0416	0.386	0.0874
Finch	1113	54/19	3	0.08138	0.0851	0.0931	0.101	0.0436	0.380	0.0867
Bunting	1192.5	45/7	3	0.07619	0.0805	0.0880	0.0954	0.0431	0.382	0.0864
Grackle	1192.5	54/19	3	0.07600	0.0798	0.0872	0.0947	0.0451	0.376	0.0856
Bittern	1272	45/7	3	0.07146	0.0759	0.0828	0.0898	0.0445	0.378	0.0855
Pheasant	1272	54/19	3	0.07122	0.0751	0.0820	0.0890	0.0466	0.372	0.0847
Dipper	1351.5	45/7	3	0.06724	0.0717	0.0783	0.0848	0.0459	0.374	0.0846
Martin	1351.5	54/19	3	0.06706	0.0710	0.0775	0.0840	0.0480	0.368	0.0838
Bobolink	1431	45/7	3	0.06352	0.0681	0.0742	0.0804	0.0472	0.371	0.0837
Plover	1431	54/19	3	0.06332	0.0673	0.0734	0.0796	0.0495	0.365	0.0829
Nuthatch	1510.5	45/7	3	0.06017	0.0649	0.0706	0.0765	0.0485	0.367	0.0829
Parrot	1510.5	54/19	3	0.06003	0.0641	0.0699	0.0757	0.0508	0.362	0.0821
Lapwing	1590	45/7	3	0.05714	0.0620	0.0674	0.0729	0.0498	0.364	0.0822
Falcon	1590	54/19	3	0.05699	0.0611	0.0666	0.0721	0.0521	0.358	0.0814
Chukar	1780	84/19	4	0.05119	0.0561	0.0609	0.0658	0.0534	0.355	0.0803
Mockingbird	2034.5	72/7	4	0.04488	0.0507	0.0549	0.0591	0.0553	0.348	0.0788
Bluebird	2156	84/19	4	0.04229	0.0477	0.0516	0.0555	0.0588	0.344	0.0775
Kiwi	2167	72/7	4	0.04228	0.0484	0.0522	0.0562	0.0570	0.348	0.0779
Thrasher	2312	76/19	4	0.03960	0.0454	0.0486	0.0528	0.0600	0.343	0.0767
Joree	2515	76/19	4	0.03643	0.0428	0.0459	0.0491	0.0621	0.338	0.0756

1 Direct current (dc) resistance is based on 16 946 ohm-cmil/ft. (61.2% IACS) at 20°C for nominal aluminum area of the conductors, and 129.64 ohm-cmil/ft (8% IACS) at 20°C for the nominal steel area, with standard increments for stranding. ASTM B 232.

2 Alternating current (ac) resistance is based on the resistance corrected for temperature using 0.00404 as temperature coefficient of resistivity per degree C for aluminum and 0.0029 per degree C for steel, and for skin effect.

3 The effective ac resistance of 3-layer ACSR increases with current density due to core magnetization.

Source: Aluminum Association, *Aluminum Electrical Conductor Handbook*, 3rd ed., 1989, Table 4.16.

Table A.7 Shaped-Wire Concentric-Lay Compact Aluminum Conductors Steel-Reinforced (ACSR/TW)

Area Equal to Stranded ACSR Sizes

Code Word	Size kcmil	Type No.	Stranding Al./St.	No. of Aluminum Layers	Resistance dc 20°C Ohms/Mile	Resistance ac—60 Hz 25°C Ohms/Mile	Resistance ac—60 Hz 50°C Ohms/Mile	Resistance ac—60 Hz 75°C Ohms/Mile	GMR ft	Phase-to-Neutral 60 Hz Resistance at One ft Spacing: Inductive Ohms/Mile X_a	Phase-to-Neutral 60 Hz Resistance at One ft Spacing: Capacitive Megohm-Miles X_a
Merlin/TW	336.4	6	14/1	2	0.2654	0.2715	0.2986	0.3258	0.0200	0.475	0.1079
Flicker/TW	477.0	13	18/7	2	0.1860	0.1904	0.2094	0.2284	0.0257	0.444	0.1017
Hawk/TW	477.0	16	18/7	2	0.1854	0.1878	0.2087	0.2277	0.0264	0.441	0.1013
Parakeet/TW	556.5	13	18/7	2	0.1593	0.1633	0.1796	0.1959	0.0277	0.435	0.0994
Dove/TW	556.5	16	20/7	2	0.1588	0.1628	0.1790	0.1953	0.0286	0.431	0.0991
Swift/TW	636.0	3	27/1	3	0.1416	0.1461	0.1605	0.1748	0.0273	0.437	0.0991
Rook/TW	636.0	13	18/7	2	0.1395	0.1432	0.1574	0.1717	0.0296	0.427	0.0978
Grosbeak/TW	636.0	16	20/7	2	0.1390	0.1426	0.1568	0.1710	0.0305	0.423	0.0971
Tern/TW	795.0	7	17/7	2	0.1123	0.1160	0.1274	0.1388	0.0312	0.4209	0.0955
Puffin/TW	795.0	10	18/7	2	0.1118	0.1152	0.1266	0.1380	0.0323	0.4165	0.0949
Condor/TW	795.0	13	20/7	2	0.1113	0.1147	0.1260	0.1373	0.0331	0.4137	0.0945
Drake/TW	795.0	16	20/7	2	0.1111	0.1144	0.1257	0.1370	0.0339	0.4105	0.0940
Phoenix/TW	954.0	5	30/7	3	0.0942	0.0982	0.1077	0.1172	0.0343	0.4094	0.0928
Rail/TW	954.0	7	32/7	3	0.0940	0.0979	0.1073	0.1160	0.0349	0.407	0.0925
Cardinal/TW	954.0	13	20/7	2	0.0931	0.0962	0.1056	0.1151	0.0362	0.403	0.0919
Snowbird/TW	1033.5	5	30/7	3	0.0868	0.0908	0.0995	0.1083	0.0356	0.405	0.0917
Ortolan/TW	1033.5	7	32/7	3	0.0867	0.0906	0.0993	0.1081	0.0363	0.402	0.0914
Curlew/TW	1033.5	13	21/7	2	0.0859	0.0389	0.0976	0.1063	0.0377	0.398	0.0906
Avocet/TW	1113.0	5	30/7	3	0.0807	0.0847	0.0928	0.1009	0.0369	0.400	0.0906
Bluejay/TW	1113.0	7	33/7	3	0.0805	0.0845	0.0925	0.1005	0.0376	0.398	0.0903
Finch/TW	1113.0	13	38/19	3	0.0802	0.0837	0.0917	0.0998	0.0399	0.391	0.0891
Oxbird/TW	1192.5	5	30/7	3	0.0753	0.0794	0.0869	0.0945	0.0382	0.396	0.0896
Bunting/TW	1192.5	7	33/7	3	0.0752	0.0791	0.0866	0.0941	0.0390	0.394	0.0893
Grackle/TW	1192.5	13	38/19	3	0.0749	0.0783	0.0859	0.0934	0.0412	0.387	0.0883
Scissortail/TW	1272.0	5	30/7	3	0.0706	0.0747	0.0817	0.0889	0.0394	0.392	0.0888
Bittern/TW	1272.0	7	35/7	3	0.0705	0.0745	0.0815	0.0885	0.0403	0.390	0.0884
Pheasant/TW	1272.0	13	39/19	3	0.0701	0.0736	0.0806	0.0876	0.0428	0.383	0.0874
Dipper/TW	1351.5	7	35/7	3	0.0664	0.0704	0.0769	0.0836	0.0415	0.386	0.0874
Martin/TW	1351.5	13	39/19	3	0.0659	0.0694	0.0760	0.0826	0.0438	0.377	0.0865
Bobolink/TW	1431.0	7	36/7	3	0.0627	0.0668	0.0730	0.0792	0.0427	0.383	0.0867
Plover/TW	1431.0	13	39/19	3	0.0624	0.0659	0.0721	0.0784	0.0451	0.376	0.0860
Lapwing/TW	1590.0	7	36/7	3	0.0564	0.0606	0.0661	0.0717	0.0449	0.377	0.0851
Falcon/TW	1590.0	13	42/19	3	0.0561	0.0598	0.0653	0.0709	0.0476	0.370	0.0841
Chukar/TW	1780.0	8	37/19	3	0.0503	0.0545	0.0594	0.0644	0.0482	0.368	0.0832
Bluebird/TW	2156.0	8	64/19	4	0.0415	0.0465	0.0504	0.0544	0.0538	0.355	0.0801

1. Direct current (dc) resistance is based on 16.727 ohm-cmil/ft (62.0% IACS) at 20°C for nominal aluminum area of the conductors, and 129.64 ohm-cmil/ft (8.0% IACS) at 20°C for the nominal steel area.
2. Alternating current (ac) resistance is based on dc resistance corrected for temperature using 0.00409 as temperature coefficient of resistivity per degree C for aluminum and 0.0029 per degree C for steel, and for skin effect.
3. The effective ac resistance of a layer ACSR/TW increases with current density due to core magnetization.

Source: Aluminum Association, *Aluminum Electrical Conductor Handbook*, 3rd ed., 1989, Table 4.20.

Table A.8 Shaped-Wire Concentric-Lay Compact Aluminum Conductors Steel-Reinforced (ACSR/TW)

Sized to Have Diameters Equal to Standard ACSR Conductors

Code Word	Size kcmil	Type No.	Stranding Al./St.	No. of Aluminum Layers	Resistance: dc 20°C Ohms/Mile	Resistance: ac—60 Hz 25°C Ohms/Mile	ac—60 Hz 50°C Ohms/Mile	ac—60 Hz 75°C Ohms/Mile	GMR ft	Phase-to-Neutral 60 Hz Reactance at One ft Spacing: Inductive Ohms/Mile X_a	Capacitive Megohm-Miles X_a'
Monongahela/TW	405.1	6	14/1	2	0.2205	0.2258	0.2483	0.2709	0.0218	0.464	0.1097
Mohawk/TW	571.7	13	18/7	2	0.1550	0.1590	0.1748	0.1907	0.0281	0.433	0.0991
Calumet/TW	565.3	16	20/7	2	0.1564	0.1603	0.1763	0.1923	0.0288	0.430	0.0988
Mystic/TW	666.6	13	20/7	2	0.1331	0.1367	0.1503	0.1638	0.0304	0.424	0.0970
Oswego TW	664.8	16	20/7	2	0.1329	0.1364	0.1500	0.1635	0.0310	0.421	0.0964
Nechako TW	768.9	3	27/1	3	0.1171	0.1214	0.1332	0.1451	0.0300	0.425	0.0965
Maumee TW	768.2	13	20/7	2	0.1155	0.1189	0.1306	0.1424	0.0325	0.416	0.0949
Wabash TW	762.8	16	20/7	2	0.1159	0.1191	0.1309	0.1428	0.0330	0.413	0.0946
Kettle TW	957.2	7	32/7	3	0.0938	0.0976	0.1071	0.1166	0.0350	0.407	0.0925
Fraser TW	946.7	10	35/7	3	0.0945	0.0982	0.1077	0.1173	0.0358	0.404	0.0919
Columbia/TW	966.2	13	21/7	2	0.0918	0.0949	0.1042	0.1136	0.0364	0.402	0.0917
Suwanee TW	959.6	16	22/7	2	0.0922	0.0951	0.1045	0.1138	0.0373	0.399	0.0913
Cheyenne/TW	1168.1	5	30/7	3	0.0769	0.0810	0.0886	0.0963	0.0378	0.397	0.0901
Genesee TW	1158.0	7	33/7	3	0.0774	0.0813	0.0891	0.0968	0.0384	0.395	0.0897
Hudson TW	1158.4	13	25/7	2	0.0764	0.0794	0.0871	0.0948	0.0400	0.391	0.0889
Catawba/TW	1272.0	5	30/7	3	0.0706	0.0747	0.0817	0.0888	0.0394	0.392	0.0889
Nelson TW	1257.1	7	35/7	3	0.0713	0.0753	0.0824	0.0895	0.0400	0.390	0.0886
Yukon TW	1233.6	13	38/19	3	0.0723	0.0758	0.0830	0.0903	0.0420	0.385	0.0877
Truckee/TW	1372.5	5	30/7	3	0.0654	0.0697	0.0761	0.0826	0.0409	0.388	0.0877
Mackenzie/TW	1359.7	7	36/7	3	0.0658	0.0698	0.0764	0.0829	0.0420	0.386	0.0874
Thames/TW	1334.6	13	39/19	3	0.0668	0.0703	0.0770	0.0837	0.0436	0.380	0.0866
St Croix/TW	1467.8	5	33/7	3	0.0612	0.0655	0.0716	0.0776	0.0424	0.840	0.0867
Miramichi/TW	1455.3	7	36/7	3	0.0616	0.0658	0.0718	0.0780	0.0431	0.382	0.0867
Merrimac/TW	1433.6	13	39/19	3	0.0622	0.0658	0.0720	0.0782	0.0450	0.376	0.0856
Platte TW	1569.0	5	33/7	3	0.0573	0.0617	0.0673	0.0730	0.0439	0.379	0.0858
Potomac TW	1557.4	7	36/7	3	0.0575	0.0617	0.0674	0.0731	0.0445	0.378	0.0853
Rio Grande/TW	1533.3	13	39/19	3	0.0582	0.0618	0.0676	0.0734	0.0466	0.372	0.0847
Schuylkill TW	1657.4	7	36/7	3	0.0541	0.0584	0.0637	0.0690	0.0459	0.374	0.0845
Pecos TW	1622.0	13	39/19	3	0.0549	0.0585	0.0639	0.0690	0.0481	0.368	0.0839
Pee Dee TW	1758.6	7	37/7	3	0.0510	0.0554	0.0603	0.0653	0.0473	0.370	0.0837
James/TW	1730.6	13	39/19	3	0.0516	0.0553	0.0604	0.0654	0.0494	0.365	0.0829
Athabaska/TW	1949.6	7	42/7	3	0.0460	0.0506	0.0550	0.0595	0.0500	0.363	0.0822
Cumberland/TW	1926.9	13	42/19	3	0.0462	0.0501	0.0546	0.0591	0.0523	0.358	0.0815
Powder TW	2153.8	8	64/19	4	0.0414	0.0464	0.0503	0.0543	0.0538	0.355	0.0803
Santee/TW	2627.3	8	64/19	4	0.0341	0.0395	0.0427	0.0459	0.0594	0.343	0.0775

1 Direct current (dc) resistance is based on 16 727 ohm-cmil/ft (62.0% IACS) at 20°C for nominal aluminum area of the conductors, and 129 64 ohm-cmil/ft (8 0% IACS) at 20°C for the nominal steel area.

2 Alternating current (ac) resistance is based on dc resistance corrected for temperature using 0 00409 as temperature coefficient of resistivity per degree C for aluminum and 0 0029 per degree C for steel, and for skin effect

3 The effective ac resistance of 3 layer ACSR/TW increases with current density due to core magnetization.

Source: Aluminum Association, *Aluminum Electrical Conductor Handbook*, 3rd ed., 1989, Table 4.22.

Table A.9 Bare Aluminum Conductors, 1350-H19 Wires Stranded with Aluminum-Clad Steel Wires (Alumoweld) as Reinforcement (AWAC) in Distribution and Neutral-Messenger Sizes

							Resistances					Phase to Neutral 60 Hz Reactance at One ft Spacing	
								ac—60 Hz					
Equiv. AWG Size and Stranding	Total Aluminum Area cmil	Outside Diameter in.	Diameter of Each Wire in.	Total Area Sq. in.	Weight per 1000 ft lb	Rated Strength lb	dc 20°C Ohms/Mile	25°C Ohms/Mile	50°C Ohms/Mile	75°C Ohms/Mile	GMR ft	Inductive X_a Ohms/Mile X_a	Capacitive X_a Megohm-Miles X_a
No 4 6/1	41,740	0.245	0.0817	0.03671	52.4	1,710	2.165	2.218	2.455	2.697	0.006738	0.6067	0.1360
5/2	41,740	0.261	0.0871	0.04172	69.7	2,790	2.126	2.172	2.459	2.717	0.004992	0.6431	0.1341
4/3	41,740	0.281	0.0937	0.04831	92.4	4,200	2.078	2.130	2.420	2.703	0.003899	0.6720	0.1320
3/4	41,740	0.307	0.1022	0.05737	123.7	6,160	2.011	2.068	2.377	2.657	0.003402	0.6896	0.1293
2/5	41,740	0.340	0.1133	0.07061	169.4	8,990	1.930	1.992	2.294	2.572	0.002287	0.7389	0.1263
No. 3 6/1	52,620	0.275	0.0918	0.04629	66.1	2,140	1.115	1.757	1.952	1.151	0.007563	0.5927	0.1326
5/2	52,620	0.293	0.0978	0.05261	87.8	3,500	1.686	1.723	1.963	2.175	0.005604	0.6291	0.1307
4/3	52,600	0.316	0.1053	0.06092	116.5	5,260	1.645	1.686	1.940	2.173	0.004385	0.6588	0.1285
3/4	52,620	0.344	0.1147	0.07234	155.9	7,700	1.597	1.642	1.908	2.147	0.003812	0.6758	0.1260
2/5	52,620	0.382	0.1273	0.08903	213.5	11,300	1.529	1.559	1.840	2.076	0.002547	0.7248	0.1228
No 2 6/1	66,360	0.309	0.1030	0.05837	83.3	2,650	1.363	1.395	1.557	1.721	0.008498	0.5766	0.1291
5/2	66,360	0.330	0.1099	0.06634	110.8	4,370	1.335	1.364	1.565	1.740	0.006311	0.6147	0.1272
4/3	66,360	0.355	0.1182	0.07682	146.9	6,600	1.306	1.338	1.555	1.754	0.004926	0.6447	0.1250
3/4	66,360	0.388	0.1288	0.09122	196.6	9,690	1.266	1.302	1.525	1.735	0.004277	0.6618	0.1225
2/5	66,360	0.429	0.1429	0.1123	269.3	13,500	1.213	1.237	1.479	1.681	0.002880	0.7107	0.1194
No 1 6/1	83,690	0.347	0.1157	0.07362	105.1	3,310	1.080	1.108	1.240	1.376	0.009543	0.5645	0.1257
5/2	83,690	0.370	0.1234	0.08366	139.7	5,450	1.059	1.082	1.250	1.395	0.007076	0.6008	0.1238
4/3	83,690	0.398	0.1327	0.09687	185.3	8,100	1.036	1.062	1.248	1.417	0.005522	0.6309	0.1216
3/4	83,690	0.434	0.1446	0.1150	247.9	11,200	1.005	1.033	1.230	1.406	0.004809	0.6479	0.1191
2/5	83,690	0.482	0.1605	0.1416	339.6	16,500	0.9620	0.9808	1.187	1.360	0.003212	0.6966	0.1159
No 1/0 6/1	105,600	0.390	0.1300	0.09289	132.6	4,080	0.8554	0.8760	0.9900	1.106	0.001073	0.5503	0.1222
5/2	105,600	0.416	0.1385	0.1055	176.1	6,580	0.8406	0.8590	1.003	1.126	0.007956	0.5865	0.1203
4/3	105,600	0.447	0.1490	0.1221	233.5	9,680	0.8216	0.8422	1.008	1.158	0.006202	0.6167	0.1182
3/4	105,600	0.487	0.1624	0.1450	312.6	13,800	0.7968	0.8191	0.9957	1.152	0.005396	0.6337	0.1156
2/5	105,600	0.541	0.1802	0.1785	428.0	19,500	0.7630	0.7781	0.9700	1.120	0.003607	0.6825	0.1125
No 2/0 6/1	133,100	0.438	0.1459	0.1171	167.1	4,930	0.6780	0.6954	0.7960	0.8979	0.01205	0.5362	0.1188
5/2	133,100	0.467	0.1556	0.1330	222.1	8,030	0.6663	0.6806	0.8072	0.9148	0.008931	0.5725	0.1169
4/3	133,100	0.502	0.1674	0.1541	294.6	11,900	0.6510	0.6673	0.8203	0.9584	0.006965	0.6027	0.1147
3/4	133,100	0.547	0.1824	0.1829	394.3	16,400	0.6315	0.6494	0.7876	0.9587	0.006061	0.6196	0.1122
No 3/0 6/1	167,800	0.492	0.1639	0.1478	210.7	6,060	0.5380	0.5511	0.6377	0.7252	0.01353	0.5221	0.1153
5/2	167,800	0.524	0.1747	0.1677	280.6	9,660	0.5285	0.5399	0.6487	0.7403	0.01002	0.5585	0.1135
4/3	167,800	0.564	0.1880	0.1943	371.4	14,200	0.5161	0.5290	0.6650	0.7873	0.007826	0.589	0.1113
No 4/0 6/1	211,600	0.552	0.1840	0.1861	265.7	7,380	0.4270	0.4373	0.5142	0.5917	0.01518	0.5082	0.1119
15/4	211,600	0.575	0.1150	0.1974	305.4	10,600	0.4264	0.4357	0.4865	0.5346	0.01375	0.5202	0.1107

1 Data shown are subject to normal manufacturing tolerances.

2 Direct current (dc) resistance is based on 17.002 ohm-cmil/ft for aluminum wires and 51.01 ohm-cmil/ft for aluminum-clad wires in strands and cladding.

3 Alternating current (ac) resistance is based on dc resistance corrected for temperature and for skin effect.

4 Properties of the individual wires are those of ASTM B 502 and ASTM B 230.

Source: Aluminum Association, *Aluminum Electrical Conductor Handbook*, 3rd ed., 1989, Table 4.25.

Bibliography

This is a selected list from many books and papers related to the subject of this book and are the ones the author has found to be most useful. No doubt there are others!

Anderson, P. M., *Analysis of Faulted Power Systems*, Iowa State University Press, Ames, Iowa, 1973.

ANSI/IEEE C37.5, *Guide for Calculation of Fault Current for Application of AC High-Voltage Circuit Breakers Rated on a Total Current Basis*, 1979.

ANSI/IEEE C37.010, *Application Guide for AC High-Voltage Circuit Breakers Rated on a Symmetrical Current Basis.*

ANSI/IEEE C37.013, *Standard for AC High Voltage Generator Circuit Breakers Rated on a Symmetrical Current Basis*, 1990.

ANSI/IEEE C37.13, *Standard for Low Voltage AC Power*, 1990.

Blackburn, J. L., *Protective Relaying: Principles and Applications*, Marcel Dekker, Inc., New York, 1987.

Blackburn, J. L., Voltage Induction in Paralleled Transmission Circuits, *IEEE Transactions*, Vol. 81, Part III, Feb. 1963, pp. 921–929.

Clarke, Edith, *Circuit Analysis of A-C Power Systems*, Vols. I and II, General Electric Co., Schenectady, N.Y., 1943, 1950.

Electrical Transmission and Distribution Reference Book, 4th ed., Westinghouse Electric Co., Schenectady, N.Y., 1964.

Electrical Utility Engineering Reference Book, Vol. 3, *Distribution Systems*, Westinghouse Electric Co., Schenectady, N.Y., 1965.

Fitzgerald, A. E., Kingsley, C., and Umans, S. D., *Electric Machinery*, McGraw-Hill Book Company, New York, 1983.

Harder, E. L., Sequence Network Connections for Unbalanced Load and Fault Conditions, *The Electric Journal*, Dec. 1937, pp. 481–488.

IEEE Red Book, Standard 141, *Recommended Practice for Electric Power Distribution for Industrial Plants*, 1976.

Seidman, A. H., Mahrous, H., and Hicks, T. G., *Handbook of Electric Power Calculations*, McGraw-Hill Book Company, New York, 1983.

Stevenson, W. D., Jr., *Elements of Power Systems Analysis*, 4th ed., McGraw-Hill Book Company, New York, 1982.

Wagner, C. F., and Evans, R. D., *Symmetrical Components*, McGraw-Hill Book Company, New York, 1933. (Available in reprint from R. E. Kreiger Publishing Co., Inc., Melbourne, Fla., 1982.)

Weedy, B. M., *Electric Power Systems*, 3rd rev., John Wiley & Sons, Inc., New York, 1987.

ANSI and IEEE documents are available from IEEE Service Center, P.O. Box 1331, Piscataway, NJ 08855-1331.

Index